I0065950

Earth Surface Engineering and Technology

Earth Surface Engineering and Technology

Edited by **Matt Weilberg**

New York

Published by Syrawood Publishing House,
750 Third Avenue, 9th Floor,
New York, NY 10017, USA
www.syrawoodpublishinghouse.com

Earth Surface Engineering and Technology
Edited by Matt Weilberg

International Standard Book Number: 978-1-68286-081-6 (Hardback)

Printed in the United States of America.

Contents

Preface

I am honored to present to you this unique book which encompasses the most up-to-date data in the field. I was extremely pleased to get this opportunity of editing the work of experts from across the globe. I have also written papers in this field and researched the various aspects revolving around the progress of the discipline. I have tried to unify my knowledge along with that of stalwarts from every corner of the world, to produce a text which not only benefits the readers but also facilitates the growth of the field.

Earth surface engineering is an emerging field of study that incorporates concepts from different branches of earth science. This book traces the progress of this field and highlights some of its crucial aspects such as various processes shaping earth's surface, earth surface interaction with different spheres, advanced tools and instrumentation for evaluation and measurement of earth surface processes, etc. A number of latest researches have been included to keep the readers up-to-date with the emerging concepts in this area of study which will provide the readers with a thorough understanding of the subject.

Finally, I would like to thank all the contributing authors for their valuable time and contributions. This book would not have been possible without their efforts. I would also like to thank my friends and family for their constant support.

Editor

1

sedFlow – a tool for simulating fractional bedload transport and longitudinal profile evolution in mountain streams

F. U. M. Heimann[1,2], D. Rickenmann[1], J. M. Turowski[3,1], and J. W. Kirchner[2,1]

[1]WSL Swiss Federal Institute for Forest, Snow and Landscape Research, 8903 Birmensdorf, Switzerland
[2]Department of Environmental System Sciences, ETH Zurich, 8092 Zurich, Switzerland
[3]Helmholtz Centre Potsdam, GFZ German Research Centre for Geosciences, Telegrafenberg, 14473 Potsdam, Germany

Correspondence to: F. U. M. Heimann (florian.heimann@wsl.ch)

Abstract. Especially in mountainous environments, the prediction of sediment dynamics is important for managing natural hazards, assessing in-stream habitats and understanding geomorphic evolution. We present the new modelling tool sedFlow for simulating fractional bedload transport dynamics in mountain streams. sedFlow is a one-dimensional model that aims to realistically reproduce the total transport volumes and overall morphodynamic changes resulting from sediment transport events such as major floods. The model is intended for temporal scales from the individual event (several hours to few days) up to longer-term evolution of stream channels (several years). The envisaged spatial scale covers complete catchments at a spatial discretisation of several tens of metres to a few hundreds of metres. sedFlow can deal with the effects of streambeds that slope uphill in a downstream direction and uses recently proposed and tested approaches for quantifying macro-roughness effects in steep channels. sedFlow offers different options for bedload transport equations, flow-resistance relationships and other elements which can be selected to fit the current application in a particular catchment. Local grain-size distributions are dynamically adjusted according to the transport dynamics of each grain-size fraction. sedFlow features fast calculations and straightforward pre- and postprocessing of simulation data. The high simulation speed allows for simulations of several years, which can be used, e.g., to assess the long-term impact of river engineering works or climate change effects. In combination with the straightforward pre- and postprocessing, the fast calculations facilitate efficient workflows for the simulation of individual flood events, because the modeller gets the immediate results as direct feedback to the selected parameter inputs. The model is provided together with its complete source code free of charge under the terms of the GNU General Public License (GPL) (www.wsl.ch/sedFlow). Examples of the application of sedFlow are given in a companion article by Heimann et al. (2015).

1 Introduction

Environmental models typically seek to predict the future state of a system, based on information about its current state and the mechanisms that regulate its evolution through time. In the case of sediment transport by flowing water in open channels, the temporal evolution of these variables is determined by a complex interaction of multiple processes including hydraulic water routing, sediment entrainment, erosion and deposition. In recent years many numerical models have been developed for simulating sediment transport in rivers. However, most of these models are intended for, and only applicable in, lowland rivers with gentle slopes. In mountain streams the effects of macro-roughness and shear stress partitioning have to be considered. Otherwise, sediment transport rates may be overestimated by several orders of magnitude

(Rickenmann and Recking, 2011; Nitsche et al., 2011, 2012). Based on these observations and contrasting the dominantly gradient-based definition used by, e.g., Wohl (2000), we define mountain streams as streams which are located within a mountainous region and in which the effects of macro-roughness and shear stress partitioning play an important role in the sediment transport system.

Few sediment transport models have been specifically designed for mountain streams. Cui et al. (2006) developed the two Dam Removal Express Assessment Models (DREAM-1&2), based on the previous models of Cui and Parker (2005) and Cui and Wilcox (2008), to focus specifically on dam removal scenarios. The DREAM models therefore feature the simulation of (a) bank erosion during the downcutting of reservoir deposits, (b) transcritical flow conditions, (c) combined bedload and suspended load transport, (d) the details of gravel abrasion and (e) staged dam removal and partial dredging as options in the dam removal scenarios. Due to their specific focus, the wider applicability of the DREAM models is limited. Both the model of García-Martinez et al. (2006) and the model MIKE21C (DHI, 1999, with its modifications by Li and Millar, 2007) focus on a two-dimensional representation of hydraulic and sediment transport processes. Therefore, these models require more extensive input data and longer calculation times compared to one-dimensional model representations. As another example, the model of Papanicolaou et al. (2004) is intended for studying sediment transport under transcritical flow conditions by solving the unsteady form of the Saint-Venant equations, which results in long calculation times.

Other sediment transport models have been described by Mouri et al. (2011), Lopez and Falcon (1999) and Hoey and Ferguson (1994), all of which feature the one-dimensional simulation of fractional bedload transport using a simplified representation of the hydraulic processes. The model of Mouri et al. (2011) can represent a combination of debris flow, bedload and suspension load processes. In contrast, SEDROUT (Hoey and Ferguson, 1994) is designed to study the spatial and temporal evolution of local grain-size distributions. Therefore, it determines the composition of the sediment surface layer by a numeric iteration within each time step. In its latest version, SEDROUT has been also extended to deal with islands and other features of river bifurcation (Verhaar et al., 2008). However, neither the source code, the executable model binary nor a detailed description of the model implementation is available for any of the three models mentioned in this paragraph.

The model Tom^{Sed} (formerly known as SEdiment TRansport model in Alpine Catchments (SETRAC)) was developed to study the influence of different shapes of channel cross sections on bedload transport in steep streams (Chiari et al., 2010; Chiari and Rickenmann, 2011). Therefore, the user can define cross sections with laterally varying bed elevations. The shape of a particular cross section stays the same during the complete simulation. Most published model applications used bedload transport calculations for a single grain size. In such a situation, all grain sizes and their spatial distribution are constant for the complete simulation. In this set-up, Tom^{Sed} is slightly faster than real time in a typical application. A fractional transport approach with dynamic grain-size distributions is implemented in Tom^{Sed} as well. However, it is rarely used due to the long calculation times.

The TOPographic Kinematic wave APproximation and Integration (Topkapi) model was originally developed as a rainfall–runoff model providing fast hydrologic simulations (Ciarapica and Todini, 2002). Later a sediment transport module was added, and this model version is called Topkapi ETH (Konz et al., 2011). The code is intended for the study of reach-scale sediment transport in the context of large-scale hydrologic processes. Due to this scope that integrates different processes and scales, the model features a spatial as well as temporal sub-gridding approach. The hydrologic processes are simulated on a coarse two-dimensional grid with time steps that are an integer multiple of the time steps for the hydraulic and sediment transport processes. The latter two processes are simulated in a one-dimensional channel at a finer spatial resolution. This channel receives water from the hydrologic two-dimensional grid, but the morphodynamic changes due to bedload transport have no influence on the topography used for the hydrologic calculations. The channel cross section is represented by a rectangle and bedload transport is based on a single grain-size approach, in which local grain-size distributions do not change over time. In typical applications, a flood event of several days can be simulated within a few minutes of calculation time.

Currently, no model is available that combines short calculation times with easy use and up-to-date sediment transport equations for mountain catchments. The new model sedFlow (Figs. 1, A1, A2, A3) presented in this contribution has been developed to provide an efficient tool for the simulation of bedload transport in mountain streams. By efficient tool we mean a model that combines straightforward pre- and post-processing of simulation data with fast calculation speeds. sedFlow is intended for temporal scales from the individual event (several hours to a few days) to longer-term channel evolution (several years). The envisaged spatial scale covers complete catchments at a spatial discretisation of several tens to a few hundreds of metres. The following elements were important for the development of sedFlow:

1. a sediment transport model provided together with its complete source code, open source and free of charge;

2. implementation of recently proposed and tested approaches for calculating bedload transport in steep channels accounting for macro-roughness effects;

3. individual calculations for several grain diameter fractions (fractional transport) resulting in more robust simulations as compared to single-grain approaches;

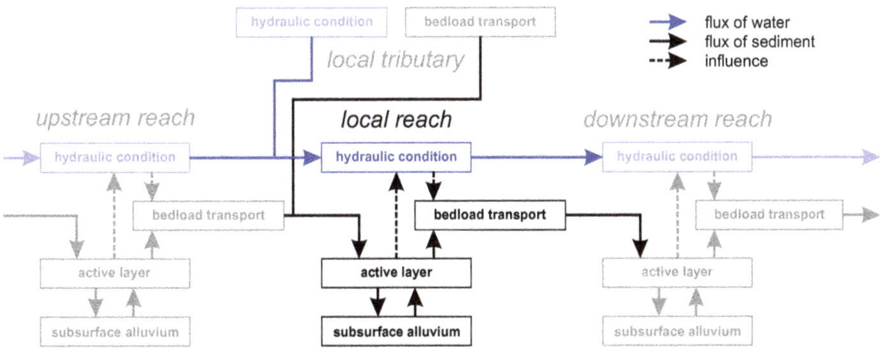

Figure 1. Simplified overview over the main process interactions within the sedFlow model.

4. consideration of ponding effects of adverse slopes (uphill slopes in the downstream direction), e.g. due to sudden sediment deposition by debris flow inputs;

5. fast calculations for modelling entire catchments, and for automated calculation of multiple scenarios exploring a range of parameter space;

6. an object-oriented code design that facilitates flexibility in model development;

7. flexibility in model application featuring different options (Figs. A1, A2), e.g. for the bedload transport equations, which can be selected to fit the current application, as well as straightforward pre- and postprocessing of simulation data.

The model sedFlow thus fills a gap in the range of existing sediment transport models for mountain streams (Table 1) and the goals outlined above have led to the implementation described in the following sections. This implementation represents the current state of the model, and may be easily extended and adjusted in the future.

Examples of sedFlow application are given in a companion article by Heimann et al. (2015).

2 Implementation of the sedFlow model

2.1 Hydraulic calculation

Hydraulic equations describe the temporal evolution of the three-dimensional flow field of the water continuum. A formalised description of the dynamics has been provided by Navier and Stokes (given in the form for incompressible flow).

$$\rho\left(\frac{\partial(\boldsymbol{v})}{\partial t}+\boldsymbol{v}\cdot\nabla\boldsymbol{v}\right)=-\nabla p+\mu\nabla^2\boldsymbol{v}+\boldsymbol{f}, \tag{1}$$

where ρ is fluid density, $\boldsymbol{v}$ is flow velocity, t is time, p is pressure and μ is dynamic viscosity. $\boldsymbol{f}$ summarises other influencing body forces. If large-scale backwater effects are not present, as is often the case in steep channels of mountain streams, the energy slope can be approximated by the bed slope (i.e. assumption of kinematic wave propagation) and the complete Navier–Stokes equation can be reduced to the following simplified, cross-sectional averaged conservation of volume (e.g. Chow et al., 1988).

$$\frac{\partial Q}{\partial x}+\frac{\partial A}{\partial t}=Q_{\mathrm{lat}}, \tag{2}$$

where Q is discharge, x is distance in flow direction, A is wetted cross-sectional area and Q_{lat} is lateral water influx.

2.1.1 Flow routing

Within sedFlow a channel network joined by confluences can be simulated. At the upstream ends of the main channel and each of the user-defined tributaries, a discharge time series is input and has to be routed through the channel system. For the following discussion of hydraulic routing schemes we will differentiate between three cases: first, in *ponding* – i.e. water collecting behind sediment obstructions – the friction slope S_{f} is approximately zero ($S_{\mathrm{f}}\approx 0$). Second, in *situations with parallel slopes*, the friction slope approximately equals the channel bed slope S_{b} ($S_{\mathrm{f}}\approx S_{\mathrm{b}}$), which is commonly true for steep S_{b}. Third, the *situations of moderate backwater effects* cover all cases between the extremes of ponding on the one hand and situations with parallel slopes on the other.

Especially in models not focussing on the details of the hydraulic routing, the kinematic wave approach (assuming the situation of parallel slopes) can be implemented using a temporally explicit Eulerian forward approach (van de Wiel et al., 2007; Chiari et al., 2010). Such an approach can be used in sedFlow as well (see Sect. A1). However, Eulerian forward approaches must assume that all parameters within one time step can be sufficiently approximated by their values at the beginning of the time step. To ensure this approximation, Eulerian forward approaches require very small time steps, especially for fast processes. This can be problematic when a relatively fast process, such as the routing of water, is combined with a relatively slow process, such as bedload

Table 1. Comparison of bedload transport models for steep mountain streams. Estimated calculation speeds refer to the simulation of a 20 km long study reach of a regular mountain river.

	Topkapi ETH	Tom^{Sed}	SEDROUT	sedFlow
Main aims	integral simulation of different processes at different scales featuring spatial and temporal sub-gridding	simulation of the effect of the shape of channel cross sections on bedload transport featuring a user-defined, detailed channel geometry	detailed simulation of the spatial and temporal evolution of local grain-size distributions; river bifurcations	fractional transport, consideration of uphill slopes, fast simulations and straightforward pre- and postprocessing of simulation data
Speed	simulation of several days within few minutes of computation time	slightly faster than real time		simulation of several years within few hours of computation time
Input format	partially MATLAB preprocessing required	XML files		mainly regular spreadsheets
Intended applications	mainly scientific	engineering and scientific	mainly scientific	mainly engineering and operational
References	Konz et al. (2011); Carpentier et al. (2012)	Chiari et al. (2010); Chiari and Rickenmann (2011); Kaitna et al. (2011)	Hoey and Ferguson (1994); Ferguson et al. (2001); Talbot and Lapointe (2002); Hoey et al. (2003); Verhaar et al. (2008); Boyer et al. (2010)	Junker et al. (2014); Heimann et al. (2015)

transport including bed level adjustments. The water routing requires small time steps and thus calculation times that may be orders of magnitude too long and slow from the perspective of bedload dynamics. Therefore, as an alternative, an implicit discharge routing is implemented in sedFlow. The implicit routing is unconditionally stable, and thus has no requirements concerning the length of time steps. In sedFlow the approach of Liu and Todini (2002) is used, which omits time-consuming iterations and finds the solution for the kinematic wave analytically via a Taylor series approximation. However, the approach depends on a power-law representation of discharge as a function of water volume in a reach. This means that it can only be applied to the specific cross-sectional shapes of infinitely deep rectangular or V-shaped channels in combination with a power-law flow-resistance equation.

The kinematic wave assumption of parallel slopes is usually valid for steep channel gradients, which are typical of mountain catchments. Nevertheless, the kinematic wave assumption can be problematic, especially in mountain streams, when tributaries deliver large amounts of sediment to the main channel within a relatively short time, e.g. during debris flow events. This may result in adverse channel slopes (uphill slopes in the downstream direction) and backwaters in the main channel, violating the assumptions of a kinematic wave. If configured for kinematic wave routing, sedFlow will abort simulations whenever adverse channel slopes occur.

When it is necessary to deal with adverse channel slopes, one has to drop the kinematic wave approximation and use a backwater calculation instead. Unfortunately, the backwater calculation is numerically intensive. Therefore, within sedFlow, a pragmatic approach can be selected to deal with adverse channel slopes: discharge is assumed to be uniform and thus equal along the entire channel for a given time step only increasing at confluences. This assumption of uniform discharge is reasonable because the temporal scale for water routing is orders of magnitude smaller than the temporal scale for morphodynamic adjustments. In the case of positive slopes, flow depth and velocity are commonly calculated using the bed slope as proxy for the friction slope (thus assuming the bed slope and water surface are parallel). However, the flow resistance may be adjusted such that a maximum Froude number is not exceeded. In cases of adverse channel slopes, the formation of ponding is simulated. That is, flow depth and velocity are selected to ensure a minimum gradient of hydraulic head, which is positive but close to zero, and thus ensures numeric stability, and approximately corresponds to the hydraulic gradient of ponding water. For bedload transport calculations the gradient of the hydraulic head is used, which by definition can only have positive slopes. Thus, the energy slope for bedload transport estimation is *not* the result of a backwater calculation, but it is the gradient between individual hydraulic head values, which under non-ponding conditions have been calculated independently from each other using the local bed slope as a proxy for friction slope. This approach is based on the assumption that the simulated system only consists of the two extreme cases of ponding on the one hand and parallel slopes on the other. At a spatial discretisation of several tens of metres, the assumption of the two extreme cases is valid for many mountain streams and it allows efficient simulation of ponding, by omitting numerically intensive backwater calculations. However, it has to be noted that this approach will produce large errors when intermediate cases of moderate backwater effects are part of the simulated system. In such systems, the first approach, which uses bed slope both as friction slope for

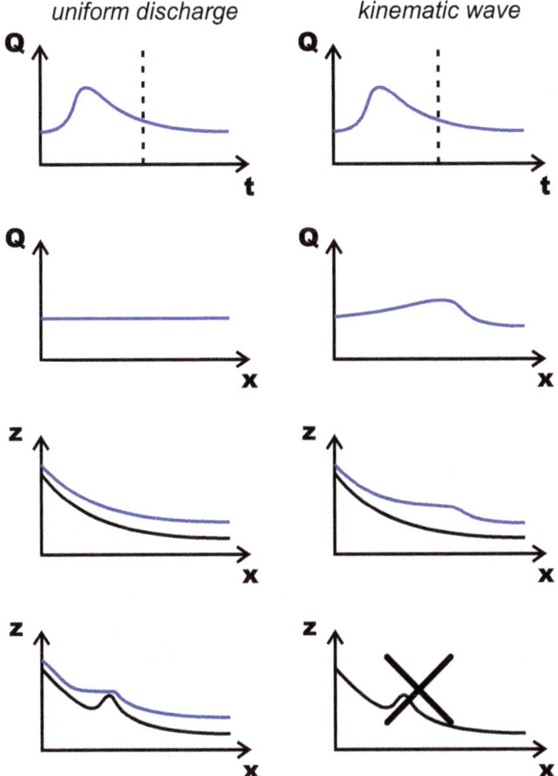

Figure 2. Qualitative comparison of uniform discharge (left) and kinematic wave (right) flow routing approaches. The top row shows a hypothetical discharge time series at the upstream boundary, which can be used in both approaches. The point in time for the following rows is indicated by the dashed vertical line. In the uniform discharge approach, the current discharge value of the time series defines the discharge for all reaches of the simulated system at the current point in time (second row left). In contrast, for the kinematic wave approach, the temporal variability of discharge is reflected in a spatial variability as well (second row right). Therefore, in the uniform discharge approach, the spatial variation of flow depth (third row), as well as water surface (blue curve), is mainly a function of roughness and slope, which is determined by the river bed (black curve). For the kinematic wave approach, flow depth may also vary due to the spatial variation of discharge (third row right). In cases of uphill channel slopes in the downstream direction, the uniform discharge approach will reproduce the effects of ponding (fourth row left), while the kinematic wave approach cannot deal with such situations (fourth row right).

the hydraulic calculations and as energy slope for the sediment transport calculations, will produce better estimates of the transported sediment volumes, but it cannot accommodate adverse channel gradients.

Heimann et al. (2015) have demonstrated that, despite their simplicity, the implemented hydraulic concepts (Fig. 2) appear to be sufficient for a realistic integrated representation of bedload transport processes. In that study, for the different hydraulic routing schemes, results have been obtained, which

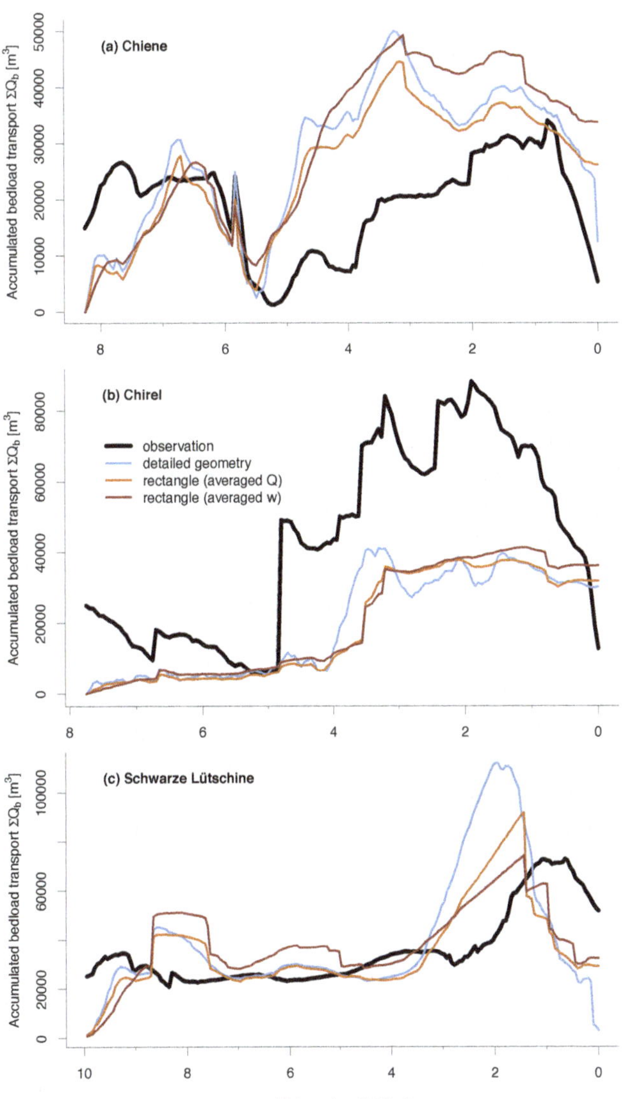

Figure 3. Comparison of the effects of different channel representations on accumulated bedload transport estimates simulated with the Tom^{Sed} model in three Swiss mountain rivers. See Table 2, text, and Stephan (2012) for details.

are close to the observed morphodynamic changes and very similar among each other.

2.1.2 Flow resistance

The interaction of flowing water with the river bed and banks determines the relation between the average downstream velocity and the wetted cross section. This interaction is summarised as flow resistance, which can be described by the following physically based relation:

$$\sqrt{\frac{8}{f}} = \frac{v}{\sqrt{g \cdot r_{\mathrm{h}} \cdot S_{\mathrm{f}}}}, \qquad (3)$$

where f is the Darcy–Weisbach friction factor, v is cross-sectional mean flow velocity, g is gravitational acceleration and r_h is hydraulic radius. The flow routing method of Liu and Todini (2002) requires a power-law relation between discharge and water volume within a reach. Therefore, the following flow-resistance law can be used in sedFlow.

$$\sqrt{\frac{8}{f}} = j_1 \cdot \left(\frac{r_\mathrm{h}}{k \cdot D_x}\right)^l \tag{4a}$$

$$\sqrt{\frac{8}{f}} = 6.5 \cdot \left(\frac{r_\mathrm{h}}{D_{84}}\right)^{\frac{1}{6}} \tag{4b}$$

Here j_1, k and l are empirical constants and D_x is the xth percentile diameter of the local grain-size distribution. Unless otherwise stated, all grain sizes refer to the surface layer. By selecting $l = \frac{1}{6}$, this formula represents a classic grain-size dependent Gauckler–Manning–Strickler relation. For the other variables, the values $j_1 = 6.5, k = 1$ and $x = 84$ (Eq. 4b) have been found to reproduce observational data well for deeper flows with $\frac{r_\mathrm{h}}{D_{84}}$ larger than about 7–10 (Rickenmann and Recking, 2011). If another flow routing is used, one can select the variable power-equation flow-resistance approach provided by Ferguson (2007) with the parameter values proposed by Rickenmann and Recking (2011), which was recommended also for applications in steep channels, including shallow flows with small relative flow depths $\frac{r_\mathrm{h}}{D_{84}}$:

$$\sqrt{\frac{8}{f}} = \frac{j_1 \cdot j_2 \cdot \frac{r_\mathrm{h}}{D_{84}}}{\sqrt{j_1^2 + j_2^2 \cdot \left(\frac{r_\mathrm{h}}{D_{84}}\right)^{\frac{5}{3}}}},$$

where $j_1 = 6.5$ and $j_2 = 2.5$. (5)

In general, flow resistance describes the effects of drag forces exerted on the bed and its structures. A part of this drag, namely skin drag, is responsible for the transport of sediment grains (e.g. Morvan et al., 2008). The drag created by larger-scale surface geometries, such as bed forms and channel shape features (e.g. bends and irregular channel width), may be summarised as macro-roughness, which reduces the energy available for the transport of sediment. If macro-roughness is not accounted for in steep channels, bedload transport capacity may be greatly overestimated (Rickenmann, 2001, 2012; Yager et al., 2007; Badoux and Rickenmann, 2008; Chiari and Rickenmann, 2011; Nitsche et al., 2011, 2012; Yager et al., 2012). To correct for macro-roughness, Nitsche et al. (2011) suggested the use of a reduced energy slope, which represents a fraction of the real gradient, and which is based on a flow-resistance partitioning approach of Rickenmann and Recking (2011) and Nitsche et al. (2011).

$$S_\mathrm{red} = S \cdot \left(\frac{f_0}{f_\mathrm{tot}}\right)^{0.5 \cdot e} = S \cdot \left[\frac{2.5 \cdot \left(\frac{r_\mathrm{h}}{D_{84}}\right)^{\frac{5}{6}}}{\sqrt{6.5^2 + 2.5^2 \cdot \left(\frac{r_\mathrm{h}}{D_{84}}\right)^{\frac{5}{3}}}}\right]^e \tag{6}$$

Here S_red is the reduced slope that accounts for macro-roughness effects, S is channel or hydraulic energy slope, f_0 is base-level flow resistance according to Eq. (4b), f_tot is total flow resistance according to Eq. (5) and e is an exponent ranging from 1 to 2, with a typical value of $e = 1.5$. Within sedFlow, one can select to use S_red based on Eq. (6) to account for macro-roughness.

2.2 Bedload transport calculation

2.2.1 Bedload transport rate

Several methods for the calculation of bedload transport capacity are implemented in sedFlow: Sects. A2–A5 describe the method of Wilcock and Crowe (2003) based on flume data, the method of Recking (2010) based on field observations and the method of Rickenmann (2001) based on flume data together with a simplified version of the Rickenmann method and another version based on discharge instead of shear stress. The method of Rickenmann (2001) was tested together with Eq. (6), by comparison with bedload transport observations in steep mountain streams (Nitsche et al., 2011). The equations of Wilcock and Crowe (2003) were derived from fractional bedload transport data. The equation of Recking (2010) was developed for the estimation of total bedload transport rates.

In the same way as the equations of Meyer-Peter and Müller (1948), Fernandez Luque and van Beek (1976) and Soulsby and Damgaard (2005), the equation of Rickenmann (2001) (Sect. A4; especially its simplified version in Eq. A20) is a good example of the following generic type of bedload estimation methods:

$$\Phi_\mathrm{b} = a \cdot \theta^b \cdot (\theta - \theta_c)^d, \tag{7}$$

where $\Phi_\mathrm{b} = \frac{q_\mathrm{b}}{\sqrt{(s-1) g D^3}}$ is dimensionless bedload flux, $\theta = \frac{\tau}{(s-1)\rho g D}$ is dimensionless bed shear stress, θ_c is the dimensionless bed shear stress threshold for the initiation of motion, q_b is bedload flux per unit flow width, D is grain diameter, a, b and d are empirical constants, $\tau = (\rho \cdot g \cdot r_h \cdot S)$ is bed shear stress and $s = \frac{\rho_\mathrm{s}}{\rho}$ is the density ratio of solids ρ_s and fluids ρ. To account for macro-roughness, S can be replaced by S_red in the calculation of the bed shear stress τ. In the case of the Rickenmann (2001) equation, a is the product of an empirical constant and the Froude number Fr. In equations like Eq. (7), bedload transport is mainly a power law of the dimensionless bed shear stress that exceeds some threshold for the initiation of bedload motion. This threshold is known as the Shields criterion (Shields, 1936) with typical values ranging from 0.03 to 0.05. In natural channels, geometric complexity and thus energy losses increase at steep bed slopes S_b. Therefore, Lamb et al. (2008) suggested the following empirical relation to account for increasing θ_c values with increasing S_b:

$$\theta_c = 0.15 \cdot S_\mathrm{b}^{0.25}. \tag{8}$$

The main objective of Lamb et al. (2008) is a theoretical explanation for the observed increase of θ_c with increasing bed slopes S_b. However, for very small values of S_b, the power law of Eq. (8) predicts values of θ_c that are approaching zero. This contrasts with the results of other studies (e.g. Recking, 2009) that observed roughly constant values of θ_c larger than zero for small slopes. sedFlow can be configured to use either a constant threshold or a slope-dependent threshold according to Eq. (8) combined with a minimum value θ_{cMin}.

The estimated bedload flux can be corrected for gravel abrasion according to the classic equation of Sternberg (1875), in which $q_{b_{abr}}$ is bedload flux per unit flow width corrected for abrasion, λ is an empirical abrasion coefficient and ΔX is the travel distance of the grains. Here the material loss due to abrasion is regarded as suspension throughput load:

$$q_{b_{abr}} = q_b \cdot \exp(-\lambda \cdot \Delta X). \tag{9}$$

If grain-size fractions are treated individually, the calculated bedload capacity Φ_b needs to be normalised with F_i, the relative volumetric portion of bed surface material of a grain-size fraction i, compared to the total surface material with $D > 2\,\text{mm}$ (e.g. Parker, 1990). Here F_i can be interpreted as the availability of a certain grain-size fraction in the bed; for an example, see Eq. (A23) in Sect. A6 compared to Eq. (A20) in Sect. A4. Further details also have to be accounted for, such as the varying exposure of different grain-size fractions, grain-size-dependent grain–grain interactions, etc. This is commonly done using some sort of hiding function. Hiding functions not only focus mainly on grain exposure but also integrate all kinds of grain-size-dependent effects which are not covered by the capacity estimation methods. Within sedFlow a relatively simple power-law hiding function can be used (Parker, 2008):

$$\theta_{ci} = \theta_c \cdot \left(\frac{D_i}{D_x}\right)^m, \tag{10}$$

as well as the hiding function of Wilcock and Crowe (2003):

$$\theta_{ci} = \theta_c \cdot \left(\frac{D_i}{D_m}\right)^{m_{wc}},$$

$$\text{where} \quad m_{wc} = \frac{0.67}{1+\exp\left(1.5 - \frac{D_i}{D_m}\right)} - 1. \tag{11}$$

Here θ_{ci} is the θ_c for the ith grain-size fraction, D_i is the mean grain diameter for ith grain-size fraction, m is an empirical hiding exponent, D_m is the geometric mean diameter of the local grain-size distribution and m_{wc} is the hiding exponent according to Wilcock and Crowe (2003). The empirical exponent m ranges from 0 to -1, where $m = -1$ corresponds to the so-called "equal mobility" case in which all grains start moving at the same bed shear stress τ, and $m = 0$ corresponds to no influence by hiding at all. For $x = 50$, the values for m, which have been derived from various field observations, typically vary within a range from -0.60 to -1.00 (Recking, 2009), and unfortunately there are only few data points for $D_i > D_{50}$ (Bathurst, 2013; Bunte et al., 2013).

For consistency, the following $\theta_{ci,r}$ is used in bedload transport calculations.

$$\theta_{ci,r} = \theta_{ci} \cdot \gamma \tag{12}$$

Within sedFlow two alternatives are implemented for the calculation of the correction factor γ:

$$\gamma = \frac{S_{red}}{S} \tag{13a}$$

and

$$\gamma = \frac{S_c}{S}. \tag{13b}$$

In Eq. (13a), $\theta_{ci,r}$ varies with discharge, as it depends on S_{red}, which in turn is a function of the hydraulic radius r_h. In Eq. (13b), suggested by Nitsche et al. (2011), $\theta_{ci,r}$ is independent of discharge. The value of S_c is calculated using Eq. (6), with the value of r_h replaced by the critical hydraulic radius $r_{h,c}$:

$$r_{h,c} = \theta_c \cdot \left(\frac{\rho_s}{\rho} - 1\right) \cdot D_{50} \cdot \frac{1}{S}. \tag{14}$$

Good arguments can be found for both approaches (Eqs. 13a and 13b). Due to the lack of suitable data, it is unclear which approach is more plausible.

2.2.2 Evolution of channel bed elevation and slope

The temporal evolution of the longitudinal profile is simulated in sedFlow based on a finite-difference version of the general Exner equation (e.g. Parker, 2008).

$$(1 - \eta_{pore}) \cdot \frac{\partial z}{\partial t} = q_{b_{lat}} - \frac{\partial q_b}{\partial x} \tag{15}$$

Here η_{pore} is pore volume fraction, z is elevation of the channel bed and $q_{b_{lat}}$ is lateral bedload influx per unit flow width. Eq. (15) allows for the calculation of the new channel slope $\frac{\Delta z}{\Delta x}$ after each time step. Heretofore, infinitely deep rectangles are used within sedFlow as the shape of the cross-sectional profiles, with the complete width defined as active width (i.e. sediment transport takes place over the complete width).

All three elements (the cross-sectional channel geometry, its alteration due to morphodynamics and the determination of the active width) are implemented as abstract classes. Thus, the presented realisations just represent the current state of the code and any programmer can easily extend the code to deal with more complex cross-sectional geometries. However, the implicit flow routing by Liu and Todini (2002), with its advantages in terms of simulation efficiency, requires

Table 2. Relative root mean square deviations (rRMSDs) determined according to Eq. (16) for the simulations and observations displayed in Fig. 3: obs denotes the reference data derived from observations, orig denotes the simulation results using the detailed channel geometry, and rect(Q) and rect(w) denote simulation results using a rectangular substitute channel based on averaged representative discharges (Q) and widths (w), respectively. The numerical values in the table are the rRMSDs between the two data sets determined by the row and column. The labels for the rows and columns (bold type face) are given in the diagonal of each panel. For example, at the Chirel the simulations using a detailed channel geometry and using a substitute rectangle based on averaged widths differ from each other by a rRMSD value of 0.28.

obs 0.65	0.60	0.71
orig	0.13	0.18
	rect(Q)	0.22
(a) Chiene		**rect(w)**
obs 0.90	0.94	0.89
orig	0.24	0.28
	rect(Q)	0.12
(b) Chirel		**rect(w)**
obs 0.69	0.50	0.46
orig	0.30	0.40
	rect(Q)	0.20
(c) Schwarze Lütschine		**rect(w)**

infinitely deep rectangular or V-shaped channels (together with a simple power-equation flow-resistance law such as Eq. 4a).

Additionally, Stephan (2012) has studied the impact of the rectangular-shaped approximation and found that, at least during major flood events, it is negligible compared to the other uncertainties. Stephan (2012) recalculated bedload transport for the August 2005 transport event in the catchments of the Chiene, Chirel and Schwarze Lütschine. For details on the catchments and event characteristics see Chiari and Rickenmann (2011). For the simulations, Stephan (2012) used the one-dimensional model Tom^{Sed} (Chiari et al., 2010), which allows for the definition of cross sections with laterally varying bed elevations. The simulations were repeated with a detailed channel geometry as presented in Chiari and Rickenmann (2011) and with two different rectangular substitute channels. The widths of the rectangular substitute channels w were determined based on a discharge Q_{rep} that is representative of the simulation period. The channel width was selected to produce the same wetted cross-sectional area and hydraulic radius for the representative discharge as was simulated for the detailed channel geometry. In one approach the threshold discharge for the initiation of bedload motion Q_c, as well as the maximum discharge of the simulation period Q_{max}, were averaged to find Q_{rep}, for which the representative channel width was determined. In the second approach, one width w was determined for both Q_c and Q_{max} and then the two widths were averaged to find the representative channel width. The detailed channel geometry produced results that were broadly similar to the rectangular substitutes when compared with the field observations on bedload transport (Fig. 3). To quantify the deviations between the different simulations and between the simulations and the observations, Table 2 summarises the relative root mean square deviations (rRMSD) which have been calculated according to the following equation:

$$\mathrm{rRMSD} = \frac{\sqrt{\frac{\sum_{i=1}^{n}(\chi_i-\psi_i)^2}{n}}}{\frac{\overline{\chi}+\overline{\psi}}{2}}. \tag{16}$$

Here χ and ψ are two arbitrary data sets of length n, with $\overline{\chi}$ and $\overline{\psi}$ being their average values, and i is a running index. As can be seen from Table 2, the simulations on average differ from each other by a rRMSD value of only 0.23, while the simulations differ from the observations on average by a rRMSD value of 0.70. However, it has to be noted that these results are for a flood event with high discharge values. For low discharges close to the initiation of bedload motion, a detailed representation of channel geometry may play a more important role than the one suggested by Fig. 3 and Table 2. For further details see Stephan (2012).

Finally, the introduction of more complex cross-sectional shapes raises the question of how these shapes are influenced and altered through morphodynamics. As far as we know, no generally accepted concepts are available for this problem.

2.3 Grain-size distribution changes

In sedFlow the alluvial substrate of the river is represented by a stack of horizontal layers with homogeneous grain-size characteristics. The topmost layer of the bed interacts with the flow and is typically called the active surface layer. The grain-size distribution of the active surface layer is used for the determination of the flow resistance, hiding processes and bedload transport capacity (Fig. 1). All deposited material is added to this layer; all eroded material is taken from it. The thickness of this layer determines the inertia of its evolving grain-size distribution. Therefore, the active surface layer and especially its thickness play an important role in the numeric representation of bedload transport systems (e.g. Belleudy and Sogreah, 2000). When the alluvium thickness is smaller than the expected usual active layer thickness, sedFlow makes use of the shape properties of bedrock to determine flow resistance and hiding. The thickness of the active surface layer may be set constant or dynamic as a multiple of some grain-size percentile. Three different approaches are available within sedFlow for the interaction between the active surface layer and the underlying subsurface alluvium.

The first method (Fig. A4) has been adapted from the one described by van de Wiel et al. (2007). Lower and upper thresholds are defined for the thickness of the active surface layer. Whenever these thresholds are exceeded, sediment increments are incorporated from, or released to, the subsurface alluvium underneath until the active surface layer thickness again takes a value within the given thresholds. The sediment increments are stored as bed strata underneath the active surface layer. In this way the simulated river bed is able to remember its history. In contrast to the procedure of van de Wiel et al. (2007) the thickness of the sediment increments can be defined independently of the thickness of the active surface layer. Trivially, the thickness of the increments defines the minimum distance between the thresholds for the active surface layer thickness.[1] The smaller this distance between the thresholds is, the more intense the interaction is between the active surface layer and the underlying subsurface alluvium.

The second approach (Fig. A5), which has been applied in various models (e.g. Hunziker, 1995), can be described as an extreme case of the first one, in which the two thresholds collapse to a single target thickness for the active surface layer. In this case, any addition or removal of material to or from the active surface layer is instantaneously balanced against the underlying subsurface alluvium. In this case of maximum interaction between active surface layer and subsurface alluvium, the subsurface alluvium is represented by one homogenised volume without any internal structure. The target thickness is usually determined at the start of a simulation and then kept constant. Alternatively, it can be dynamically adjusted based on a Eulerian forward approach, in which the thickness is updated at the end of each time step.

The third approach (Fig. A6) is a variation of the second one. When sediment is eroded, only the volume of the active surface layer is instantaneously replaced from the subsurface, while the grain-size distribution stays the same. The sediment volume that is transported from the subsurface alluvium to the active surface layer shares the grain-size distribution of the subsurface alluvium only if a condition for the break-up of an armouring layer is fulfilled:

$$\theta_{50} \geq \theta_{c,s}. \tag{17}$$

Here θ_{50} is a representative dimensionless shear stress θ for the median diameter D_{50} of the active surface layer and $\theta_{c,s}$ is a representative θ_c for the active surface layer. To avoid artefacts due to a hard threshold, some fraction i_{sub} of the sediment transported from the subsurface alluvium to the active surface layer has the grain-size distribution of the subsurface alluvium already before the break-up condition is fulfilled. The rest $(1 - i_{\text{sub}})$ has the grain-size distribution of the active surface layer:

$$i_{\text{sub}} = \frac{\theta_{50} - \theta_{c,\text{sub}}}{\theta_{c,s} - \theta_{c,\text{sub}}} \text{ with } \quad 0 \leq i_{\text{sub}} \leq 1$$
$$\text{and} \quad 0 < \theta_{c,\text{sub}} < \theta_{c,s}. \tag{18}$$

Here i_{sub} is the relative grain-size influence from the subsurface alluvium and $\theta_{c,\text{sub}}$ is a representative θ_c for the subsurface alluvium. The value of $\theta_{c,\text{sub}}$ can be estimated, e.g. according to Eq. (8), while the value of $\theta_{c,s}$ can be estimated using the following relation of Jäggi (1992):

$$\theta_{c,s} = \theta_{c,\text{sub}} \cdot \left(\frac{D_{\text{mArith}_s}}{D_{\text{mArith}_{\text{sub}}}} \right)^{\frac{2}{3}}. \tag{19}$$

Here D_{mArith_s} and $D_{\text{mArith}_{\text{sub}}}$ are the arithmetic mean diameters of the grain-size distribution of the active surface layer (s) and of the subsurface alluvium (sub).

For non-fractional studies, the active surface layer concept can be turned off. In that case, the complete alluvium is represented by a single homogeneous layer, which directly interacts with the flow.

3 Discussion

3.1 Comparison of sedFlow implementations with similar models

In the following subsections, various details of the implementation of sedFlow are discussed and compared to implementations in similar models. Differences between the individual models are explained in the context of their differing objectives.

[1] Here h_s is the active surface layer thickness, within a time step, after erosion or deposition but before the interaction between the active surface layer and the subsurface alluvium; accordingly, $h_{s,\text{post}}$ is the active surface layer thickness after the layer interaction, Δh_s is the thickness of sediment increments, $|y|$ is the number of sediment increments that are incorporated from or released to the subsurface alluvium, $\text{thresh}_{hs,\text{low}}$ and $\text{thresh}_{hs,\text{high}}$ are the thresholds for the thickness of the active surface layer with $\text{thresh}_{hs,\text{low}} < \text{thresh}_{hs,\text{high}}$ and $\text{thresh}_{hs,\text{prox}}$ is an alias for the threshold that is closer to h_s. It is the objective of the layer interaction to reach a state in which $h_{s,\text{post}}$ is as close as possible to the middle between the thresholds and

$$\text{thresh}_{hs,\text{low}} \leq h_{s,\text{post}} \leq \text{thresh}_{hs,\text{high}}$$
$$\text{with } h_{s,\text{post}} = h_s + (y \cdot \Delta h_s) \text{ and } y \in \mathbb{Z}. \tag{F1}$$

However, the state of Eq. (F1) cannot be reached if

$$\operatorname{mod} \frac{|h_s - \text{thresh}_{hs,\text{prox}}|}{\Delta h_s} < \Delta h_s$$
$$- (\text{thresh}_{hs,\text{high}} - \text{thresh}_{hs,\text{low}}). \tag{F2}$$

This might cause instability. In simple terms, if Δh_s was larger than the distance between the thresholds, situations might occur in which the addition or subtraction of another Δh_s will cause $h_{s,\text{post}}$ to jump over both thresholds, such that it can never reach a value between the thresholds. Therefore, Δh_s is defined as the minimum distance between the thresholds. With this minimum distance, the right-hand side of Eq. (F2) cannot become larger than zero, while the left-hand side of Eq. (F2) cannot become smaller than zero anyway.

3.1.1 Fractional transport and grain-size distributions

In streams, local grain-size distributions and bedload transport influence each other in various ways. For example, flow resistance is often determined as a function of a characteristic bed surface grain diameter such as in Eqs. (4b) and (5) that use D_{84}. Therefore, local grain-size distributions have a high influence on the local hydraulic radius r_h, which in turn determines bed shear stress and thus bedload transport capacity. In addition, the shear stress partitioning in Eq. (6) is based on D_{84} as well. Finally, the stream can only mobilise and transport the grains which are available in the bed and, as transport capacity is inversely related to the transported grain diameter, fine-grained bed material will yield high transport rates. These mechanisms let the bed surface grain-size distribution influence the transport capacity, which in turn alters the local grain-size distributions. At high transport capacity, the stream will erode and deplete particles with transportable diameters and will leave only coarse grains in the bed. At low transport capacity, the stream cannot transport and will, thus, even deposit grains with small diameters. This will accumulate fine-grained material in the bed. Therefore, the spatial pattern of bed surface grain-size distributions at the end of a simulation can be interpreted as a function of bed slope (coarse grains in steep sections), channel width (coarse grains in narrow sections) and channel network effects (coarse grains at confluences with steep tributaries).

In order to study dynamically evolving grain-size distributions and their effects on hydraulics and bedload transport, together with the effects of an evolving channel slope, sedFlow is optimised for the simulation of fractional bedload transport. In this context, different concepts are provided for the interaction between the active surface layer and the underlying alluvium. As grain-size distributions dynamically adjust to be consistent with the local circumstances (channel width, slope, etc.), this numeric concept might partially compensate the uncertainty in local grain-size distribution data. For a more detailed discussion of this topic see Heimann et al. (2015). To the authors' knowledge there are no in-depth studies assessing the influence of different representations of active surface layer dynamics on simulated bedload volumes. As sedFlow contains three different formulations of active surface layer dynamics in the same modelling tool, it provides the base for a future study on the effects of different active surface layer algorithms.

Within Topkapi ETH (Konz et al., 2011) fractional transport is not implemented, and within Tom^{Sed} (Chiari et al., 2010) it is rarely used due to long calculation times. These models have a different application objective, for which dynamic grain-size distributions are of lesser relevance.

The threshold-based layer interaction (Fig. A4) implemented in sedFlow is similar to the approach of Lopez and Falcon (1999) for the evolution of the local grain-size distribution. However, Lopez and Falcon (1999) only introduced a lower threshold for the thickness of the active surface layer. This means that the subsurface alluvium remains constant even in cases of massive aggradation and that the active surface layer may reach unrealistically high thickness values, especially in cases of intense aggradation.

In SEDROUT (Hoey and Ferguson, 1994), the surface layer grain-size distribution is determined as a function of the spatial derivative of fractional transport rates and the thickness of the surface layer, which in turn is a function of the surface layer grain-size distribution. This set of equations is solved by numeric iteration within each time step. In sedFlow, this numerically extensive procedure is replaced by a constant surface layer thickness or an Eulerian forward approach, in which the layer thickness is updated at the end of each time step. These more pragmatic approaches have been selected because fast simulations are one of the main aims of sedFlow.

3.1.2 Adverse channel slopes

Within sedFlow adverse channel slopes (uphill slopes in the downstream direction) and their effects in terms of ponding can be considered using uniform discharge hydraulics. This approach is based on the assumption that the simulated system only consists of two extreme cases: ponding on the one hand, and parallel slopes on the other. This assumption is valid in many mountain streams and it allows for fast simulations, which have been another main objective for the development of sedFlow. However, in the intermediate case of moderate backwater effects, it may produce large errors. The implemented approach corresponds to the situation of a confined channel, in which the sudden deposition of large volumes of sediment, e.g. by debris flow inputs, may produce ponding.

Within Topkapi ETH any large volumes of deposited material, which would produce adverse channel slopes, are instantaneously distributed to downstream river reaches until all slopes are positive. This algorithm would correspond to instantaneous landslides within the channel or to debris flows with short travel distances, but such phenomena are typically not observed in mountain streams. However, instantaneous lateral sediment input is not the main focus of the Topkapi ETH model. In the context of its intended applications, the described algorithm of Topkapi ETH is an appropriate pragmatic approach to conserve mass and ensure positive bed slopes, which are used as energy slopes in the implemented kinematic wave approach.

Within Tom^{Sed} any deposited material that would produce adverse channel slopes is fed to a virtual sediment storage, which does not contribute to elevation changes in the main channel. As long as there is sediment in this virtual storage, any erosion or deposition is applied to this storage keeping the main channel untouched. That means that the elevation of the main channel is frozen as long as there is material in the virtual storage. In some way this algorithm corresponds to a lateral displacement of the river channel due to large volumes of deposited material that are stored next to the new channel. However, in mountain rivers the amount of material that is fed from the deposits to the main channel depends on the stability of the deposit slopes. Therefore, the new channel may well lower its elevation due to erosion, even if there is still some material in storage next to the channel. However, the situation of lateral channel displacement is close to the limits of a one-dimensional simulation and the described algorithm of Tom^{Sed} ensures positive bed slopes (used as proxy for the energy slopes) in a way which corresponds to some extent to a natural process, even though it violates conservation of mass within the main channel.

3.1.3 Simulation speed

sedFlow has been designed for optimal computation speeds. Besides the selection of the coding language C++, for which there are powerful compilers available, we have implemented a spatially uniform discharge (within each segment of the channel network), as well as an implicit kinematic wave flow routing approach, both aimed at providing a modelling tool for fast simulations. Both hydraulic approaches allow for coarse temporal discretisations and the implemented algorithm of Liu and Todini (2002) omits computationally demanding iterations. The ideal temporal discretisation (as fine as necessary and as coarse as possible) can be obtained from the Courant–Friedrichs–Lewy (CFL) criterion based on the speed of bedload multiplied by a user-defined safety factor. As a result, several years of bedload transport and resulting slope and grain-size distribution adjustment can be simulated with sedFlow within only few hours of calculation time on a regular 2.8 GHz central processing unit (CPU) core.

In Topkapi ETH the implicit kinematic wave flow routing is also implemented using the algorithms of Liu and Todini (2002). However, the length of the time steps used for the bedload transport simulations may differ from the ideal length, as it is defined as an integer multiple of the time steps used for the hydrologic simulations. This implementation is due to Topkapi ETH's aim to simulate different processes at different scales.

Within Tom^{Sed}, explicit kinematic wave flow routing is implemented. Thus, time steps need to be determined using the CFL criterion based on water flow velocity. Therefore, time steps are shorter than in sedFlow or Topkapi ETH and slow down the simulations considerably. The choice of the explicit water flow routing is due to Tom^{Sed}'s aim to simulate the effects of the shape of channel cross sections. The implicit flow routing based on the algorithms of Liu and Todini (2002) requires simple rectangular or V-shaped channels and would therefore prevent any detailed study of channel geometry.

3.1.4 Flexibility

To ensure flexibility of application, we selected regular spreadsheets as the file format for the data input to sedFlow. Thus, preprocessing can be done with common software applications, which are familiar to most users, so that data from any study catchment can be quickly and easily prepared for a sedFlow simulation. This contrasts with Tom^{Sed}, which uses the extensible markup language (XML) file format for data input, as well as with Topkapi ETH, which partially requires MATLAB preprocessing. In sedFlow different equation sets can be selected and combined by the user for the main process representations (flow routing, flow resistance, initiation of bedload motion, transport capacity, etc.). The number, content and format of the output files can be defined by the user as well, in order to get the best solution for the respective study objectives.

To ensure flexibility of model development we selected an object-oriented design for the internal structure of the sedFlow code. In such a code, succeeding programmers just create new realisations for predefined code interfaces without revising the model core. It is not necessary to know (virtually) anything about the model itself. The only piece of code that the programmer will have to read is the specification of the relevant code interface, which is typically not longer than two printout pages.

3.2 Advantages and limits of the sedFlow modelling approach

The limits of the sedFlow modelling approach are defined by the implemented process representations. Hydraulics are considered in a cross-sectionally averaged way. The effects of turbulence and eddies, such as the formation of dunes and pool-riffle sequences, are not considered explicitly. Therefore, sedFlow should not be applied at spatial discretisations small enough for these processes to become relevant. Within sedFlow, discharge is assumed to be generally subcritical. It should not be applied in systems where Froude numbers larger than one occur at the scale of spatial discretisation.

As sedFlow generally assumes kinematic wave propagation, bed slopes S_b should be steep enough for friction slopes S_f to approximately equal bed slopes ($S_f \approx S_b$). Slight violations of this assumption will not cause major errors. However, if one uses the flow routing approach that allows for adverse channel slopes (uphill slopes in the downstream direction), any occurrence of moderate backwater effects ($S_f \not\approx 0 \wedge S_f \not\approx S_b$) will produce large errors (cf. Sect. 2.1.1).

sedFlow considers sediment transport only in terms of bedload. Therefore, it should not be applied if suspension load and wash load play a major role in the studied system.

However, if the limits of applicability are respected, sedFlow provides several benefits for the user interested in the simulation of bedload transport in steep mountain streams. The model is optimised for the calculation of fractional bedload transport resulting in more robust simulations as compared to single-grain approaches. Based on an assumption that is valid for many mountain streams, sedFlow provides for the efficient simulation of adverse slopes and ponding. The selected coding language and two implemented hydraulics representations are all aimed at providing a modelling tool for fast simulations. Finally, with the user-defined structure of the model outputs, with the straightforward preprocessing of the inputs, and with the provision of different selectable options, e.g., for the flow resistance and bedload transport equations, sedFlow offers the flexibility to be easily adjusted to fit a specific application in a particular catchment.

4 Conclusions

The new model sedFlow complements the range of existing tools for the simulation of bedload transport in steep mountain streams. It is an appropriate tool if (i) grain-size distributions need to be dynamically adjusted in the course of a simulation, (ii) the effects of ponding, e.g., due to debris flow inputs might play a role in the study catchment or (iii) one simply needs a fast simulation of bedload transport in a mountain stream with quick and easy pre- and postprocessing. Examples of the application of sedFlow are presented in a companion article by Heimann et al. (2015).

The current version of the sedFlow code and model can be downloaded under the terms of the GNU General Public License (GPL) at the following web page: www.wsl.ch/sedFlow.

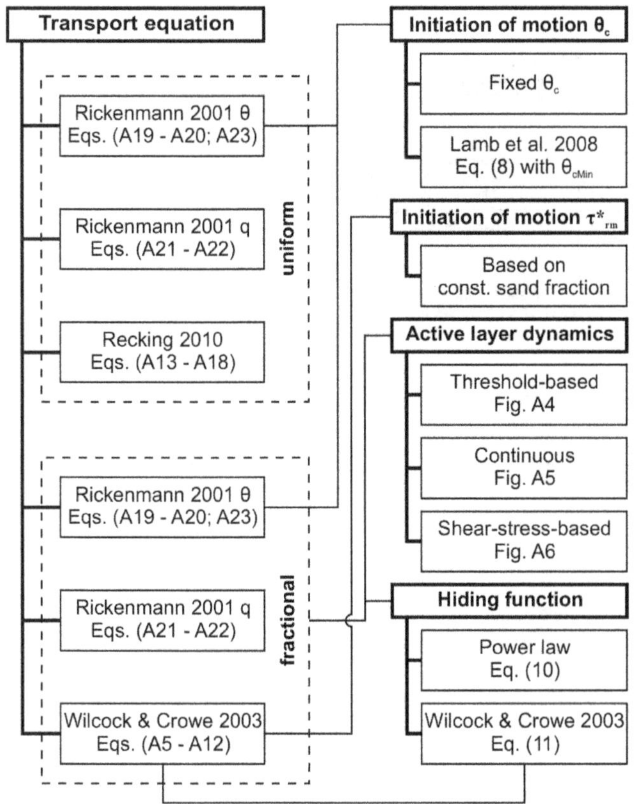

Figure A1. Simplified summary diagram of the bedload transport part of sedFlow, showing a selection of the most important general types of functions (bold type face) and their selectable realisations (regular type face). For the transport equations of Rickenmann (2001), versions for both fractional and uniform grain sizes are implemented in sedFlow. Among the implemented equations, only the θ-based ones of Rickenmann (2001) use θ_c to determine the initiation of bedload motion. The selection of a hiding function and active surface layer dynamics is only necessary if a fractional transport equation is used. The transport equation of Wilcock and Crowe (2003) is always combined with their proposed hiding function and uses τ^*_{rm} to determine the initiation of bedload motion.

Appendix A: Supplementary methods

A1 Explicit hydraulics

$$\frac{\partial V}{\partial t} = Q^{\mathrm{u}}_{T-1} - Q_{T-1} \tag{A1}$$

$$V_T = V_{T-1} + \left(\frac{\partial V}{\partial t} \cdot \Delta t\right) \tag{A2}$$

$$r_{\mathrm{h}T} = \mathrm{geom}(V_T) \tag{A3}$$

$$Q_T = \mathrm{fr}(r_{\mathrm{h}T}) \tag{A4}$$

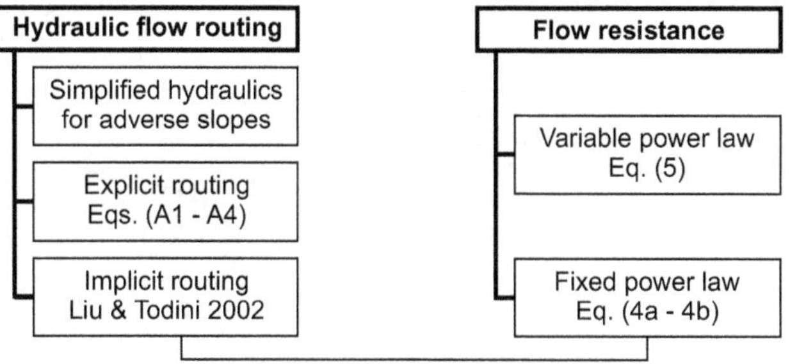

Figure A2. Simplified summary diagram of the hydraulic part of sedFlow, showing a selection of the most important general types of functions (bold type face) and their selectable realisations (regular type face). The implicit flow routing of Liu and Todini (2002) is always combined with a fixed power law flow resistance.

A2 Bedload capacity estimation according to Wilcock and Crowe (2003)

$$v^* = \sqrt{\frac{\tau}{\rho}} \tag{A5}$$

$$m_{\mathrm{wc}} = \frac{0.67}{1 + \exp\left(1.5 - \frac{D_i}{D_{\mathrm{m}}}\right)} - 1 \tag{A6}$$

$$\frac{\tau_{\mathrm{r}}}{\tau_{\mathrm{rm}}} = \left(\frac{D_i}{D_{\mathrm{m}}}\right)^{m_{\mathrm{wc}}+1} \tag{A7}$$

$$\tau^*_{\mathrm{rm}} = 0.021 + \left[0.015 \cdot \exp\left(-20F_{\mathrm{s}}\right)\right] \tag{A8}$$

$$\tau_{\mathrm{rm}} = \tau^*_{\mathrm{rm}} \cdot \rho \cdot g \cdot D_{\mathrm{m}} \cdot \left(\frac{\rho_{\mathrm{s}}}{\rho} - 1\right) \tag{A9}$$

$$W^* = 0.002 \cdot \left(\frac{\tau}{\tau_{\mathrm{r}}}\right)^{7.5} \quad \text{for} \quad \frac{\tau}{\tau_{\mathrm{r}}} < 1.35 \tag{A10}$$

$$W^* = 14 \cdot \left[1 - \frac{0.894}{\sqrt{\frac{\tau}{\tau_{\mathrm{r}}}}}\right]^{4.5} \quad \text{for} \quad \frac{\tau}{\tau_{\mathrm{r}}} \geq 1.35 \tag{A11}$$

$$q_{\mathrm{b}} = F_i \cdot \frac{W^* \cdot v^{*3}}{\left(\frac{\rho_{\mathrm{s}}}{\rho} - 1\right) \cdot g} \tag{A12}$$

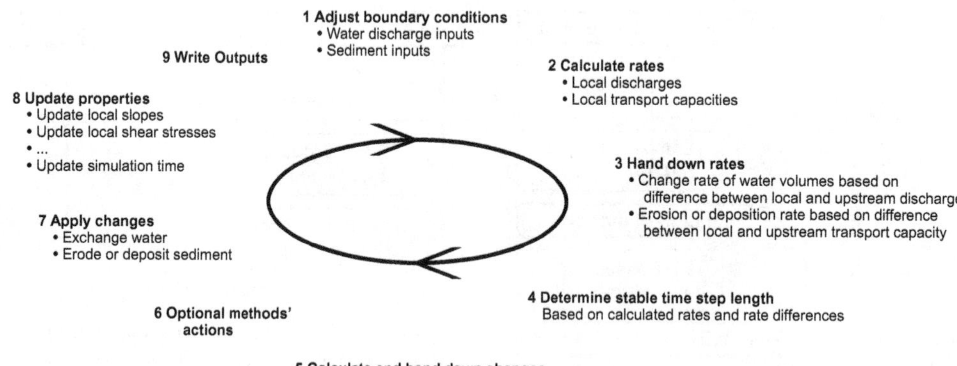

Figure A3. Simplified summary diagram of a time step in sedFlow. The timing of the optional methods' actions has been selected in a way that all changes, such as, e.g., erosion or deposition volumes, are readily calculated and available for possible modifications before the changes, such as, e.g., erosion or deposition, are then actually applied.

A3 Bedload capacity estimation according to Recking (2010)

$$\theta_{c84} = (1.32 \cdot S + 0.037) \cdot \left(\frac{D_{84}}{D_{50}}\right)^{-0.93} \quad \text{(A13)}$$

$$L = 12.53 \cdot \left(\frac{D_{84}}{D_{50}}\right)^{4.445\sqrt{S}} \cdot \theta_{c84}^{1.605} \quad \text{(A14)}$$

$$\theta_{84} = \frac{\tau}{(\rho_s - \rho) \cdot g \cdot D_{84}} \quad \text{(A15)}$$

$$\Phi_b = 0.0005 \cdot \left(\frac{D_{84}}{D_{50}}\right)^{-18\sqrt{S}} \cdot \left(\frac{\theta_{84}}{\theta_{c84}}\right)^{6.5} \quad \text{for} \quad \theta_{84} < L; \quad \text{(A16)}$$

$$\Phi_b = 14 \cdot \theta_{84}^{2.45} \quad \text{for} \quad \theta_{84} \geq L \quad \text{(A17)}$$

$$q_b = \Phi_b \cdot \sqrt{\left(\frac{\rho_s}{\rho} - 1\right) \cdot g \cdot D_{84}^3} \quad \text{(A18)}$$

A4 Bedload capacity estimation according to Rickenmann (2001) based on θ

$$\Phi_b = 3.1 \cdot \left(\frac{D_{90}}{D_{30}}\right)^{0.2} \cdot \sqrt{\theta} \cdot (\theta - \theta_c) \cdot Fr \cdot \frac{1}{\sqrt{\frac{\rho_s}{\rho} - 1}} \quad \text{(A19)}$$

In this equation, Φ_b and θ are based on D_{50}. Equation (A19) may be simplified using the mean experimental value of $\left(\frac{D_{90}}{D_{30}}\right)^{0.2} = 1.05$ and a common value of $\frac{\rho_s}{\rho} = 2.65$:

$$\Phi_b = 2.5 \cdot \sqrt{\theta} \cdot (\theta - \theta_c) \cdot Fr \quad \text{(A20)}$$

A5 Bedload capacity estimation according to Rickenmann (2001) based on q

$$q_b = 3.1 \cdot \left(\frac{\rho_s}{\rho} - 1\right)^{-1.5} \cdot \left(\frac{D_{90}}{D_{30}}\right)^{0.2} \cdot (q - q_c) \cdot S^{1.5} \quad \text{(A21)}$$

$$q_c = 0.065 \cdot \left(\frac{\rho_s}{\rho} - 1\right)^{1.67} \cdot \sqrt{g} \cdot D_{50}^{1.5} \cdot S^{-1.12} \quad \text{(A22)}$$

A6 Fractional bedload capacity estimation according to Rickenmann (2001) based on θ

$$\Phi_{bi} = 2.5 \cdot \sqrt{\theta_{i,r}} \cdot (\theta_{i,r} - \theta_{ci,r}) \cdot Fr \quad \text{(A23)}$$

with $\Phi_{bi} = \frac{q_{bi}}{F_i\sqrt{(\frac{\rho_s}{\rho} - 1) g D_i^3}}$ and $q_b = \Sigma q_{bi}$.

In the simulations of the companion manuscript of Heimann et al. (2015), a fractional version of Eq. (A19) was used instead of Eq. (A23).

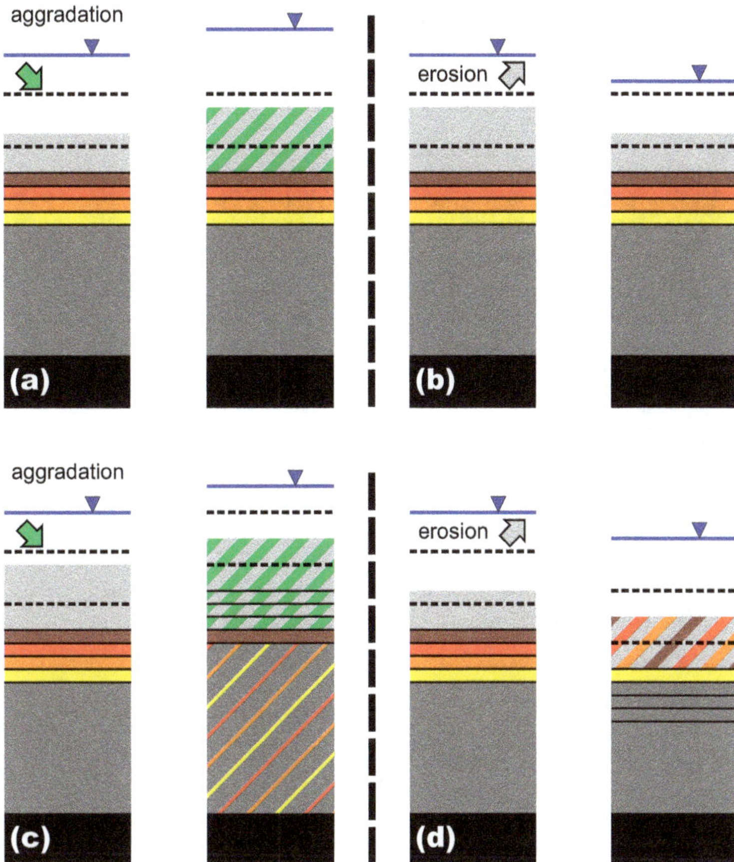

Figure A4. Qualitative sketch of threshold-based interaction between the active surface layer and subsurface alluvium. The water level is displayed in blue, the active surface layer with its variable thickness is light grey and bedrock is black. The subsurface alluvium consists of several strata layers with user-defined constant thickness displayed in reddish colours and one base layer with variable thickness displayed in dark grey. The horizontal dashed lines indicate the thresholds for the thickness of the active surface layer. In case of aggradation (**a** and **c**), the material input from upstream, which is displayed in green, is added to the active surface layer, which is instantaneously homogenised. Any homogenisation process is displayed by diagonal stripes. If the thickness of the active surface layer does not exceed its thresholds after aggradation or erosion (**a** and **b**), the thresholds of the active surface layer and the layers of the subsurface alluvium remain constant. If the active surface layer thickness exceeds its upper threshold after aggradation (**c**), the complete system of layers and thresholds is shifted upwards so that the upper boundary of the active surface layer is in the middle between its thresholds. The upper strata layers are populated with material from the homogenised active surface layer, while the material of the lower strata layers is added to the base layer, which is instantaneously homogenised. If the active surface layer thickness exceeds its lower threshold after erosion (**d**), the complete system of layers and thresholds is shifted downwards so that the upper boundary of the active surface layer is in the middle between its thresholds. The lower strata layers are populated with material from the base layer, while the material of the upper strata layers is added to the active surface layer, which is instantaneously homogenised.

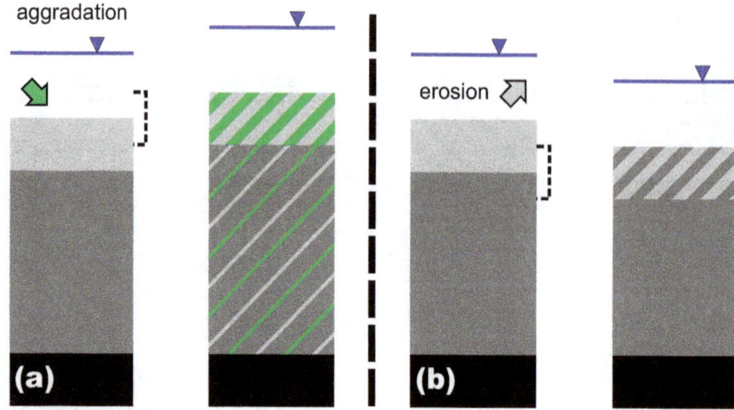

Figure A5. Qualitative sketch of continuous interaction between the active surface layer and subsurface alluvium. The water level is displayed in blue, the active surface layer with its constant thickness is light grey, the subsurface alluvium consisting of one layer with variable thickness is dark grey and bedrock is black. The dashed bracket indicates the position of the active surface layer after aggradation or erosion. In case of aggradation **(a)**, the material input from upstream, which is displayed in green, is added to the active surface layer, which is instantaneously homogenised. Any material of the homogenised active surface layer, which exceeds the constant thickness, is transferred to the subsurface alluvium, which is instantaneously homogenised as well. Any homogenisation process is displayed by diagonal stripes. In case of erosion **(b)**, the sediment deficit of the active surface layer is replaced by material from the subsurface alluvium and the active surface layer is instantaneously homogenised.

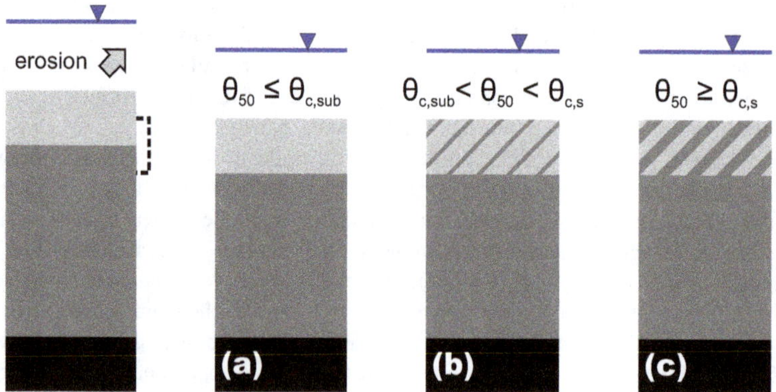

Figure A6. Qualitative sketch of shear-stress-based interaction between the active surface layer and subsurface alluvium. For an explanation of the symbols see the caption of Fig. A5. In this approach, the aggradation case is treated identically to the continuous update approach displayed in Fig. A5a. For erosion, three cases are differentiated, with the representative dimensionless shear stress θ_{50} increasing from case **(a)** to case **(c)**. If θ_{50} equals or exceeds the threshold for the active surface layer $\theta_{c,s}$ **(c)**, the layers interact in the same way as in the continuous update approach displayed in Fig. A5b. If θ_{50} does not exceed the threshold for the subsurface alluvium $\theta_{c,\mathrm{sub}}$ **(a)**, the volume deficit of the active surface layer is replaced by material from the subsurface alluvium, but the grain-size distribution of the active surface layer remains constant. For the intermediate case with θ_{50} greater than $\theta_{c,\mathrm{sub}}$ and smaller than $\theta_{c,s}$ **(b)**, the influence of the grain-size distribution of the subsurface alluvium on the grain-size distribution of the active surface layer is interpolated linearly between cases **(a)** and **(c)**.

Table A1. Notation.

The following symbols are used in this article.	
∇	vector differential operator
β	an empirical constant
γ	correction factor for θ_{ci}
Δh_s	thickness of sediment increments for the interaction between the active surface layer and the subsurface alluvium
Δt	temporal discretisation (time step)
Δx	spatial discretisation (reach length)
ΔX	travel distance of grains
η_{pore}	pore volume fraction
θ	dimensionless bed shear stress
θ_{50}	representative θ for D_{50}
θ_c	dimensionless bed shear stress threshold for initiation of bedload motion
θ_{cMin}	minimum value for θ_c
θ_{ci}	θ_c for ith grain size fraction
$\theta_{ci,\text{r}}$	θ_{ci} corrected for macro-roughness
$\theta_{c,\text{s}}$	representative θ_c for the active surface layer
$\theta_{c,\text{sub}}$	representative θ_c for the subsurface alluvium
θ_{c84}	θ_c for D_{84}
θ_i	θ for ith grain-size fraction
$\theta_{i,\text{r}}$	θ_i corrected for macro-roughness
λ	empirical abrasion coefficient
μ	dynamic viscosity
ρ	fluid density
ρ_{s}	sediment density
τ	bed shear stress
τ_{r}	reference bed shear stress
τ_{rm}	reference bed shear stress of mean bed surface grain size
τ^*_{rm}	reference dimensionless Shields stress for mean bed surface grain size
Φ_{b}	Einstein parameter of dimensionless bedload transport rate
Φ_{bi}	Φ_{b} for ith grain size fraction
χ, ψ	two arbitrary data sets
$\overline{\chi}, \overline{\psi}$	the average values of χ and ψ
A	wetted cross-sectional area
a,b,d	empirical constants
D	grain diameter
D_i	mean grain diameter for ith grain-size fraction
D_{m}	geometric mean for grain diameters
D_{mArith}	arithmetic mean for grain diameters
$D_{\text{mArith}_\text{s}}$	D_{mArith} for the active surface layer
$D_{\text{mArith}_\text{sub}}$	D_{mArith} for the subsurface alluvium
D_x	xth percentile for grain diameters
D_{50}	median grain diameter
e	empirical constant ranging from 1 to 2
exp	exponential function
$\boldsymbol{f}$	body forces
f	Darcy–Weisbach friction factor
F_i	volumetric proportion of ith grain-size fraction
fr	flow resistance
Fr	Froude number
g	gravitational acceleration
geom	channel geometry
h_s	active surface layer thickness after erosion or deposition but before the interaction between the active surface layer and the subsurface alluvium
$h_{s,\text{post}}$	active surface layer thickness after the interaction between the active surface layer and the subsurface alluvium

Table A1. Continued.

i	a running index
i_{sub}	grain-size influence from the subsurface alluvium
j_1, j_2, k, l	empirical constants
L	process break point
m	empirical hiding exponent
m_{wc}	hiding exponent according to Wilcock and Crowe (2003)
mod	modulo function
n	the length of χ and ψ
p	pressure
q	discharge per unit flow width
q_b	bedload flux per unit flow width
$q_{b_{abr}}$	q_b corrected for gravel abrasion
$q_{b_{lat}}$	lateral bedload influx per unit flow width
q_c	threshold q for initiation of bedload motion
Q	discharge
Q_c	threshold Q for initiation of bedload motion
Q_{max}	maximum Q for the simulation period
Q_{rep}	representative Q for the simulation period
Q_{lat}	lateral water influx
r_h	hydraulic radius
$r_{h,c}$	r_h for $[\theta_{50} = \theta_c]$
rRMSD	relative root mean square deviation
s	density ratio of solids and fluids
S	slope
S_b	channel bed slope
S_c	virtual slope for the correction of θ_{ci} based on $r_{h,c}$
S_f	friction slope
S_{red}	slope reduced for macro-roughness
t	time
T	of current time step
$T-1$	of previous time step
$\text{thresh}_{hs,\text{low}}, \text{thresh}_{hs,\text{high}}$	thresholds for the thickness of the active surface layer with $[\text{thresh}_{hs,\text{low}} < \text{thresh}_{hs,\text{high}}]$
$\text{thresh}_{hs,\text{prox}}$	an alias for the threshold (either $\text{thresh}_{hs,\text{low}}$ or $\text{thresh}_{hs,\text{high}}$) that is closer to h_s
u	of upstream river reach
$\boldsymbol{v}$	flow velocity vector
v	flow velocity scalar
v^*	shear velocity
V	water volume in reach
w	channel width
W^*	dimensionless bedload transport rate according to Parker et al. (1982)
x	distance in flow direction
$\lvert y \rvert$	number of sediment increments that are incorporated from or released to the subsurface alluvium
z	elevation of channel bed

Acknowledgements. We are especially grateful to Christa Stephan (project thesis ETH/WSL), Lynn Burkhard (MSc thesis ETH/WSL) and Martin Böckli (WSL) for their contributions to the development of sedFlow, and to Alexandre Badoux (WSL) for his support. We thank the Swiss National Science Foundation for funding this work in the framework of the NRP 61 project "Sedriver" (SNF grant no. 4061-125975/1/2). Jeff Warburton and two anonymous referees provided thoughtful and constructive suggestions to improve this manuscript.

Edited by: D. Parsons

References

Badoux, A. and Rickenmann, D.: Berechnungen zum Geschiebetransport während der Hochwasser 1993 und 2000 im Wallis, Wasser Energie Luft, 100, 217–226, 2008.

Bathurst, J. C.: Critical conditions for particle motion in coarse bed materials of nonuniform size distribution, Geomorphology, 197, 170–184, doi:10.1016/j.geomorph.2013.05.008, 2013

Belleudy, P., and SOGREAH: Numerical simulation of sediment mixture deposition part 1: analysis of a flume experiment, J. Hydraul. Res., 38, 417-425, doi:10.1080/00221680009498295, 2000

Boyer, C., Verhaar, P. M., Roy, A. G., Biron, P. M., and Morin, J.: Impacts of environmental changes on the hydrology and sedimentary processes at the confluence of St. Lawrence tributaries: potential effects on fluvial ecosystems, Hydrobiologia, 647, 163–183, doi:10.1007/s10750-009-9927-1, 2010.

Bunte, K. B., Abt, S. R., Swingle, K. W., Cenderelli, D. A., and Schneider, J. M.: Critical Shields values in coarse-bedded steep streams, Water Resour. Res., 49, 1–21, doi:10.1002/2012WR012672, 2013.

Carpentier, S., Konz, M., Fischer, R., Anagnostopoulos, G., Meusburger, K., and Schoeck, K.: Geophysical imaging of shallow subsurface topography and its implication for shallow landslide susceptibility in the Urseren Valley, Switzerland, J. Appl. Geophys., 83, 46–56, doi:10.1016/j.jappgeo.2012.05.001, 2012.

Chiari, M. and Rickenmann, D.: Back-calculation of bedload transport in steep channel with a numerical model, Earth Surf. Proc. Land., 36, 805–815, doi:10.1002/esp.2108, 2011.

Chiari, M., Friedl, K., and Rickenmann, D.: A one-dimensional bedload transport model for steep slopes, J. Hydraul. Res., 48, 152–160, doi:10.1080/00221681003704087, 2010.

Chow, V. T., Maidment, D. R., and Mays, L. W.: Applied Hydrology, McGraw-Hill, New York, 1988.

Ciarapica, L. and Todini, E.: TOPKAPI: a model for the representation of the rainfall–runoff process at different scales, Hydrol. Process., 16, 207–229, doi:10.1002/hyp.342, 2002.

Cui, Y. and Parker, G.: Numerical Model of Sediment Pulses and Sediment-Supply Disturbances in Mountain Rivers, J. Hydraul. Eng.-ASCE, 131, 646–656, doi:10.1061/(ASCE)0733-9429(2005)131:8(646), 2005.

Cui, Y. and Wilcox, A.: Development and application of numerical models of sediment transport associated with dam removal, in: Sedimentation Engineering – Processes, Measurements, Modeling and Practice, edited by: García, M. H., vol. 110 of ASCE Manual and Reports on Engineering Practice, American Society of Civil Engineers (ASCE), Reston, USA, chap. 23, 995–1020, doi:10.1061/9780784408148.ch23, 2008.

Cui, Y., Parker, G., Braudrick, C., Dietrich, W. E., and Cluer, B.: Dam Removal Express Assessment Models (DREAM). Part 1: Model development and validation, J. Hydraul. Res., 44, 291–307, doi:10.1080/00221686.2006.9521683, 2006.

Danish Hydraulic Institute (DHI): MIKE21C user's guide and scientific documentation, Tech. rep., DHI, Horsholm, Denmark, 1999.

Ferguson, R. I.: Flow resistance equations for gravel- and boulder-bed streams, Water Resour. Res., 43, W05427, doi:10.1029/2006WR005422, 2007.

Ferguson, R. I., Church, M., and Weatherly, H.: Fluvial aggradation in Vedder River: testing a one-dimensional sedimentation model, Water Resour. Res., 37, 3331–3347, doi:10.1029/2001WR000225, 2001.

Fernandez Luque, R. and van Beek, R.: Erosion and transport of bed-load sediment, J. Hydraul. Res., 14, 127–144, 1976.

García-Martinez, R., Espinoza, R., Valeraa, E., and González, M.: An explicit two-dimensional finite element model to simulate short- and long-term bed evolution in alluvial rivers, J. Hydraul. Res., 44, 755–766, doi:10.1080/00221686.2006.9521726, 2006.

Heimann, F. U. M., Rickenmann, D., Böckli, M., Badoux, A., Turowski, J. M., and Kirchner, J. W.: Calculation of bedload transport in Swiss mountain rivers using the model sedFlow: proof of concept, Earth Surf. Dynam., 3, 37–56, doi:10.5194/esurf-3-37-2015, 2015.

Hoey, T. B. and Ferguson, R. I.: Numerical simulation of downstream fining by selective transport in gravel bed rivers: model development and illustration, Water Resour. Res., 30, 2251–2260, doi:10.1029/94WR00556, 1994.

Hoey, T. B., Bishop, P., and Ferguson, R. I.: Testing numerical models in geomorphology: how can we ensure critical use of model predictions?, in: Prediction in Geomorphology, edited by: Wilcock, P. R. and Iverson, R. M., Vol. 135 of Geophysical Monograph, American Geophysical Union, Washington DC, USA, 241–256, doi:10.1029/135GM17, 2003.

Hunziker, R. P.: Fraktionsweiser Geschiebetransport, in: Mitteilung der Versuchsanstalt für Wasserbau, Hydrologie und Glaziologie, edited: by Vischer, D., 138, ETH, Zurich, Switzerland, 1–209, 1995.

Jäggi, M. N. R.: Sedimenthaushalt und Stabilität von Flussbauten, in: Mitteilung der Versuchsanstalt für Wasserbau, Hydrologie und Glaziologie, edited by: Vischer, D., 119, ETH, Zurich, Switzerland, 1–105, 1992.

Junker, J., Heimann, F. U. M., Hauer, C., Turowski, J. M., Rickenmann, D., Zappa, M., and Peter, A.: Assessing the impact of climate change on brown trout (Salmo trutta fario) recruitment, Hydrobiologia, doi:10.1007/s10750-014-2073-4, 2014.

Kaitna, R., Chiari, M., Kerschbaumer, M., Kapeller, H., Zlatic-Jugovic, J., Hengl, M., and Huebl, J.: Physical and numerical modelling of a bedload deposition area for an Alpine torrent, Nat. Hazards Earth Syst. Sci., 11, 1589–1597, doi:10.5194/nhess-11-1589-2011, 2011.

Konz, M., Chiari, M., Rimkus, S., Turowski, J. M., Molnar, P., Rickenmann, D., and Burlando, P.: Sediment transport modelling in a distributed physically based hydrological catchment model, Hydrol. Earth Syst. Sci., 15, 2821–2837, doi:10.5194/hess-15-2821-2011, 2011.

Lamb, M. P., Dietrich, W. E., and Venditti, J. G.: Is the critical Shields stress for incipient sediment motion dependent on channel-bed slope?, J. Geophys. Res., 113, F02008, doi:10.1029/2007JF000831, 2008.

Li, S. S. and Millar, R. G.: Simulating bed-load transport in a complex gravel-bed river, J. Hydraul. Eng.-ASCE, 133, 323–328, doi:10.1061/(ASCE)0733-9429(2007)133:3(323), 2007.

Liu, Z. and Todini, E.: Towards a comprehensive physically-based rainfall-runoff model, Hydrol. Earth Syst. Sci., 6, 859–881, doi:10.5194/hess-6-859-2002, 2002.

Lopez, J. L. and Falcon, M. A.: Calculation of bed changes in mountain streams, J. Hydraul. Eng.-ASCE, 125, 263–270, doi:10.1061/(ASCE)0733-9429(1999)125:3(263), 1999.

Meyer-Peter, E. and Müller, R.: Formulas for bed-load transport, in: Proceedings of the 2nd Meeting of the International Association for Hydraulic Structures Research, Appendix 2, Stockholm, Sweden, 7–9 June 1948, 1948.

Morvan, H., Knight, D., Wright, N., Tang, X., and Crossley, A.: The concept of roughness in fluvial hydraulics and its formulation in 1D, 2D and 3D numerical simulation models, J. Hydraul. Res., 46, 191–208, doi:10.1080/00221686.2008.9521855, 2008.

Mouri, G., Shiiba, M., Hori, T., and Oki, T.: Modeling reservoir sedimentation associated with an extreme flood and sediment flux in a mountainous granitoid catchment, Japan, Geomorphology, 125, 263–270, doi:10.1016/j.geomorph.2010.09.026, 2011.

Nitsche, M., Rickenmann, D., Turowski, J. M., Badoux, A., and Kirchner, J. W.: Evaluation of bedload transport predictions using flow resistance equations to account for macro-roughness in steep mountain streams, Water Resour. Res., 47, W08513, doi:10.1029/2011WR010645, 2011.

Nitsche, M., Rickenmann, D., Kirchner, J. W., Turowski, J. M., and Badoux, A.: Macroroughness and variations in reach-averaged flow resistance in steep mountain streams, Water Resour. Res., 48, W12518, doi:10.1029/2012WR012091, 2012.

Papanicolaou, A. N., Bdour, A., and Wicklein, E.: One-dimensional hydrodynamic/sediment transport model applicable to steep mountain streams, J. Hydraul. Res., 42, 357–375, doi:10.1080/00221686.2004.9641204, 2004.

Parker, G.: Surface-based bedload transport relation for gravel rivers, J. Hydraul. Res., 28, 417–436, doi:10.1080/00221689009499058, 1990.

Parker, G.: Transport of gravel and sediment mixtures, in: Sedimentation Engineering: Processes, Measurements, Modeling, and Practice, Vol. 110 of ASCE Manuals and Reports on Engineering Practice, American Society of Civil Engineers (ASCE), Chap. 3, 165–252, Reston, VA, USA, 2008.

Parker, G., Klingeman, P. C., and McLean, D. G.: Bedload and size distribution in paved gravel-bed streams, J. Hydraul. Eng., 108, 544–571, 1982.

Recking, A.: Theoretical development on the effects of changing flow hydraulics on incipient bed load motion, Water Resour. Res., 45, W04401, doi:10.1029/2008WR006826, 2009.

Recking, A.: A comparison between flume and field bed load transport data and consequences for surface-based bed load transport prediction, Water Resour. Res., 46, W03518, doi:10.1029/2009WR008007, 2010.

Rickenmann, D.: Comparison of bed load transport in torrent and gravel bed streams, Water Resour. Res., 37, 3295–3305, doi:10.1029/2001WR000319, 2001.

Rickenmann, D.: Alluvial steep channels: flow resistance, bedload transport prediction, and transition to debris flows, in: Gravel-bed Rivers: Processes, Tools, Environments, edited by: Church, M., Biron, P. M., and Roy, A. G., John Wiley & Sons, 386–397, Chichester, UK, 2012.

Rickenmann, D. and Recking, A.: Evaluation of flow resistance in gravel-bed rivers through a large field data set, Water Resour. Res., 47, W07538, doi:10.1029/2010WR009793, 2011.

Shields, A.: Anwendung der Aehnlichkeitsmechanik und der Turbulenzforschung auf die Geschiebebewegung, Tech. rep., Mitteilungen der Preussischen Versuchsanstalt für Wasserbau und Schiffbau, Berlin, 1936.

Soulsby, R. L. and Damgaard, J. S.: Bedload sediment transport in coastal waters, Coast. Eng., 52, 673–689, doi:10.1016/j.coastaleng.2005.04.003, 2005.

Stephan, C.: Sensitivity of bedload transport simulations to different transport formulae and cross-sectional geometry with the model Tom^{Sed}, Tech. rep., Swiss Federal Research Institute WSL & ETH, Zurich, Switzerland, 2012.

Sternberg, H.: Untersuchungen über Längen- und Querprofil geschiebeführender Flüsse, Zeitschrift für Bauwesen, 25, 483–506, 1875.

Talbot, T. and Lapointe, M.: Numerical modeling of gravel bed river response to meander straightening: the coupling between the evolution of bed pavement and long profile, Water Resour. Res., 38, 10-1–10-10, doi:10.1029/2001WR000330, 2002.

van de Wiel, M. J., Coulthard, T. J., Macklin, M. G., and Lewin, J.: Embedding reach-scale fluvial dynamics within the CAESAR cellular automaton landscape evolution model, Geomorphology, 90, 283–301, doi:10.1016/j.geomorph.2006.10.024, 2007.

Verhaar, P. M., Biron, P. M., Ferguson, R. I., and Hoey, T. B.: A modified morphodynamic model for investigating the response of rivers to short-term climate change, Geomorphology, 101, 674–682, doi:10.1016/j.geomorph.2008.03.010, 2008.

Wohl, E.: Mountain rivers, vol. 14 of *Water Resources Monograph*, American Geophysical Union, Washington, DC, USA, 2000.

Wilcock, P. R. and Crowe, J. C.: Surface-based transport model for mixed-size sediment, J. Hydraul. Eng.-ASCE, 129, 120–128, doi:10.1061/(ASCE)0733-9429(2003)129:2(120), 2003.

Yager, E. M., Kirchner, J. W., and Dietrich, W. E.: Calculating bed load transport in steep boulder bed channels, Water Resour. Res., 43, W07418, doi:10.1029/2006WR005432, 2007.

Yager, E. M., Dietrich, W. E., Kirchner, J. W., and McArdell, B. W.: Prediction of sediment transport in step-pool channels, Water Resour. Res., 48, W01541, doi:10.1029/2011WR010829, 2012.

2

Calculation of bedload transport in Swiss mountain rivers using the model sedFlow: proof of concept

F. U. M. Heimann[1,2], D. Rickenmann[1], M. Böckli[1], A. Badoux[1], J. M. Turowski[3,1], and J. W. Kirchner[2,1]

[1]WSL Swiss Federal Institute for Forest, Snow and Landscape Research, 8903 Birmensdorf, Switzerland
[2]Department of Environmental System Sciences, ETH Zurich, 8092 Zurich, Switzerland
[3]Helmholtz Centre Potsdam, GFZ German Research Centre for Geosciences, Telegrafenberg, 14473 Potsdam, Germany

Correspondence to: F. U. M. Heimann (florian.heimann@wsl.ch)

Abstract. Fully validated numerical models specifically designed for simulating bedload transport dynamics in mountain streams are rare. In this study, the recently developed modelling tool sedFlow has been applied to simulate bedload transport in the Swiss mountain rivers Kleine Emme and Brenno. It is shown that sedFlow can be used to successfully reproduce observations from historic bedload transport events with plausible parameter set-ups, meaning that calibration parameters are only varied within ranges of uncertainty that have been pre-determined either by previous research or by field observations in the simulated study reaches. In the Brenno river, the spatial distribution of total transport volumes has been reproduced with a Nash–Sutcliffe goodness of fit of 0.733; this relatively low value is partially due to anthropogenic extraction of sediment that was not considered. In the Kleine Emme river, the spatial distribution of total transport volumes has been reproduced with a goodness of fit of 0.949. The simulation results shed light on the difficulties that arise with traditional flow-resistance estimation methods when macro-roughness is present. In addition, our results demonstrate that greatly simplified hydraulic routing schemes, such as kinematic wave or uniform discharge approaches, are probably sufficient for a good representation of bedload transport processes in reach-scale simulations of steep mountain streams. The influence of different parameters on simulation results is semi-quantitatively evaluated in a simple sensitivity study. This proof-of-concept study demonstrates the usefulness of sedFlow for a range of practical applications in alpine mountain streams.

1 Introduction

The rolling, sliding or saltating transport of sediment grains along river beds, which is summarised as bedload transport, represents one of the main morphodynamic processes in mountain streams. Bedload transport has implications which go beyond mere morphodynamics. It exerts considerable ecological influence by reorganising the bed and thus potential spawning grounds (e.g. Unfer et al., 2011). In mixed alluvial–bedrock channels, the bedload flux is one of the dominant controls on bedrock erosion (e.g. Turowski, 2012). Frequently, bedload fluxes are also responsible for damage to engineering structures (e.g. Jaeggi, 2008; Totschnig et al., 2011). Because bedload transport can amplify the impact of severe floods, it is also important in natural hazard management (e.g. Badoux et al., 2014). This wide range of implications is reflected in numerous applied engineering projects which evaluate potential bedload transport using one- or two-dimensional simulation models. A summary of the applied aspects of bedload transport assessment has been given by Habersack et al. (2011).

The available models for simulating sediment transport may be divided into two groups. The first group of models does not focus on process details. It rather sees fluvial sediment transport as a part of a network of interacting processes within the landscape. Therefore, such models use simplified representations of river hydraulics and are often combined with hydrologic or soil erosion model components. Large-scale spatial resolutions and fast calculations are common in this group of models. The SHE-TRANsport model SHETRAN with SHE standing for Système Hydrologique Européen (Lukey et al., 2000; Bathurst et al., 2010), the Distributed Hydrology-Soil-Vegetation Model (DHSVM) (Doten et al., 2006) and others (e.g. Mouri et al., 2011) fall in this group. The SHE SEDiment component SHESED (Wicks and Bathurst, 1996) for the Système Hydrologique Européen also combines sediment transport routines with hydrologic and soil erosion routines, but without the strong simplifications (and associated efficiency gains) of the models mentioned above.

The second group of models concentrates on hydraulic processes as the main driving factor of sediment transport. Therefore, such models commonly solve the full Saint-Venant equations, but neglect any processes outside the channel. Small-scale spatial resolutions and slow calculations are common in this group of models. The Steep Stream Sediment Transport 1-D model (3ST1D) (Papanicolaou et al., 2004), the Hydrologic Engineering Center model no. 6 (HEC-6) (Bhowmik et al., 2008), the model SEDROUT (Ferguson et al., 2001), the Generalized Stream Tube Alluvial River Simulation model (GSTARS) (Hall and Cratchley, 2006), the FLUvial Modelling ENgine (FLUMEN) (Beffa, 2005), the BASic EnvironMENT for simulation of environmental flow and natural hazard simulation (BASEMENT) (Faeh et al., 2011) and others (e.g. Lopez and Falcon, 1999; García-Martinez et al., 2006; Li et al., 2008) fall in this group.

Similar to Tom^{Sed} (formerly known as SEdiment TRansport model in Alpine Catchments (SETRAC)) (Chiari et al., 2010), the model sedFlow (Heimann et al., 2015) is intended to bridge the gap between these two groups of models by providing good representation of fluvial bedload transport processes at intermediate spatial scales and high calculation speeds. Here the focus of modelling is not on the details of the temporal evolution of sediment transport, but rather on a realistic reproduction of the total transport volumes and overall morphodynamic changes resulting from sediment transport events such as major floods.

In spite of the considerable need for modelling tools in scientific and engineering applications and in spite of the interest in the relevant physical processes, bedload transport in mountain streams is not entirely understood. This is partly due to the complex measurement conditions in gravel-bed rivers (Bunte et al., 2008; Gray et al., 2010). Because of these difficulties, there are relatively few data sets available for deriving conceptual models or for validating and testing numeric models.

Based on the available field observations, it has become clear that river bed morphology and thus hydraulic processes become increasingly complex as channel gradients become steeper. The range of observed grain diameters becomes larger, which entails more complex grain–grain and grain–flow interactions as well. Summarising available field data on flow velocity, Rickenmann and Recking (2011) showed that a considerable part of the river's shear stress is consumed by turbulence due to complex bed morphology, summarised as macro-roughness. They also suggested an approach to quantify the impact of macro-roughness based on the relative flow depth compared to a characteristic grain diameter. Lamb et al. (2008) and Bunte et al. (2013) have noted that in steep channels higher energies are needed for the initiation of bedload motion, compared to channels with gentle slopes. Turowski et al. (2011) have shown that the conditions for the initiation of bedload motion vary in time and are strongly linked to the conditions at the end of the last bedload transport event. Parker (2008) and Wilcock and Crowe (2003) have discussed and proposed approaches for quantifying grain–grain interactions in so-called hiding functions. Finally, several methods have been suggested for predicting bedload transport in mountain streams. Some of these methods are based on flume experiments, such as those of Rickenmann (2001) and Wilcock and Crowe (2003), and some are based on field observations, such as those of Recking (2010; 2013a). For recent applications and discussions of the conceptual models and methods mentioned in this paragraph see Chiari and Rickenmann (2011), Nitsche et al. (2011) and Rickenmann (2012). A selection of such methods related to the estimation of bedload transport in steep channels has been implemented in the modelling tool sedFlow. For the bedload transport equation, the flow-resistance relation and several other elements, sedFlow offers different options which can be selected to fit the current application in a particular catchment. The model architecture and implementation are described in detail in a companion article (Heimann et al., 2015), and are only briefly reviewed here. The program is intended for quantitatively simulating bedload transport processes in mountain streams at temporal scales from the individual event (several hours to few days) to longer-term evolution of stream channels (several years). It is designed for spatial scales covering complete catchments at a spatial discretisation of several tens of metres to a few hundred metres. sedFlow has been developed to provide a tool which combines recently proposed and tested process representations with fast computational algorithms and user-friendly file formats for easy pre- and postprocessing of simulation data.

In this article, we show that sedFlow can reproduce observations from historical bedload transport events, using plausible parameter set-ups. Here by plausible parameter set-ups we mean that calibration parameters are only varied within ranges of uncertainty that have been pre-determined either by previous research or by field observations in the simulated study reaches. The main aim of this proof-of-concept study is defined by the objective of the sedFlow model, namely the realistic simulation of total transport volumes and overall morphodynamic effects of sediment transport events such as major floods. The results of this study may help to interpret simulation results produced with sedFlow in applied engineering projects. Experiences with the simulation tool are dis-

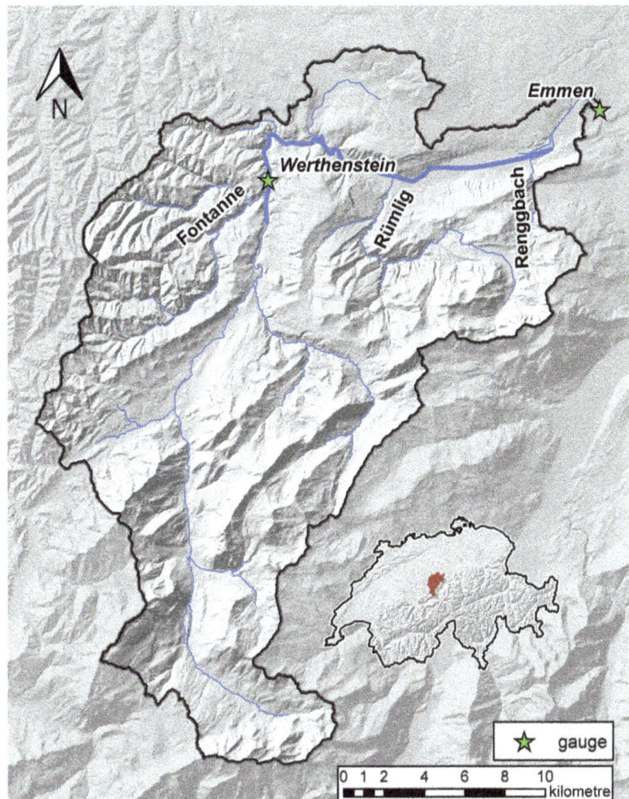

Figure 1. The Kleine Emme catchment in central Switzerland. The study reach from Doppleschwand to the confluence with the Renggbach is indicated by the bold blue line.

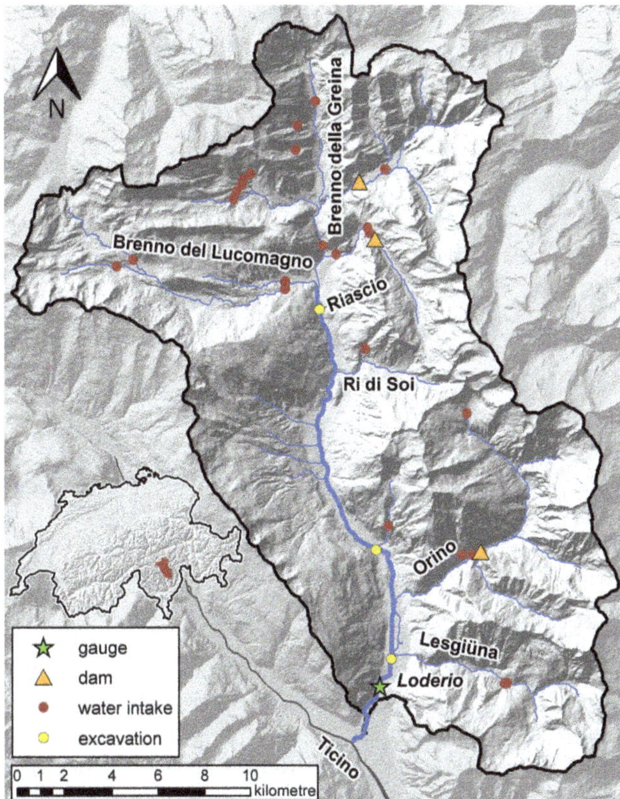

Figure 2. The Brenno catchment in southern Switzerland. The study reach from Olivone to Biasca is indicated by the bold blue line.

cussed with respect to the problems of quantifying the influence of macro-roughness within traditional flow-resistance equations. In addition, the uncertainties introduced by common graphical representations of bedload transport reconstructions are highlighted based on the results of a simple sensitivity study.

2 Material and methods

For our study we selected two Swiss rivers, the Kleine Emme and the Brenno (Figs. 1 and 2). The Kleine Emme was chosen because extensive data are available to validate and test the sedFlow model in this catchment. The Brenno river was selected as a complementary case study to cover a wider range of channel gradients and streambed morphology.

In this article we differentiate between net and gross channel gradients in the context of sills. Net channel gradients are defined as gross channel gradients corrected for the elevation differences attributable to sills or other drop-down structures.

2.1 General catchment characteristics

The Kleine Emme is a mountain river in central Switzerland (Fig. 1). It drains an area of 477 km^2 and flows into the Reuss at Reussegg. The Kleine Emme's net channel gradient averages 0.8 % with a maximum of 3.5 %. Near Doppleschwand the in situ bedrock is close to the surface, limiting the alluvium that can potentially be eroded. Further downstream the river was channelised in the late 19th and early 20th century. To mitigate the subsequent erosion, the bed was stabilised in the early 20th centuries with numerous bottom sills (documented by Geoportal Kanton Luzern, 2013). The Kleine Emme is an alpine mountain river catchment with gentle slopes, without glaciers or debris flow inputs and with only very moderate influence from hydropower installations, but with intensive modifications by fluvial engineering.

The Brenno is situated in southern Switzerland (Fig. 2) and drains into the river Ticino. Its drainage area is 397 km^2 and its channel gradient averages 2.6 %, with a maximum of 17 %. There are no sills in the Brenno, so the net and gross gradients are the same. About 1 % of the catchment area is glaciated. Especially in the northern and eastern part of the catchment, its hydrology is substantially influenced by hydropower (Fig. 2). The water used for hydropower production is returned to the Ticino river downstream of Biasca. The tributaries Riale Riascio and Ri di Soi are currently the most important sediment sources to the Brenno river (Table 1). The sediment input from the Riale Riascio is dominated by debris

Table 1. Estimated sediment input yields from tributaries to the Brenno (based on Flussbau AG, 2003, 2005; Stricker, 2010) (Process types: DF = debris flow, FT = fluvial bedload transport).

Tributary	Type	Per year [$m^3 a^{-1}$]			Calibration period [m^3]		
		Min.	Mean	Max.	Min.	Mean	Max.
Brenno della Greina	FT	2500		7500	25 000		75 000
Brenno del Lucomagno/Ri di Piera	FT	1500		5000	15 000		50 000
Riale Riascio	DF	4000	10 000	22 000	40 000	100 000	220 000
Ri di Soi	DF + FT	10 000	20 000	30 000	100 000	200 000	300 000
Lesgiüna	FT	1000	2000	5000	10 000	20 000	50 000
Crenone (Vallone)	DF	1000	1500	4000	10 000	15 000	40 000

flows, while the larger subcatchment Ri di Soi delivers sediment both as debris flows and as fluvial bedload transport. Downstream of the confluences with these tributaries, the bed of the Brenno is stabilised by large blocks and the main channel shows pronounced knickpoints at these positions. Other tributaries on the western side of the Brenno catchment were very active in the decades from 1970 to 1990, but their sediment delivery to the Brenno is much reduced at the time of writing due to intense torrent control works and sediment retention basins. The course of the Brenno is partially channelised and partially near natural. The Brenno represents a moderately steep mountain river influenced by glaciation, hydropower production and debris flow inputs.

The two catchments are impacted and show a range of engineering interventions typical of many mountain catchments. The Kleine Emme is marked by river training works, including numerous bottom sills as well as riprap and groynes in some locations. The Brenno is strongly influenced by controls on water and sediment delivery to the channel. The Brenno's hydrology is substantially influenced by hydropower production and lateral sediment input is limited by torrent control works and sediment retention basins in the tributaries. Along a few kilometres of the Brenno river gravel extraction occurred during the calibration period. The two catchments contrast with each other not only by their different management histories. Even though the two catchment areas are of similar size, channel gradients are steeper in the Brenno river than in the Kleine Emme river. While in the Kleine Emme channel bank erosion played a dominant role in feeding sediment to the transport system, in the Brenno lateral sediment input due to debris flows from tributaries was important during the calibration period. In summary, the two study catchments differ substantially and present a range of characteristics common to many mountain catchments.

Several channel cross sections are periodically surveyed for both rivers. In the case of the Kleine Emme, they are measured by the Swiss Federal Office for the Environment (FOEN) and in the case of the Brenno, they are measured by the authorities of the canton of Ticino. Cross-sectional profiles are recorded at 200 m intervals in the Kleine Emme and at about 150 m intervals in the Brenno. For the Kleine Emme we used measurements from September 2000 to November 2005. For the Brenno we used measurements from April 1999 to June/July 2009. We selected our study reaches to overlap with these surveyed cross sections.

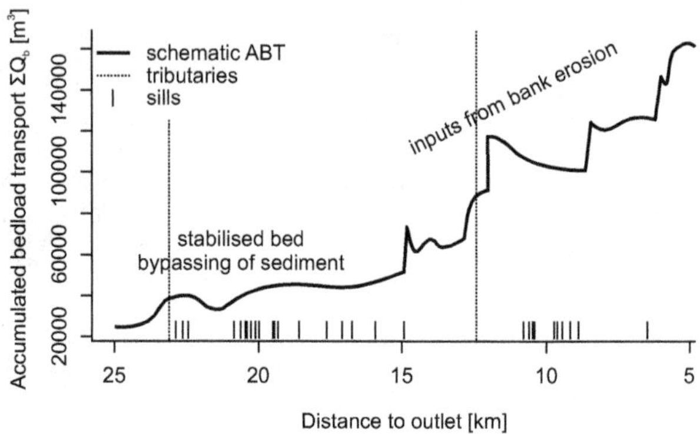

Figure 3. Schematic representation of accumulated bedload transport (ABT) in the Kleine Emme with locations of tributaries and sills (tributaries from up- to downstream: Fontanne, Rümlig).

Doppleschwand, about 25 km upstream from the Kleine Emme mouth, represents the upper boundary of our simulation reach. A large, long-duration flood event occurred in August 2005, with a return period of around 50 years for the peak discharge. During this event, widespread flooding occurred along the lowermost 5 km of the river in the area of Littau. Therefore, the lower boundary of our one-dimensional model simulations is the confluence of the Kleine Emme and the Renggbach (Fig. 1). At the Brenno, our study reach extends from Olivone at the upper end to Biasca at the confluence with the Ticino river (Fig. 2).

2.2 Hydrology

The discharge of the Kleine Emme has been measured at Werthenstein since 1985 and at Littau–Emmen since 1978 (Fig. 1). Peak discharge at Littau–Emmen during the August 2005 flood was 650 $m^3 s^{-1}$. To account for the reduced catchment area of the Kleine Emme upstream of the Renggbach at the simulation outlet, the discharge at Littau–Emmen is reduced by 5 % as suggested by VAW (1997). The discharge of the Rümlig tributary is estimated by the difference between the values of Werthenstein and the simulation outlet. The discharge of the Fontanne is simulated using the fully-distributed version of the Precipitation-Runoff-EVApotranspiration HRU model PREVAH (Viviroli et al., 2009; Schattan et al., 2013) in which HRU stands for hydrological response units. The discharge of the headwater is estimated by the difference between the measured discharge at Werthenstein and the simulated Fontanne discharge.

Discharge of the Brenno has been measured at Loderio ever since the establishment of the hydropower reservoirs in the catchment in 1962. A peak discharge of $515\,m^3\,s^{-1}$ was recorded during the July 1987 flood, corresponding to a return period of about 150 years. For the simulations, the discharge at Loderio has been distributed among the subcatchments according to rainfall–runoff simulations using the PREVAH model. The discharge is assumed to be zero at dams and reduced by the intake capacity at water intakes. In this reduction, we accounted for the regulations that specify the minimum residual discharge in the river channel downstream of a water intake. The values of this minimum residual discharge were defined based on ecological aspects and vary with intake location and time of the year.

2.3 Channel morphology and bedload observations

2.3.1 Rectangular channels

For use in the sedFlow model, the cross-sectional profiles were transformed into the equivalent width of a simple rectangular substitute channel. For this transformation a representative discharge was defined as the mean of the peak discharge of the simulation period and the discharge at the initiation of bedload motion, as these two values define the range of discharges relevant for bedload transport. The variable power equation flow-resistance relation was used to translate discharge into flow depth based on the same grain-size distributions (GSDs) that were used in the simulations. Then, a rectangular channel was found which has the same cross-sectional flow area and hydraulic radius as the original cross-sectional profile at this flow depth. The channel of the Kleine Emme has been regulated in the past and its geometry is well defined by a trapezoidal profile with steep banks. In contrast, the Brenno study reach is in a natural condition over most parts, including both more incised reaches with a well-defined width and depositional reaches in flatter areas with riparian forest. The latter reaches are characterised by river banks with gentle slopes. In such channels, a slight change of the representative discharge may result in a substantial change in the width of the rectangular substitute channel. Therefore, in the depositional reaches of the Brenno, the uncertainty in representative discharge entails a considerable uncertainty in substitute channel widths, which contrasts with the better-constrained substitute channel widths in the incised Brenno reaches and Kleine Emme reaches, due to their steeper banks.

2.3.2 Reference data

To test the sedFlow model, a reference is needed, to which the simulation results can be compared. Therefore, the bedload transport during the calibration period, which was not observed by itself, needs to be reconstructed from available observations. To volumetrically quantify the reconstructed bedload transport, the change in average bed level between each pair of cross-sectional surveys is multiplied with the mean of the substitute channel width of both profile measurements and with the distance to the next profile. These bed volume changes give an integrated value of the minimum bed material transported over the observation period. However, to obtain a complete sediment budget, data on bank erosion, lateral sediment input from tributaries and the material that leaves the catchment at the outflow have to be considered. At the Kleine Emme, bank erosion volumes were estimated from the difference between the FOEN cross-sectional profiles in 2000 and 2005 and from field assessments of the erosion scars (Flussbau AG, 2009; Hunzinger and Krähenbühl, 2008), and the sediment outflow was quantified based on data of regular gravel extraction at the confluence of the Kleine Emme with the Reuss (Hunzinger and Krähenbühl, 2008; Hunziker, Zarn and Partner AG, 2009). For the Brenno the lateral inputs by debris flows or fluvial bedload transport were estimated based on data from a number of previous studies (Flussbau AG, 2003, 2005; Stricker, 2010), as listed in Table 1. The spatial pattern of changes in sediment transport, as well as the absolute value of sediment transport, is greatly influenced by sediment input from the tributaries. Thus, the uncertainty in the estimates of tributary sediment inputs largely determines the overall uncertainty in sediment transport in the Brenno. The sediment outflow at the mouth of the Brenno and thus the volume of the throughput load of the complete system is unknown. Therefore, we used the result of the sedFlow simulations as a best guess for this parameter, since no other proxies are available. Of course, this approach for the determination of sediment outflow at the mouth of the Brenno partially compromises the independence of the evaluation of model performance, regarding the overall transport rate. However, this approach still allows for an independent evaluation of the along-channel changes in transport rates as well as any other variables such as erosion and deposition rates or characteristic bed surface grain diameters.

2.3.3 Accumulated bedload transport

All volumetric data related to the sediment budget are summarised in accumulated bedload transport (ABT) diagrams, e.g. Figs. 3 and 4. ABT represents the net bedload amount which has been transported through a given stream section during the period of interest (Chiari et al., 2010). It is a tem-

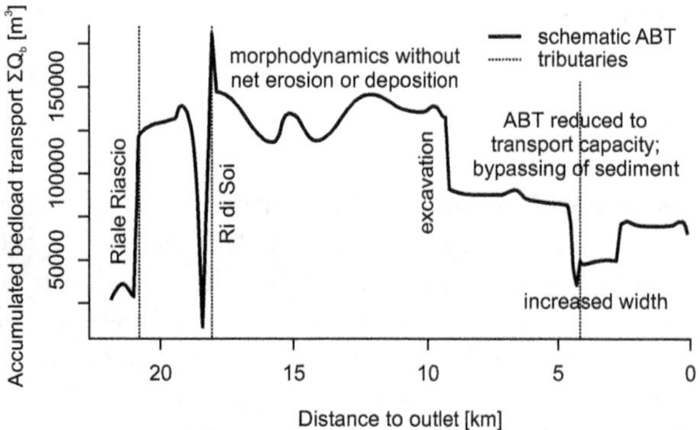

Figure 4. Schematic representation of accumulated bedload transport (ABT) in the Brenno, with labels indicating major sediment sources and sinks (tributaries from up- to downstream: Riale Riascio, Ri di Soi, Lesgiüna).

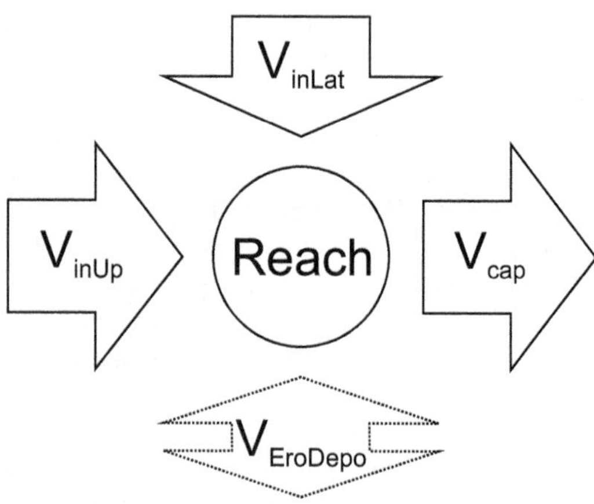

Figure 5. Schematic visualisation of Eqs. (2) to (4).

poral integral of the transport rates and it is a spatial integral of the volumetric changes including bed net erosion and deposition, lateral inputs and sediment outflow. In this article, all ABT values include an assumed pore volume fraction of 30 %. The ABT can be derived from the morphodynamic relation, which has been described by Exner in its continuous form (e.g. Parker, 2008):

$$\left(1-\eta_{\text{pore}}\right)\cdot\frac{\partial z}{\partial t}=q_{\text{b}_{\text{lat}}}-\frac{\partial q_{\text{b}}}{\partial x}. \tag{1}$$

Here η_{pore} is the pore volume fraction, z is elevation of channel bed, t is time, q_{b} is sediment flux per unit flow width, x is distance in flow direction and $q_{\text{b}_{\text{lat}}}$ is lateral sediment influx per unit flow width. Equation (1) represents a balance of input and output volumes and it can be rewritten in a discretised form for a finite reach and period of time as

$$V_{\text{in}}-V_{\text{out}}-V_{\text{EroDepo}}=0, \tag{2}$$

$$V_{\text{in}}=V_{\text{inUp}}+V_{\text{inLat}}, \tag{3}$$

$$V_{\text{out}}=V_{\text{cap}}. \tag{4}$$

Here V_{in} designates the volume of sediment that enters a reach, subdivided into the volume V_{inUp} coming from upstream and the volume V_{inLat} introduced laterally, e.g. by tributaries or bank erosion. V_{EroDepo} is the volume eroded or deposited in the reach, with positive values indicating deposition. V_{out} is the volume that exits the reach, which in the case of unlimited (or at least sufficient) supply of material (Eq. 4) equals the volume V_{cap} corresponding to the transport capacity within the reach, multiplied by the considered time interval. Equation (2) constitutes the difference between inputs and outputs is counterbalanced by erosion or deposition (Fig. 5). For erosion, the local V_{in} will always be smaller than V_{cap} and will result in ABT increasing downstream. In the same way, deposition will result in a decreasing ABT, while a roughly constant ABT reflects throughflow of sediment without net erosion or deposition.

GSDs have been estimated for different reaches, based on transect pebble counts using the method of Fehr (1986, 1987). To determine the subsurface GSD, the pebble count was transformed into a full GSD by assuming an average proportion of 25 % fine material with $D < 10$ mm according to Fehr (1987). To determine the surface GSD, the pebble count was transformed into a full GSD by assuming an average proportion of 10 % fine material with $D < 10$ mm according to observations reported in Recking (2013b) and Anastasi (1984). In some cases at the Brenno, coarser sediment portions were added to the recorded GSDs, because coarse blocks have been underrepresented in the transect counts and thus the original transect GSDs partially led to unrealistic model behaviour. The measured GSDs were assumed to be representative for entire reaches, which are separated from each other by features such as confluences or considerable changes in channel gradient. This spatial extrapolation entails some uncertainty. The current GSD measurements, which were obtained after the end of the calibration period, are used as proxy estimates for the initial GSDs at the beginning of the calibration period. This time shift introduces additional uncertainty.

The bedload transport system of the Kleine Emme can be subdivided into two regimes (Fig. 3). In the upper part from 25 to ~ 15 km, the bed is stabilised by in situ bedrock and numerous sills. Therefore, the system is dominated by throughflow of sediment without considerable trends or jumps in the along-channel evolution of the ABT. In the lower part from ~ 15 to 5 km, the bedload transport system is mainly influenced by sediment inputs from bank erosion during the 2005 flood event, which increase the downstream ABT in a step-like way.

The Brenno bedload transport system is mainly influenced by local elements (Fig. 4). The Riale Riascio at 20.8 km introduced a considerable amount of sediment to the system, resulting in a step-like downstream increase in the ABT. Large amounts of the material delivered by the Ri di Soi at 18.1 km have been deposited at the confluence. These deposits reduced upstream channel gradients and thus transport capacity. The lack of material coming from upstream is overcompensated by the input from the Ri di Soi. However, the excess material has been deposited shortly after the confluence. All processes around the confluence with the Ri di Soi are reflected in a pronounced negative peak and small positive peak in the along-channel evolution of the ABT. The following stretch down to 10 km exhibits erosion and deposition corresponding to the interaction of GSD, channel gradient and width, but without any overall erosion or deposition trend. At 10 km sediment has been anthropogenically extracted from the riverbed by excavation, which results in a step-like downstream decrease of ABT. Because the excavation reduces the amount of transported material down to the transport capacity of the river, sediment bypasses the following reaches. At 4.5 km, the deposits at the confluence with the Lesgiüna decrease the upstream slope and thus cause a drop in transport capacity. In the stretch from 4.5 to 3 km, an increased channel width keeps the ABT at low values.

2.4 The model sedFlow

The bedload transport modelling tool sedFlow has been designed especially for application to mountain rivers. Consistent with this objective, it exhibits the following main features: (i) it uses recently proposed and tested approaches for calculating bedload transport in steep channels accounting for macro-roughness, (ii) it calculates several grain diameter fractions individually, i.e. fractional transport, (iii) it uses fast algorithms and thus can be used for modelling complete catchments and for scenario studies with automated calculations over many variations in the input data or parameter set-up. Here we give a short overview of the essential components of sedFlow. For a detailed account of the model structure and implementation see Heimann et al. (2015). The current version of the sedFlow code and model can be downloaded at the following web page: www.wsl.ch/sedFlow.

Flow resistance is either calculated with the variable power equation of Ferguson (2007) according to Eq. (5) or with a grain-size-dependent Manning–Strickler equation (Eq. 6):

$$\frac{v_m}{v^*} = \frac{a_1 a_2 \left(\frac{r_\mathrm{h}}{D_{84}}\right)}{\sqrt{a_1^2 + a_2^2 \left(\frac{r_\mathrm{h}}{D_{84}}\right)^{\frac{5}{3}}}}, \quad (5)$$

$$\frac{v_m}{v^*} = a_1 \left(\frac{r_\mathrm{h}}{D_{84}}\right)^{\frac{1}{6}}. \quad (6)$$

Here v_m is the average flow velocity, $v^* = \sqrt{g r_\mathrm{h} S}$ is the shear velocity, r_h is the hydraulic radius, S is the gradient of hydraulic head, which may be approximated by the gradient of the water surface or channel bed, D_{84} is the characteristic grain diameter of the surface material, for which 84 % of the material is finer, and g is gravitational acceleration. Equation (5) has been tested by Rickenmann and Recking (2011) based on nearly 3000 field data points. With the coefficients $a_1 = 6.5$ and $a_2 = 2.5$, it shows very good agreement with the average trend of observations, especially including small relative flow depths that are characterised by high flow resistance. Rickenmann and Recking (2011) also rewrote Eq. (5) in an alternative version, in which flow velocity is written as a direct function of q, the discharge per unit flow width.

sedFlow allows three methods for the calculation of channel hydraulics: an explicit kinematic wave routing, an implicit kinematic wave routing and a uniform discharge approach.

The explicit flow routing corresponds to a Eulerian forward approach. In such an approach, all relevant variables are assumed constant for the duration of one time step. For numeric stability, time steps have to be short enough for this approximation to be valid. For morphodynamic simulations this may be impractical. The fast process of running water defines the short time step lengths, even though it is not the process of interest and the relatively slower morphodynamic changes would allow for much longer time steps and thus faster calculations. Apart from this disadvantage, the explicit flow routing provides a routing of discharge without any restrictions concerning other concepts or parameters.

To overcome the short time steps, sedFlow also provides capabilities for implicit flow routing. Because they are unconditionally stable, implicit methods impose no requirements concerning the length of time steps. However, in implicit methods the unknown variables usually have to be found via computationally demanding iterations. In sedFlow, the algorithm of Liu and Todini (2002) is implemented for solving the implicit flow routing. It avoids time-consuming iterations by analytically finding the solution using Taylor series approximations. However, this algorithm requires a power-law representation of discharge as a function of water volume in a reach. That means it can only be applied to infinitely deep rectangular or V-shaped channels in combination with a power-law flow resistance such as Eq. (6). Except for this restriction, the implicit flow-routing algorithm provides a routing of discharge with fast computational performance.

The explicit and implicit flow routings use the bed slope as proxy for energy slope for all hydraulic and bedload transport computations. This approximation, which corresponds to the assumption of a kinematic wave, is acceptable for most mountain channels, as river bed gradients are commonly steep there. However, problems arise when tributaries deposit debris flow material in the main channel, producing adverse slopes (uphill slopes in the downstream direction). A pragmatic solution to deal with adverse slopes is the uniform discharge approach. Discharge is assumed to be equal along the entire channel, only increasing at confluences for a given time step. This procedure can be justified keeping in mind that the temporal scale of hydraulic processes is very small compared to the temporal scale of morphodynamic processes. Hydraulic calculations are performed using the bed slope proxy for the hydraulic gradient. In cases of adverse slopes, ponding is simulated. That is, flow depth and velocity are selected to ensure a minimum gradient of hydraulic head, which is positive and close to zero. For bedload transport calculations the gradient of the hydraulic head is used, which by definition can only exhibit positive slopes. Thus, the energy slope for bedload transport estimation is *not* the result of a backwater calculation, but it is the gradient between individual hydraulic head values, which under normal conditions have been calculated independently from each other using the local bed slope as a proxy for friction slope. It has to be noted that this approach will produce large errors if moderate backwater effects are part of the simulated system. In such systems, the other approach, which uses bed slope both as the friction slope for the hydraulic calculations and as the energy slope for the sediment transport calculations, will produce better estimates of the transported sediment volumes, but it cannot accommodate adverse channel gradients.

Partially due to the simple and efficient hydraulic schemes, several years of bedload transport and resulting slope and GSD adjustment can be simulated with sedFlow within only few hours of calculation time on a regular 2.8 GHz central processing unit (CPU) core.

For optimising calculation speed, amongst others the time steps should be as long as possible. However, there are stability concerns that limit the potential time step lengths. Within sedFlow, the time step length used for the current time step is obtained from three different methods of calculation. When explicit or implicit kinematic-wave flow routing is used, the first method ensures that local slope changes do not exceed a user-defined fraction. When explicit kinematic-wave flow routing is used, the first method further calculates another time step length based on the Courant–Friedrichs–Lewy (CFL) criterion (Courant et al., 1928) for the water flow velocity multiplied by a user-defined safety factor[1]. The second method is based on the CFL criterion for the estimated bedload grain velocity multiplied by a user-defined safety factor. The third method ensures that erosion of the active layer is always less than a user-defined maximum fraction. The actual time step length is the minimum of the values obtained for each simulated reach from the three methods described in this paragraph, provided that this minimum is smaller than a user-defined maximum time step length.

Different formulas can be used for the estimation of bedload transport capacity. The approaches of Rickenmann (2001), Wilcock and Crowe (2003) and Recking (2010) are implemented in sedFlow. The formula of Rickenmann (2001) modified for fractional transport was used here:

$$\Phi_{\mathrm{bi}} = 3.1 \cdot \left(\frac{D_{90}}{D_{30}}\right)^{0.2} \cdot \sqrt{\theta_{i,\mathrm{r}}} \cdot \left(\theta_{i,\mathrm{r}} - \theta_{ci,\mathrm{r}}\right) \cdot Fr \cdot \frac{1}{\sqrt{s-1}},$$
$$\text{with } q_{\mathrm{b}} = \Sigma q_{\mathrm{bi}}. \tag{7}$$

Here $\Phi_{\mathrm{bi}} = \frac{q_{\mathrm{bi}}}{F_i\sqrt{(s-1)gD_i^3}}$ is the dimensionless bedload transport rate per grain-size fraction, F_i is the relative portion compared to the total surface material with $D > 2$ mm of a grain-size fraction i with D_i as its mean diameter, q_{bi} is the volumetric bedload transport per grain-size fraction and unit channel width, $s = \frac{\rho_{\mathrm{s}}}{\rho}$ is the density ratio of solids ρ_{s} and the fluid ρ, Fr is the Froude number, $\theta_{i,\mathrm{r}} = \frac{r_{\mathrm{h}} S_{\mathrm{red}}}{(s-1)D_i}$ is the dimensionless bed shear stress and S_{red} is the reduced energy slope according to Rickenmann and Recking (2011) and Nitsche et al. (2011). Here D_{90} and D_{30} are characteristic grain diameters, for which 90 or 30 % of the local GSD is finer, and q_{b} is the volumetric bedload transport rate per unit channel width. The critical dimensionless bed shear stress at the initiation of transport θ_{ci} is modified by the so-called hiding function either in the form of a relatively simple power-law relation (Parker, 2008):

$$\theta_{ci} = \theta_{c50}\left(\frac{D_i}{D_{50}}\right)^{m} \tag{8}$$

or in the form proposed by Wilcock and Crowe (2003):

$$\theta_{ci} = \theta_{c50} \cdot \left(\frac{D_i}{D_{\mathrm{m}}}\right)^{m_{\mathrm{wc}}}$$
$$\text{with } m_{\mathrm{wc}} = \frac{0.67}{1+\exp\left(1.5 - \frac{D_i}{D_{\mathrm{m}}}\right)} - 1. \tag{9}$$

Here D_{50} and D_{m} are the median and geometric mean grain diameter of surface material, m is an empirical hiding exponent and m_{wc} is the hiding exponent according to Wilcock and Crowe (2003). The empirical exponent m ranges from 0 to -1, where $m = -1$ corresponds to the so-called "equal mobility" case in which all grains start moving at the same dimensionful bed shear stress τ, and $m = 0$ corresponds to no influence by hiding at all. The critical dimensionless bed shear stress at initiation of transport θ_{c50} is estimated based on the bed slope S_{b} with the empirical relation of Lamb et al. (2008) according to Eq. (10):

$$\theta_{c50} = 0.15 \cdot S_{\mathrm{b}}^{0.25}. \tag{10}$$

Within sedFlow a minimum value $\theta_{c50,\mathrm{Min}}$ can be defined for θ_{c50}, as Eq. (10) results in unrealistically low θ_{c50} values for small channel gradients. For consistency of calculations, $\theta_{ci,\mathrm{r}} = \theta_{ci}\left(\frac{S_{\mathrm{red}}}{S}\right)$ is used in Eq. (7).

2.5 Model calibration and sensitivity calculations

Using the data on channel geometry, GSD, and hydrology from the Brenno and Kleine Emme catchments, we ran

[1] When explicit kinematic-wave flow routing is used, the model does not check whether the calculated time step length is smaller than a user-defined maximum length, because the CFL criterion for the water flow velocity usually produces time step lengths which are considerably smaller than commonly used maxima.

the model sedFlow aiming to reproduce the observed bedload transport. The following criteria were applied to assess the agreement between simulation results and observations, which are stated in order of decreasing importance: (i) the input values, such as the local GSDs, should generally remain within the uncertainty range of observations. (ii) The input parameters, such as the threshold bed shear stress at the beginning of bedload motion, should vary within a plausible range. (iii) The simulated erosion and deposition should be as close as possible to the observed pattern. (iv) The simulated ABT should be as close as possible to the one reconstructed from field observations. (v) The GSDs at the end of the simulation should vary within a plausible range. In the calibration of this study, we examined these criteria (i–v) by visual inspection.

The calibration process consists of five steps. First, a hydraulic routing scheme is selected. Second, a bedload transport relation is selected. Third, the threshold for the initiation of motion is adjusted. Fourth, if the simple power-law hiding function of Eq. (8) is used, the exponent m is adjusted as well. Fifth, some fine-tuning is made via local reach-scale adjustments. In general, the calibration parameters for bedload transport can be divided into three groups. The selection of the transport equation and the threshold for the initiation of motion θ_{c50} (or $\theta_{c50,\mathrm{Min}}$ in combination with the relation of Lamb et al., 2008) are *global* calibration parameters, which determine the overall level of transport rate. The local GSDs and representative channel widths are *local* calibration parameters, which can be used to locally modify the transport rates and thus the along-channel distribution pattern of transport rates. Finally, the selection of the hiding function and the hiding exponent m, the method for the interaction between the active surface layer and the subsurface alluvium, and the thickness of the active surface layers form the *remaining* calibration parameters. To the authors' knowledge, there are no in-depth studies assessing the effects of these remaining parameters, which are hard to predict for a natural river system without a systematic sensitivity study.

In the first step of the calibration process of the presented study, the implicit kinematic wave hydraulic routing scheme was selected for the Kleine Emme, because the gentle slopes preclude the uniform discharge approach and the long simulated time period requires fast simulations. For the Brenno, the uniform discharge approach was selected, because the intense sediment inputs from the tributaries require the consideration of adverse slopes. In the fifth step of the calibration process, reach-scale adjustments have been made to the GSD in the Kleine Emme and to the representative channel width in the Brenno river. For the Kleine Emme, the representative channel width was well constrained, while measured GSD's were relatively poorly constrained because the riverbed is accessible only at a limited number of gravel bars. For the Brenno, the uncertainty about the effective channel width is relatively large along the depositional reaches in flatter areas, and for the calibration of the sedFlow simulations the mean channel width was adjusted primarily in these reaches. The corresponding simulation set-ups are summarised in Table 2. For the sediment exchange mechanism between the active surface layer and the subsurface alluvium, in the Brenno, we used a threshold-based interaction approach with 20 and 70 cm as thresholds for the active surface layer thickness. In the Kleine Emme, we used a shear-stress-based interaction approach in which the constant active surface layer thickness equals twice the local surface D_{84} at the beginning of the simulation.

For the Brenno, the simulation of the calibration period was repeated using all three different hydraulic schemes and two flow-resistance relations, which are implemented in sedFlow. Comparing these simulation results allows us to study the influence of the hydraulic algorithm on the simulated bedload transport.

To explicitly study the influence of different time step lengths, we used a set-up in which the actual time step generally equals the user-defined maximum time step value[2]. We compared the simulation results for different maximum time steps ranging from 1 min to 1 h. For any other simulation outside this time step comparison, we used a maximum time step of 15 min for the Kleine Emme and a maximum time step of 1 h for the Brenno. These two values have been selected in order to achieve reasonably short calculation times.

After the calibration exercise, the best-fit parameter set was used as a base for two sensitivity studies. For the first study, in each simulation, all parameters but one are set to their original best-fit values and the remaining parameter is increased and decreased by a certain fraction. In the following we will call this procedure a one-at-a-time range sensitivity study. We varied the parameters discharge, minimum threshold for the initiation of bedload motion $\theta_{c50,\mathrm{Min}}$, grain size and channel width by either plus or minus 10, 20 and 30 %. The maximum variation of 30 % fits the order of magnitude of the different uncertainties typically involved in bedload transport simulations. For example, discharge values are affected by the uncertainties of the rainfall–runoff simulations. The GSD of river reaches is measured at individual and accessible points and therefore cannot sufficiently capture the spatial variability of this parameter. The value of the minimal threshold for the initiation of bedload motion $\theta_{c50,\mathrm{Min}}$ may vary along the river length (we assumed a constant value for the best-case simulation) and, as described before, the effective channel width exhibits considerable uncertainty in depositional reaches. However, considering the more detailed knowledge of the system of the Kleine Emme, a reduced uncertainty of only plus or minus 20 % is more appropriate than an uncertainty of 30 % for discharge and channel width in this catchment.

[2]However, it cannot be excluded that in a few time steps another of the conditions for temporal discretisation (listed in Sect. 2.4) caused a different time step length.

Table 2. Summary of calibration period simulations with different equation sets.

Figure	River	Flow resistance	Bedload transport	Threshold for transport	ABT-RMSE	ABT-Nash–Sutcliffe
6	Kleine Emme	Manning–Strickler-type	Rickenmann (Eq. 7) W and C hiding*	Lamb et al. (2008) $\theta_{c50,\mathrm{Min}} = 0.06$	$7.83 \times 10^3\ \mathrm{m}^3$	0.949
7	Brenno	Variable power-law	Rickenmann (Eq. 7) no hiding	Lamb et al. (2008) $\theta_{c50,\mathrm{Min}} = 0.1$	$18.0 \times 10^3\ \mathrm{m}^3$	0.733

* Wilcock and Crowe (2003) hiding (Eq. 9).

For the second sensitivity study, all possible combinations of maximum decreased (−30 %), best-fit and maximum increased (+30 %) values for all treated parameters were simulated[3]. In the following we will call this a complete range sensitivity study. In this complete range sensitivity analysis, the sediment input volumes from the tributaries to the Brenno were varied as well by plus or minus 30 %.

Some model parameters described in the companion paper by Heimann et al. (2015) have not been included in the sensitivity analyses for the following reasons:

- For the exponent e of the flow-resistance partitioning approach of Rickenmann and Recking (2011) and Nitsche et al. (2011), previous studies have shown that for various cases and conditions the value of 1.5 performed well in reproducing available observations (Nitsche et al., 2011). Therefore, we have not included e in our sensitivity study and instead recommended the use of a default value of 1.5.
- The abrasion coefficient λ of the equation of Sternberg (1875) is commonly only used in simulations of test reaches longer than 30 km, as this is the minimum distance for λ to have considerable influence.
- The hiding exponents m_{wc} and m (Eqs. 8 and 9) do not fit in the concept of the presented sensitivity analysis, which is the variation of a best-fit value by a certain percentage. In addition, there are almost no field data providing guidance for suitable values of the hiding function for the coarser part of the GSD.

3 Results

3.1 Simulations for the calibration period

At the Kleine Emme, the simulated ABT shows agreement with the observed sediment budget (Fig. 6). Locally, how-

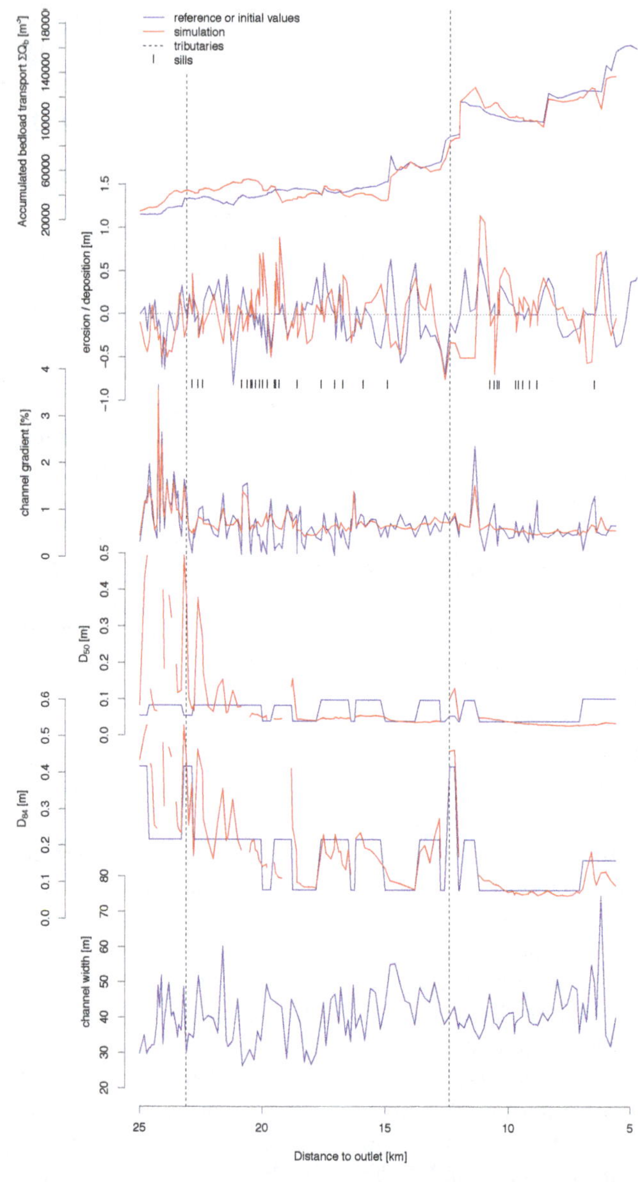

Figure 6. Comparison of predictions and observations related to bedload transport in the Kleine Emme for the period 2000–2005 (tributaries from up- to downstream: Fontanne, Rümlig).

[3]In three simulations at the Kleine Emme with high $\theta_{c50,\mathrm{Min}}$, low discharge, coarse GSD and narrow, mean or wide channel widths, the river could not transport the bank erosion sediment inputs near 12 km. This resulted in the creation of adverse slopes. Therefore, these three simulations have been excluded from the complete range sensitivity study.

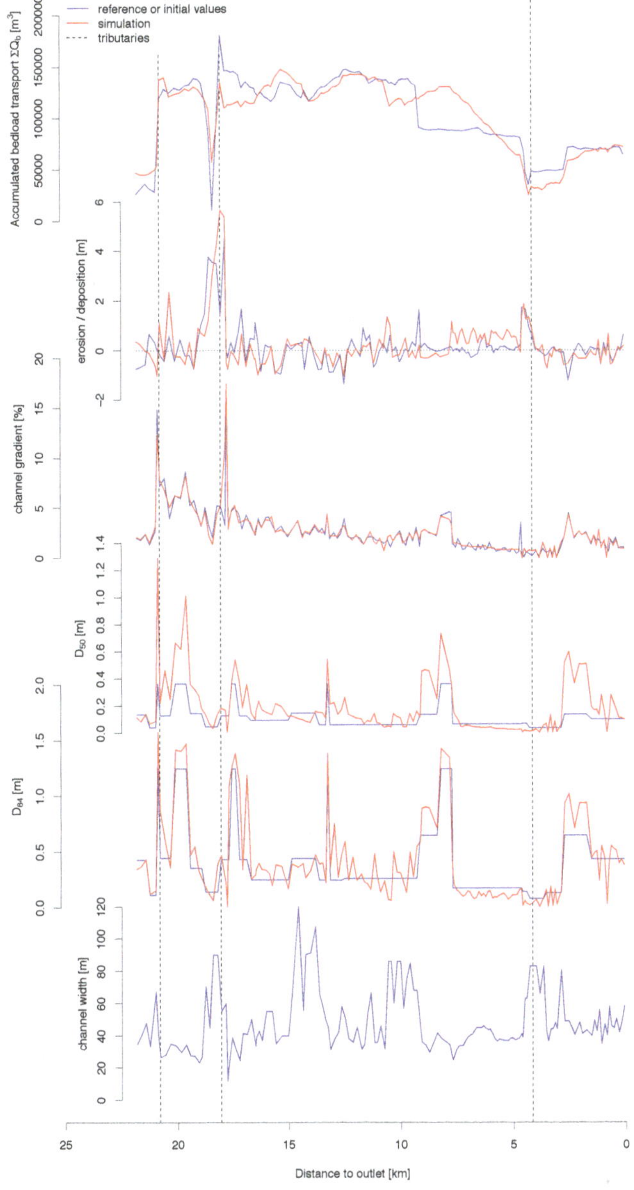

Figure 7. Comparison of predictions and observations related to bedload transport in the Brenno for the period 1999–2009 (tributaries from up- to downstream: Riale Riascio, Ri di Soi, Lesgiüna).

ever, simulations and observations of erosion and deposition can differ considerably. In the uppermost part down to ~ 17 km, peaks of very coarse GSDs are simulated. The gaps in the simulated GSD represent reaches in which the alluvial cover is washed out completely and the river runs over bedrock. Downstream of ~ 17 km, simulated final GSDs are close to the initial values.

At the Brenno, the simulation depicts well the interactions and qualitative transport behaviour in the vicinity of the tributaries Riale Riascio and Ri di Soi (Fig. 7). Downstream of the anthropogenic excavation at 10 km, which is not considered in the simulation, the model exhibits an overall depositional trend. The low sediment transport from 4.5 to 3 km due to a locally increased channel width is well reflected in the simulations. Except for the depositional trend from ~ 8 to ~ 4.5 km (which did not occur in reality because substantial sediment volume was anthropogenically excavated from this reach) the simulated erosion and deposition show good agreement with the observations. In reaches with larger channel gradients the model produces a coarsening of GSDs. Apart from these reaches, simulated final GSDs are close to their initial values.

In both rivers, the model tends to smoothen spatially varying channel gradients (Figs. 6 and 7). Furthermore, in both rivers, the GSDs evolve over the course of a model run such that the final GSDs can be interpreted as a function of bed slope (coarse grains in steep sections), channel width (coarse grains in narrow sections) and channel network (coarse grains at confluences with steep tributaries). The channel width is not modified during the simulations.

The simulations of the Brenno suggest an intense backward migrating erosion of the knickpoints at the confluences with debris flow tributaries, but this is not observed in the field. This erosion can be prevented in the simulations either by limiting the alluvium thickness and thus potential erosion depth, or by adding coarse blocks to the local GSD, which have not been captured in the transect pebble count, or by introducing a maximum Froude number limit in the flow resistance and drag-force partitioning calculations.

In both rivers, early in the course of a simulation the model tends to adjust surface GSDs, which stay roughly the same for the rest of the simulation and which therefore seem to be stable under the local conditions (i.e. local slope, channel width, subsurface GSD and discharge pattern).

In the Kleine Emme, the variation of maximum time step length caused differences in the modelled erosion and deposition only at a few locations. This results in small differences in modelled ABT along the complete river length (Fig. 10). In the Brenno, long maximum time step lengths caused an underestimation of the depositional trend from 6 to 5 km. Downstream of this position, the underestimation of deposition resulted in an overestimation of simulated ABT (Fig. 11).

3.2 Sensitivity analyses

The local sensitivity analysis (Fig. 8) shows that variations in input discharge and GSDs have a large influence on the resulting ABT in both rivers. The impact of variations of the minimum value for the threshold dimensionless shear stress at the initiation of bedload motion ($\theta_{c50,\mathrm{Min}}$) ranges from low in the Brenno to high in the Kleine Emme. In general, relative output variations are larger in the Kleine Emme than in the Brenno. However, this statement is only true when uncertainties of 30 % are assumed for both rivers. The difference in trend is less pronounced when the smaller uncertainties of

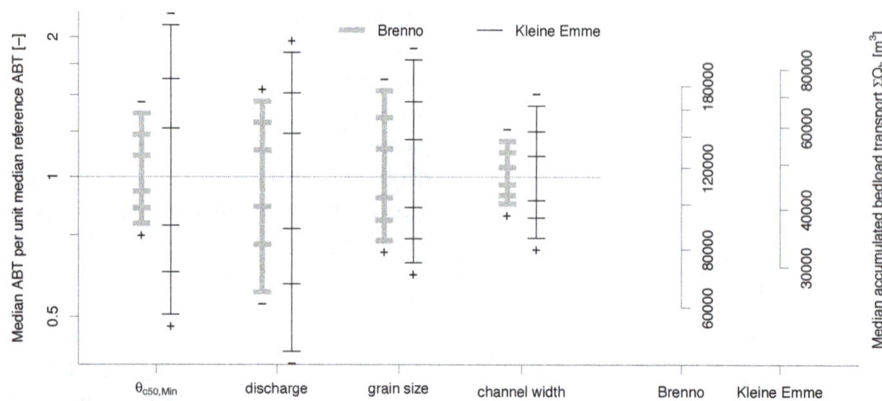

Figure 8. Simulation sensitivity with respect to simulated accumulated bedload transport (ABT) for different input parameters. The horizontal line represents the reference best-fit simulations of Figs. 6 and 7. The tick marks at the vertical bars display the simulation results for input parameter variations of either plus or minus 10, 20, and 30 %. Plus or minus signs at the end of the bars indicate whether the input parameter was increased (plus) or decreased (minus).

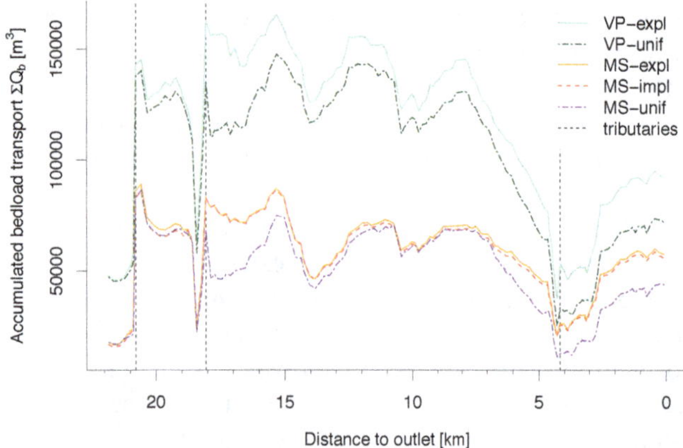

Figure 9. Comparison of simulated accumulated bedload transport in the Brenno for different combinations of flow-routing schemes and flow-resistance relations. Two flow-resistance relations are shown: the variable power relation given in Eq. (5) (denoted VP) and the grain-size-dependent Manning–Strickler relation given in Eq. (6) (denoted MS). Three flow-routing schemes are shown: explicit kinematic wave (denoted expl), implicit kinematic wave (denoted impl) and uniform discharge (denoted unif). The VP-unif curve (green dot-dashed line) is the same as the red line in the top panel of Fig. 7 and displays the reference best-fit simulation.

20 % for discharge and channel width at the Kleine Emme are taken into account.

Comparing the three implemented hydraulic schemes, the explicit and implicit hydraulic flow routing produce practically identical results and the differences to using an uniform discharge approach are small in the Brenno catchment (Fig. 9). In contrast, there is a considerable difference in ABT between the simulations based on the two different flow-resistance relations (Fig. 9).

As a main result of the complete range sensitivity study, the variation of input values caused considerable variation in the simulated ABT, but caused very little variability in the simulated erosion and deposition (Figs. 12 and 13).

4 Discussion

4.1 Simulations for the calibration period

Bedload transport and morphodynamic observation of both rivers can be reproduced with plausible parameter set-ups (Table 2 and Figs. 6 and 7). At the Kleine Emme the simulated absolute values of net erosion and net deposition at the end of the calibration period are small and thus close to the noise of the measurements. Therefore, the differences between observed and simulated morphodynamics may be partly explained as noise. The simulated peaks of very coarse GSD in the upper part of the Kleine Emme are due to the small alluvium thickness, which is in some places washed out completely (or nearly so). If only a few coarse grains are left in a reach, they will produce extremely coarse grain-size percentiles. At the Brenno, the deposition from ~ 8 to ~ 4.5 km (Fig. 7), which substitutes for the unconsidered excavation, appears as a plausible behaviour of the river without any anthropogenic interventions. Coarsening at reaches with increased channel gradient is plausible as well. At the Brenno, the minimum threshold dimensionless shear stress $\theta_{c50,\mathrm{Min}}$ for the initiation of bedload motion has been calibrated to a value of 0.1 (Table 2). This corresponds to the findings of Lamb et al. (2008) and Bunte et al. (2013), who showed that in mountain rivers θ_c may well assume values in this order of magnitude.

The good agreement of bedload transport simulations and observations may be surprising, given that the natural system is complex and the model representation is relatively simple, with only a few parameters for calibration. The selected transport equation and threshold for the initiation of motion determine the average level of transport volumes.

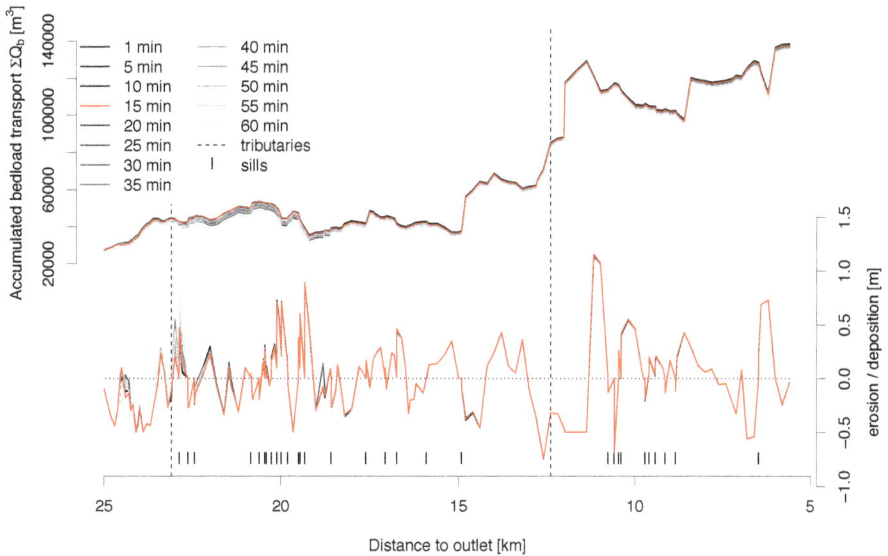

Figure 10. Comparison of simulated accumulated bedload transport and erosion and deposition in the Kleine Emme for different maximum time step lengths denoted in the plot legend. The maximum time step length value, which has been used for any other simulation in the Kleine Emme (e.g. Fig. 6), is displayed in red.

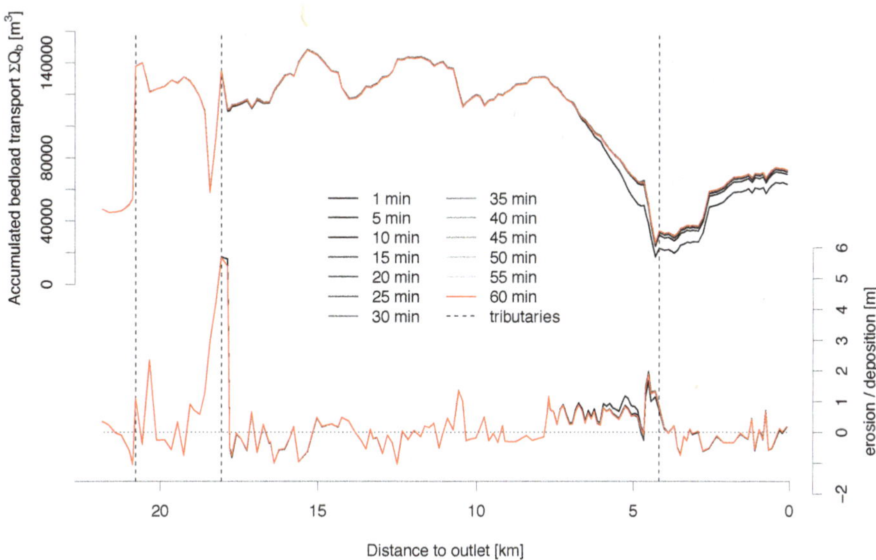

Figure 11. Comparison of simulated accumulated bedload transport and erosion and deposition in the Brenno for different maximum time step lengths denoted in the plot legend. The maximum time step length value, which has been used for any other simulation in the Brenno (e.g. Fig. 7), is displayed in red.

The selected hiding function locally modulates the calculated volumes and in particular influences the evolution of the GSD. Despite its simplicity, the described modelling framework appears to be adequate for a quantitative description of bedload transport processes, as suggested by the reasonable agreement of simulation and observation.

The better agreement of simulated and reference ABT at the Kleine Emme compared to the Brenno is not surprising. At the Kleine Emme, there are no debris flow inputs, the influence of tributaries is limited and the sediment outflow is known. The Kleine Emme is a well-defined system with low uncertainties and thus is ideal for simulation. In addition, spatially distributed calibration was applied more extensively to the Kleine Emme than to the Brenno. For the Brenno, spatially distributed calibration was performed by adjusting the width of the channel. This was done only at depositional reaches, which entail considerable uncertainty in the representative substitute channel width and which cor-

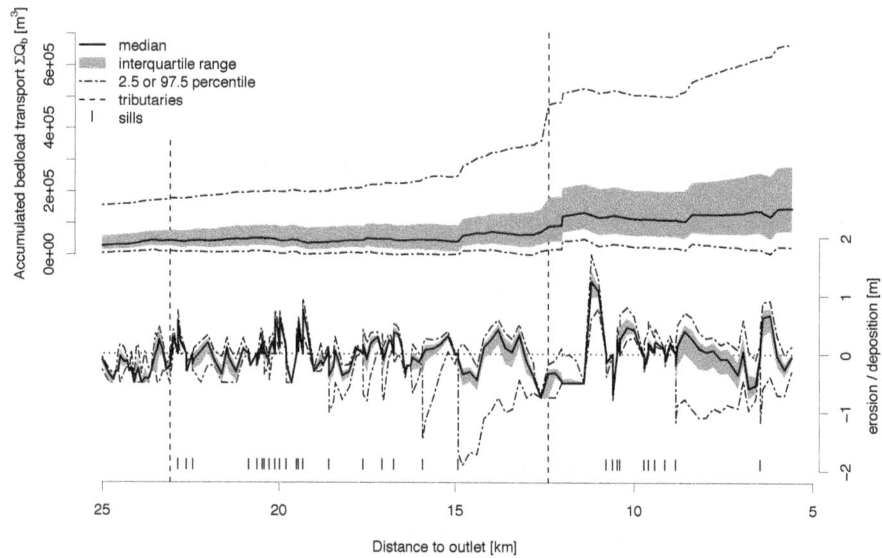

Figure 12. Output variability within the sensitivity study for the Kleine Emme.

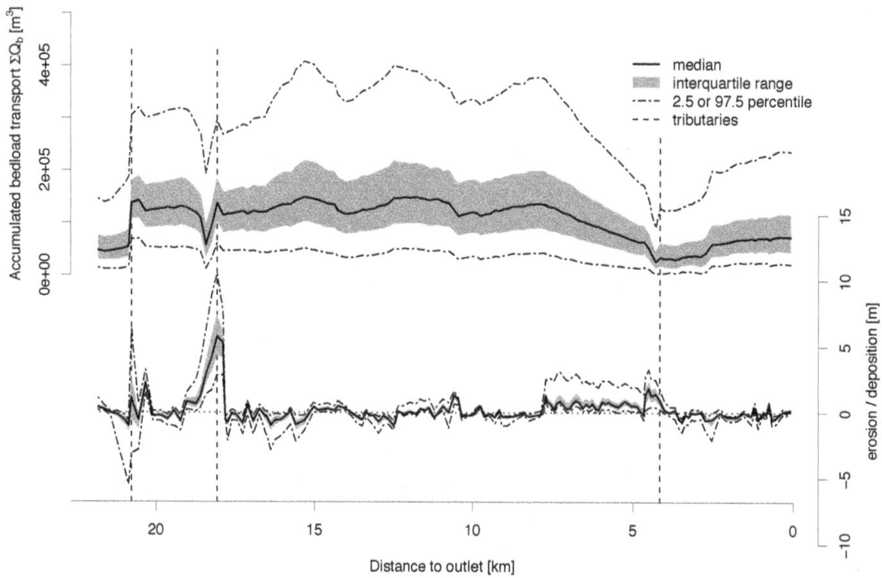

Figure 13. Output variability within the sensitivity study for the Brenno.

respond to only ca. 30 % of the total study-reach length. In contrast, at the Kleine Emme, spatially distributed calibration was performed by adjusting local GSDs along the complete study reach. This more extensive, spatially distributed calibration at the Kleine Emme partly also explains the better agreement of simulated and reference ABT at the Kleine Emme compared to the Brenno.

Few studies (Lopez and Falcon, 1999; Chiari and Rickenmann, 2011; Mouri et al., 2011) have performed a spatially distributed comparison of simulations and field observations, similar to what is presented in this article. However, these studies focused on shorter river lengths than the Brenno and the Kleine Emme. Lopez and Falcon (1999) performed a lumped calibration by simply multiplying calculated transport rates by four. In all aforementioned studies, the models have been calibrated but not independently validated (similarly to the present investigation). This contrasts with approaches used in other research fields such as hydrology (Beven and Young, 2013), where it is common practice to perform a calibration and a validation separately. The lack of independent validation is mainly due to the marked scarcity of available field data on bedload transport. Other studies compared simulation results against point data derived from field observations (Hall and Cratchley, 2006; Li et al., 2008),

Figure 14. River bed of the Brenno at the confluence with the Riale Riascio (at 20.8 km, Fig. 7) exhibiting blocks with diameters of up to 2 m.

against analytic considerations (García-Martinez et al., 2006) or against a combination of field data and analytical results along with additional flume experiment data and results from other models (Papanicolaou et al., 2004). Many studies discuss model behaviour without any explicit comparison between that behaviour and observational data (Lopez and Falcon, 1999; Papanicolaou et al., 2004; Hall and Cratchley, 2006; Li et al., 2008; García-Martinez et al., 2006; Radice et al., 2012).

The simulated GSDs might be seen as a proxy for GSDs which are consistent with the local slope, channel geometry and discharge pattern. This idea is rather attractive, as the model would use variables with a low uncertainty to estimate the local GSD, which is associated with a relatively large uncertainty. Unfortunately, the simulated surface GSD also depends on the subsurface GSD, and on the algorithm regulating the exchange between the surface and subsurface layers (for details see Heimann et al., 2015). In any case, the simulated surface GSD is consistent with local conditions. However, the simulated surface GSD is influenced either by an unrealistically small interaction between surface and subsurface, or by a highly uncertain and possibly incorrect subsurface GSD. Nevertheless, these simulated GSDs will be internally consistent with the other assumptions in the model and thus may have the potential to serve as input for calibration exercises and follow-up studies. A detailed investigation of this topic is beyond the scope of this article.

The simulated erosion of knickpoints in the Brenno was not observed in the field and is thus unrealistic. This suggests that large blocks, which are present in these reaches, but which have not been captured by the transect pebble counts, are important to stabilise the bed. The influence of large blocks also explains why the GSD at these positions had to be coarsened to achieve realistic model behaviour. In addition, the simulated GSDs coarsened even further. Both the unrealistic erosion and the need for coarsened GSDs point to the limitations of a volumetric percentile grain diameter to serve as a proxy for channel roughness. Flow-resistance estimation depends on the representative grain diameter D_{84} in both the Manning–Strickler and variable power equation formulations (Ferguson, 2007). However, even a few large blocks, possibly at percentiles higher than 84, can heavily influence the properties of the flow. The problems of a single representative grain-size percentile used as a proxy for bed roughness become more severe in the case of a discontinuous GSD, for example if the coarse blocks originate from rock fall and thus from a different source than the alluvial gravel. In such cases, any percentile diameter will considerably over- or underestimate the roughness, if its value falls in the gap of the discontinuous GSD. Coarse blocks are also a problem for the general concept of a volumetric percentile. Only a small fraction of the volume of a large block belongs to the surface layer of the river bed, which is assumed to define its roughness. Large parts of such blocks protrude into the deeper alluvium or into the water flow not belonging to the surface layer (or even into the air above the flow). Therefore, the volumetric contribution of such blocks to the surface layer is hard to determine. These issues are reflected in conceptual models for flow resistance, such as the ones of Yager et al. (2007) or Nitsche et al. (2012), which consider large blocks explicitly, e.g., in terms of a surface block density. In a recent study, Ghilardi (2013) suggested that the protrusion height of large blocks into the flow could be used as a potential proxy for flow resistance. Based on this approach, the visual appearance of the Brenno river bed (Fig. 14) suggests a roughness of about 1–2 m. This value is of the same order of magnitude as the D_{84} of the coarsened GSDs (Fig. 7), which we used as a roughness proxy in our simulations. It is further supported by additional area block counts in the Brenno, which showed that grains with a diameter smaller than 1 m only make up 90 % of the surface layer's sediment volume or even only 75 % at the confluence with the Riale Riascio (Fig. 14). These blocks observed in the field dominate the macro-roughness. Since D_{84} is selected to represent macro-roughness, the block counts support the D_{84} values which are used in the simulations, and which are of the same order of magnitude as the observed block diameters.

To assess the influence of time step length, the user-defined maximum time step length was varied between 1 min and 1 h. In the Kleine Emme, the influence of time step length is negligible compared to the overall uncertainty of bedload transport simulations (Fig. 10). In the Brenno, the effect of large maximum time step lengths is spatially limited and well defined (Fig. 11) and thus can be easily considered in the interpretation of the simulation results.

4.2 Sensitivity analyses

The limitations of simple one-at-a-time sensitivity studies for the analysis of non-linear processes are well known (Saltelli et al., 2006). However, an adequate global sensitivity analysis, in which the complete parameter space is covered, would go beyond the scope of this article.

As shown in Fig. 8 the model reacts differently to input changes, depending on which parameter is modified. The model's reaction to input changes also depends on the current river setting. For example, the relative variability and thus uncertainty of model outputs is generally larger in the Kleine Emme as compared to the Brenno. This may be partially due to the fact that the volumes of transported sediment are generally smaller in the Kleine Emme as compared to the Brenno. However, the output uncertainty can be partially compensated by better-supported knowledge and thus higher confidence in the inputs (reduced uncertainty of only plus or minus 20 % for discharge and channel width at the Kleine Emme). Interestingly, even the order of parameter sensitivities may change depending on the current river setting. For example, the reaction to changes in the minimum threshold for the initiation of bedload transport $\theta_{c50,\mathrm{Min}}$ differs considerably for the two rivers. In the Kleine Emme, the uncertainty of this parameter seems to be responsible for a large part of the model output uncertainty. In contrast, in the Brenno $\theta_{c50,\mathrm{Min}}$ plays a rather subordinate role.

In the complete range sensitivity study (Figs. 12 and 13) all input variations have been applied to the complete length of the river. This may explain why the simulated erosion and deposition show only limited variation compared to the simulated ABT. Erosion and deposition are a function of changes of channel properties (gradient, width, GSD, inputs) along the river. Applying the input variation to the complete length of the river keeps the relative changes of channel properties the same. Even though bedload transport is not a linear system, the input variation on the complete length of the river did not cause considerable variation of simulated erosion and deposition. Nevertheless, the sensitivity study with its highly variable ABT and almost constant morphodynamics stresses the uncertainty of ABT estimates that are only derived from morphologic changes. These simulation results support previous studies that have discussed this issue (Kondolf and Matthews, 1991; Reid and Dunne, 2003; Erwin et al., 2012). This is especially important because ABT plots are very common for the description of bedload transport in applied engineering practice and are even recommended by authorities (e.g. Schälchli and Kirchhofer, 2012).

As is illustrated in Fig. 9 for the Brenno river, the two different flow-resistance relations produce considerably different values of simulated ABT. This further stresses the limitations of Manning–Strickler-type flow-resistance relations in steep mountain streams, as discussed in Rickenmann and Recking (2011). In contrast, the three different flow-routing schemes predict similar transported bedload volumes in the Brenno river (Fig. 9). Differences can be neglected when compared to the overall uncertainties of bedload transport simulations. Therefore, the influence on the model outputs does not constitute a preference for any of the hydraulic schemes and any scheme can be selected based on its characteristics. If adverse slopes occur or if the variable power equation flow resistance, which is more suitable for shallow flow in steeper channels, is to be used without slowing down the calculations, one may select the uniform discharge approach. If one needs neither the ability to deal with adverse slopes nor the use of the variable power equation flow resistance, one may select the implicit kinematic wave routing, as it provides a routing of discharge. If a variable power equation approach is to be combined with a routing of discharge, one may select the explicit kinematic wave routing, even though this option is not recommended due to its long calculation times.

5 Conclusions

In this article, we used the model sedFlow to calculate bedload transport in two Swiss mountain rivers. sedFlow is a tool designed for the simulation of bedload dynamics in mountain streams. Observations of bedload transport in these two rivers have been successfully reproduced with plausible parameter settings. The results of the one-at-a-time range sensitivity analysis have shown that a defined change of an input parameter produces larger relative changes of output sediment transport rates in the Kleine Emme as compared to the Brenno, which may be due to the generally smaller transport rates at the Kleine Emme. Simulation results highlighted the problems that can arise because traditional flow-resistance estimation methods fail to account for the influence of large blocks. As an important result of our study, we conclude that a very detailed and sophisticated representation of hydraulic processes is apparently not necessary for a good representation of bedload transport processes in steep mountain streams. Both uniform flow routing and kinematic wave routing performed well in simulating field observations related to bedload transport. Moreover, it has been shown that bedload transport events with widely differing accumulated bedload transport (ABT) may produce identical patterns of erosion and deposition. This highlights the uncertainty in ABT estimates that are derived only from morphologic changes. This proof-of-concept study demonstrates the usefulness of sedFlow for a range of practical applications in alpine mountain streams.

Appendix A

Table A1. Notation.

The following symbols are used in this article.	
η_{pore}	pore volume fraction
θ_i	dimensionless bed shear stress for ith grain-size fraction
$\theta_{i,\text{r}}$	θ_i using S_{red} to account for macro-roughness
θ_c	dimensionless bed shear stress at initiation of bedload motion
θ_{ci}	θ_c for ith grain-size fraction
θ_{c50}	θ_c for the median grain diameter
$\theta_{c50,\text{Min}}$	minimum value for θ_{c50}
$\theta_{ci,\text{r}}$	θ_{ci} accounting for macro-roughness
λ	abrasion coefficient of the equation of Sternberg (1875)
ρ	fluid density
ρ_{s}	sediment density
τ	dimensionful bed shear stress
Φ_{bi}	dimensionless bedload flux for ith grain-size fraction
a_1,a_2	empirical constants
D_i	mean grain diameter for ith grain-size fraction
D_{m}	geometric mean for grain diameters
D_x	xth percentile for grain diameters
D_{50}	median grain diameter
e	exponent of the flow-resistance partitioning approach of Rickenmann and Recking (2011) and Nitsche et al. (2011)
F_i	proportion of ith grain-size fraction
Fr	Froude number
g	gravitational acceleration
m	empirical hiding exponent ranging from 0 to -1
m_{wc}	hiding exponent according to Wilcock and Crowe (2003)
q	discharge per unit flow width
Q_{b}	bedload flux
q_{b}	bedload flux per unit flow width
q_{bi}	q_b for ith grain-size fraction
$q_{\text{b}_{\text{lat}}}$	lateral bedload influx per unit flow width
q_c	threshold q for initiation of bedload motion
r_{h}	hydraulic radius
s	density ratio of solids and the fluid
S	slope of hydraulic head
S_{b}	slope of river bed
S_{red}	slope reduced for macro-roughness
t	time
v_m	average flow velocity
v^*	shear velocity
V_{cap}	volume of sediment corresponding to the transport capacity in a reach
V_{EroDepo}	volume of sediment that is eroded or deposited in a reach
V_{in}	volume of sediment that enters a reach
V_{inUp}	volume of sediment that enters a reach from upstream
V_{inLat}	volume of sediment that is introduced laterally to a reach
V_{out}	volume of sediment that exits a reach
x	distance in flow direction
z	elevation of channel bed

Acknowledgements. We are grateful to Christa Stephan (project thesis ETH/WSL), Lynn Burkhard (MSc thesis ETH/WSL), Anna Pöhlmann (WSL), Claudia Bieler (MSc thesis ETH/WSL) and Christian Greber (MSc thesis ETH/WSL) for their contributions to the development and application of sedFlow. Special thanks to Massimiliano Zappa for his PREVAH support and the hydrologic input data. We thank the Swiss National Science Foundation for funding this work in the framework of the NRP 61 project "Sedriver" (SNF grant no. 4061-125975/1/2). The simulations in the Brenno river were also supported by the BAFU (GHO) project "Feststofftransport in Gebirgs-Einzugsgebieten" (contract no. 11.0026.PJ/K154-7241) of the Swiss Federal Office for the Environment. Jeff Warburton and an anonymous referee provided thoughtful and constructive suggestions to improve this manuscript.

Edited by: D. Parsons

References

Anastasi, G.: Geschiebeanalysen im Felde unter Berücksichtigung von Grobkomponenten. Mitteilung Nr. 70. Versuchsanstalt für Wasserbau, Hydrologie und Glaziologie, ETH Zurich, Zurich, 99p., 1984.

Badoux, A., Andres, N., and Turowski, J. M.: Damage costs due to bedload transport processes in Switzerland, Nat. Hazards Earth Syst. Sci., 14, 279–294, doi:10.5194/nhess-14-279-2014, 2014.

Bathurst, J. C., Bovolo, C. I., and Cisneros, F.: Modelling the effect of forest cover on shallow landslides at the river basin scale, Ecol. Eng., 36, 317–327, doi:10.1016/j.ecoleng.2009.05.001, 2010.

Beffa, C.: 2D-Simulation der Sohlenreaktion in einer Flussverzweigung, Österreichische Wasser- und Abfallwirtschaft, 42, 1–6, 2005.

Beven, K. and Young, P.: A guide to good practice in modeling semantics for authors and referees, Water Resour. Res., 49, 5092–5098, doi:10.1002/wrcr.20393, 2013.

Bhowmik, N. G., Tsai, C., Parmar, P., and Demissie, M.: Case study: application of the HEC-6 model for the main stem of the Kankakee River in Illinois, J. Hydraul. Eng.-ASCE, 134, 355–366, doi:10.1061/(ASCE)0733-9429(2008)134:4(355), 2008.

Bunte, K., Abt, S. R., Potyondy, J. P., and Swingle, K. W.: A comparison of coarse bedload transport measured with bedload traps and Helley–Smith samplers, Geodin. Acta, 21, 53–66, doi:10.3166/ga.21.53-66, 2008.

Bunte, K. B., Abt, S. R., Swingle, K. W., Cenderelli, D. A., and Schneider, J. M.: Critical Shields values in coarse-bedded steep streams, Water Resour. Res., 49, 1–21, doi:10.1002/2012WR012672, 2013.

Chiari, M. and Rickenmann, D.: Back-calculation of bedload transport in steep channels with a numerical model, Earth Surf. Proc. Land., 36, 805–815, doi:10.1002/esp.2108, 2011.

Chiari, M., Friedl, K., and Rickenmann, D.: A one-dimensional bedload transport model for steep slopes, J. Hydraul. Res., 48, 152–160, doi:10.1080/00221681003704087, 2010.

Courant, R., Friedrichs, K., and Lewy, H.: Über die partiellen Differenzengleichungen der mathematischen Physik, Math. Ann., 100, 32–74, doi:10.1007/BF01448839, 1928.

Doten, C. O., Bowling, L. C., Lanini, J. S., Maurer, E. P., and Lettenmaier, D. P.: A spatially distributed model for the dynamic prediction of sediment erosion and transport in mountainous forested watersheds, Water Resour. Res., 42, W04417, doi:10.1029/2004WR003829, 2006.

Erwin, S. O., Schmidt, J. C., Wheaton, J. M., and Wilcock, P. R.: Closing a sediment budget for a reconfigured reach of the Provo River, Utah, United States, Water Resour. Res., 48, W10512, doi:10.1029/2011WR011035, 2012.

Faeh, R., Mueller, R., Rousselot, P., Vetsch, D., Volz, C., Vonwiller, L., Veprek, R., and Farshi, D.: System manuals of BASEMENT. vol 2.2., VAW, ETH Zurich, Switzerland, available at: www.basement.ethz.ch (last access: 15 July 2014), 2011.

Fehr, R.: A method for sampling very coarse sediments in order to reduce scale effects in movable bed models, in: Proc. Symp. Scale effects in modelling sediment transport phenomena, Toronto, IAHR, Delft, 383–397, 1986.

Fehr, R.: Geschiebeanalysen in Gebirgsflüssen. Mitteilung Nr. 92. Versuchsanstalt für Wasserbau, Hydrologie und Glaziologie, ETH Zurich, Zurich, 139p., 1987.

Ferguson, R.: Flow resistance equations for gravel- and boulder-bed streams, Water Resour. Res., 43, W05427, doi:10.1029/2006WR005422, 2007.

Ferguson, R. I., Church, M., and Weatherly, H.: Fluvial aggradation in Vedder River: Testing a one-dimensional sedimentation model, Water Resour. Res., 37, 3331–3347, doi:10.1029/2001WR000225, 2001.

Flussbau AG: Geschiebetransport im Brenno. Einfluss der Murgangablagerung aus dem Ri di Soi, Tech. rep., Reppublica e Cantone Ticino – Divisione delle costruzioni – Ufficio dei corsi d'acqua, 2003 (unpublished report).

Flussbau AG: Revitalisierung von Auenökosystemen (Risanamento dei Ecosistemi Alluvionali) Fachbericht Morphologie und Geschiebe, Tech. rep., Reppublica e Cantone Ticino – Dipartimento del Territorio – Consorzio Risanamento Ecosistemi Alluvionali, 2005 (unpublished report).

Flussbau AG: Ereignisanalyse Hochwasser 2005 – Seitenerosion, Tech. rep., Swiss Federal Office for the Environment FOEN, 2009 (unpublished report).

García-Martinez, R., Espinoza, R., Valeraa, E., and González, M.: An explicit two-dimensional finite element model to simulate short- and long-term bed evolution in alluvial rivers, J. Hydraul. Res., 44, 755–766, doi:10.1080/00221686.2006.9521726, 2006.

Geoportal Kanton Luzern: available at: www.geo.lu.ch/map/gewaessernetz (last access: 27 August 2013), 2013.

Ghilardi, T.: Intense sediment transport and flow conditions in steep mountain rivers considering the large immobile boulders, Ph.D. thesis, École polytechnique fédérale de Lausanne, 2013.

Gray, J. R., Laronne, J. B., and Marr, J. D.: Bedload-Surrogate Monitoring Technologies, Tech. rep., US Geological Survey Scientific Investigations Report 2010-5091, available at: http://pubs.usgs.gov/sir/2010/5091 (last access: 15 July 2014), 2010.

Habersack, H., Hengl, M., Huber, B., Lalk, P., and Tritthart, M. (Eds.): Fließgewässermodellierung – Arbeitsbehelf Feststofftransport und Gewässermorphologie, Austrian Federal Ministry of Agriculture, Forestry, Environment and Water Management and Österreichischer Wasser- und Abfallwirtschaftsverband ÖWAV, Vienna, 2011.

Hall, G. and Cratchley, R.: Sediment erosion, transport and deposition during the July 2001 Mawddach extreme flood event, in: Sediment Dynamics and the Hydromorphology of Fluvial Systems, Dundee, UK, July, 136–147, 2006.

Heimann, F. U. M., Rickenmann, D., Turowski, J. M., and Kirchner, J. W: sedFlow – a tool for simulating fractional bedload transport and longitudinal profile evolution in mountain, Earth Surf. Dynam., 3, 15–35, doi:10.5194/esurf-3-15-2015, 2015.

Hunziker, Zarn & Partner AG: Kleine Emme, Geschiebehaushaltstudie. Abschnitt Fontanne bis Reuss, Tech. rep., by order of Kanton Luzern Dienststelle Verkehr und Infrastruktur, Aarau, Switzerland, 2009.

Hunzinger, L. and Krähenbühl, S.: Sohlenveränderungen und Geschiebefrachten, in: Ereignisanalyse Hochwasser 2005. Teil 2 – Analyse von Prozessen, Massnahmen und Gefahrengrundlagen, edited by: Bezzola, G. R. and Hegg, C., Swiss Federal Office for the Environment FOEN and Swiss Federal Research Institute WSL, Chap. 4.2, 118–125, Berne, Switzerland, 2008.

Jaeggi, M.: The floods of August 22–23, 2005, in Switzerland: some facts and challenges, in: Gravel-Bed Rivers VI: From Process Understanding to River Restoration, edited by: Habersack, H., Piégay, H., and Rinaldi, M., Elsevier, 587–604, Amsterdam, The Netherlands, 2008.

Kondolf, G. M. and Matthews, W. V. G.: Unmeasured residuals in sediment budgets: a cautionary note, Water Resour. Res., 27, 2483–2486, doi:10.1029/91WR01625, 1991.

Lamb, M. P., Dietrich, W. E., and Venditti, J. G.: Is the critical Shields stress for incipient sediment motion dependent on channel-bed slope?, J. Geophys. Res., 113, F02008, doi:10.1029/2007JF000831, 2008.

Li, S. S., Millar, R. G., and Islam, S.: Modelling gravel transport and morphology for the Fraser River Gravel Reach, British Columbia, Geomorphology, 95, 206–222, doi:10.1016/j.geomorph.2007.06.010, 2008.

Liu, Z. and Todini, E.: Towards a comprehensive physically-based rainfall-runoff model, Hydrol. Earth Syst. Sci., 6, 859–881, doi:10.5194/hess-6-859-2002, 2002.

Lopez, J. L. and Falcon, M. A.: Calculation of bed changes in mountain streams, J. Hydraul. Eng.-ASCE, 125, 263–270, doi:10.1061/(ASCE)0733-9429(1999)125:3(263), 1999.

Lukey, B., Sheffield, J., Bathurst, J., Hiley, R., and Mathys, N.: Test of the SHETRAN technology for modelling the impact of reforestation on badlands runoff and sediment yield at Draix, France, J. Hydrol., 235, 44–62, doi:10.1016/S0022-1694(00)00260-2, 2000.

Mouri, G., Shiiba, M., Hori, T., and Oki, T.: Modeling reservoir sedimentation associated with an extreme flood and sediment flux in a mountainous granitoid catchment, Japan, Geomorphology, 125, 263–270, doi:10.1016/j.geomorph.2010.09.026, 2011.

Nitsche, M., Rickenmann, D., Turowski, J. M., Badoux, A., and Kirchner, J. W.: Evaluation of bedload transport predictions using flow resistance equations to account for macro-roughness in steep mountain streams, Water Resour. Res., 47, W08513, doi:10.1029/2011WR010645, 2011.

Nitsche, M., Rickenmann, D., Kirchner, J. W., Turowski, J. M., and Badoux, A.: Macroroughness and variations in reach-averaged flow resistance in steep mountain streams, Water Resour. Res., 48, W12518, doi:10.1029/2012WR012091, 2012.

Papanicolaou, A. N., Bdour, A., and Wicklein, E.: One-dimensional hydrodynamic/sediment transport model applicable to steep mountain streams, J. Hydraul. Res., 42, 357–375, doi:10.1080/00221686.2004.9641204, 2004.

Parker, G.: Transport of gravel and sediment mixtures, in: Sedimentation Engineering: Theories, Measurements, Modeling, and Practice, Vol. 110 of ASCE Manuals and Reports on Engineering Practice, American Society of Civil Engineers (ASCE), Chap. 3, 165–252, 2008.

Radice, A., Giorgetti, E., Brambilla, D., Longoni, L., and Papini, M.: On integrated sediment transport modelling for flash events in mountain environments, Acta Geophys., 60, 191–2013, doi:10.2478/s11600-011-0063-8, 2012.

Recking, A.: A comparison between flume and field bed load transport data and consequences for surface-based bed load transport prediction, Water Resour. Res., 46, W03518, doi:10.1029/2009WR008007, 2010.

Recking, A.: Simple method for calculating reach-averaged bed-load transport, J. Hydraul. Eng.-ASCE, 139, 70–75, doi:10.1061/(ASCE)HY.1943-7900.0000653, 2013a.

Recking, A.: An analysis of nonlinearity effects on bed load transport prediction, J. Geophys. Res.-Earth Surf., 118 , 1264–1281, doi:10.1002/jgrf.20090, 2013b.

Reid, L. M. and Dunne, T.: Sediment budgets as an organizing framework in fluvial geomorphology, in: Tools in Fluvial Geomorphology, edited by: Kondolf, G. M. and Piégay, H., John Wiley and Sons, Ltd., Chap. 16, 463–500, 2003.

Rickenmann, D.: Comparison of bed load transport in torrent and gravel bed streams, Water Resour. Res., 37, 3295–3305, doi:10.1029/2001WR000319, 2001.

Rickenmann, D.: Alluvial steep channels: flow resistance, bedload transport prediction, and transition to debris flows, in: Gravel Bed Rivers: Processes, Tools, Environment, edited by: Church, M., Biron, P. M., and Roy, A. G., John Wiley & Sons, 386–397, Chichester, UK, 2012.

Rickenmann, D. and Recking, A.: Evaluation of flow resistance in gravel-bed rivers through a large field data set, Water Resour. Res., 47, W07538, doi:10.1029/2010WR009793, 2011.

Saltelli, A., Ratto, M., Tarantola, S., Campolongo, F., European Commission, and Joint Research Centre of Ispra (I): Sensitivity analysis practices: strategies for model-based inference, Reliab. Eng. Syst. Safe., 91, 1109–1125, doi:10.1016/j.ress.2005.11.014, 2006.

Schälchli, U. and Kirchhofer, A.: Sanierung Geschiebehaushalt. Strategische Planung. Ein Modul der Vollzugshilfe Renaturierung der Gewässer, Swiss Federal Office for the Environment FOEN, Berne, Switzerland, 2012.

Schattan, P., Zappa, M., Lischke, H., Bernhard, L., Thürig, E., and Diekkrüger, B.: An approach for transient consideration of forest change in hydrological impact studies, in: Climate and Land Surface Changes in Hydrology: Proceedings of H01, IAHS-IAPSO-IASPEI Assembly, Vol. 359 of IAHS Publ., 311–319, 2013.

Sternberg, H.: Untersuchungen über Längen- und Querprofil geschiebeführender Flüsse, Zeitschrift für Bauwesen, 25, 483–506, 1875.

Stricker, B.: Murgänge im Torrente Riascio (TI): Ereignisanalyse, Auslösefaktoren und Simulation von Ereignissen mit RAMMS, M.S. thesis, University of Zurich and Swiss Federal Research Institute WSL, 2010.

Totschnig, R., Sedlacek, W., and Fuchs, S.: A quantitative vulnerability function for fluvial sediment transport, Nat. Hazards, 58, 681–703, doi:10.1007/s11069-010-9623-5, 2011.

Turowski, J. M.: Semi-alluvial channels and sediment-flux-driven bedrock erosion, in: Gravel-bed Rivers: Processes, Tools, Environments, edited by: Church, M., Biron, P. M., and Roy, A. G., John Wiley & Sons, 401–418, Chichester, UK, 2012.

Turowski, J. M., Badoux, A., and Rickenmann, D.: Start and end of bedload transport in gravel-bed streams, Geophys. Res. Lett., 38, L04401, doi:10.1029/2010GL046558, 2011.

Unfer, G., Hauer, C., and Lautsch, E.: The influence of hydrology on the recruitment of brown trout in an Alpine river, the Ybbs River, Austria, Ecol. Freshw. Fish, 20, 438–448, doi:10.1111/j.1600-0633.2010.00456.x, 2011.

VAW Laboratory of Hydraulics, Hydrology and Glaciology, ETH Zurich: Geschiebehaushalt Kleine Emme, Studie über den Geschiebehaushalt der Kleinen Emme und Prognose der zukünftigen Sohlenveränderungen zwischen der Lammschlucht und der Mündung in die Reuss, Tech. rep. no. 4106, by order of Tiefbauamt des Kantons Luzern, Zurich, Switzerland, 1997 (unpublished report).

Viviroli, D., Zappa, M., Gurtz, J., and Weingartner, R.: An introduction to the hydrological modelling system PREVAH and its pre- and post-processing-tools, Environ. Modell. Softw., 24, 1209–1222, doi:10.1016/j.envsoft.2009.04.001, 2009.

Wicks, J. and Bathurst, J.: SHESED: a physically based, distributed erosion and sediment yield component for the SHE hydrological modelling system, J. Hydrol., 175, 213–238, doi:10.1016/S0022-1694(96)80012-6, 1996.

Wilcock, P. R. and Crowe, J. C.: Surface-based transport model for mixed-size sediment, J. Hydraul. Eng.-ASCE, 129, 120–128, doi:10.1061/(ASCE)0733-9429(2003)129:2(120), 2003.

Yager, E. M., Kirchner, J. W., and Dietrich, W. E.: Calculating bed load transport in steep boulder bed channels, Water Resour. Res., 43, W07418, doi:10.1029/2006WR005432, 2007.

Re-evaluating luminescence burial doses and bleaching of fluvial deposits using Bayesian computational statistics

A. C. Cunningham[1,2], **J. Wallinga**[3], **N. Hobo**[3,4,5], **A. J. Versendaal**[3], **B. Makaske**[3,4], **and H. Middelkoop**[5]

[1]School of Geosciences, University of the Witwatersrand, Johannesburg, South Africa
[2]Centre for Archaeological Science, School of Earth and Environmental Sciences, University of Wollongong, Wollongong, Australia
[3]Soil Geography and Landscape group & Netherlands Centre for Luminescence dating, Wageningen University, Wageningen, the Netherlands
[4]Alterra, Wageningen University and Research Centre, Wageningen, the Netherlands
[5]Department of Physical Geography, Utrecht University, Utrecht, the Netherlands

Correspondence to: A. C. Cunningham (acunning@uow.edu.au)

Abstract. The optically stimulated luminescence (OSL) signal from fluvial sediment often contains a remnant from the previous deposition cycle, leading to a partially bleached equivalent-dose distribution. Although identification of the burial dose is of primary concern, the degree of bleaching could potentially provide insights into sediment transport processes. However, comparison of bleaching between samples is complicated by sample-to-sample variation in aliquot size and luminescence sensitivity. Here we begin development of an age model to account for these effects. With measurement data from multi-grain aliquots, we use Bayesian computational statistics to estimate the burial dose and bleaching parameters of the single-grain dose distribution. We apply the model to 46 samples taken from fluvial sediment of Rhine branches in the Netherlands, and compare the results with environmental predictor variables (depositional environment, texture, sample depth, depth relative to mean water level, dose rate). Although obvious correlations with predictor variables are absent, there is some suggestion that the best-bleached samples are found close to the modern mean water level, and that the extent of bleaching has changed over the recent past. We hypothesise that sediment deposited near the transition of channel to overbank deposits receives the most sunlight exposure, due to local reworking after deposition. However, nearly all samples are inferred to have at least some well-bleached grains, suggesting that bleaching also occurs during fluvial transport.

1 Introduction

The use of optically stimulated luminescence (OSL) for dating Holocene fluvial deposits is widespread. However, fluvial sediments are not ideal for OSL dating because the intensity of sunlight under water may not be sufficient to reset the OSL signal in some grains prior to their deposition. The remnant OSL signal can then cause the burial dose to be overestimated, leading to an overestimate of the age. This phenomenon is referred to as poor, partial or heterogeneous bleaching (e.g. Wallinga, 2002a).

While the burial age is usually the primary consideration, there are good reasons to quantify the degree of bleaching too. Firstly, it may provide information on the robustness of an OSL age. Secondly, the degree of bleaching might yield information on the sediment source or sediment-transport processes (e.g. Reimann et al., 2015). For instance, if a tsunami deposit appears well bleached, it could indicate that shallow shore-face or intertidal deposits provided the primary sediment source (Murari et al., 2007). For fluvial deposits, poor bleaching might, for instance, reflect short transport distances or an old deposit acting as the primary source.

To compare the bleaching between samples, it is first necessary to distinguish between the part of the equivalent dose (D_e) built up since the time of deposition and the poorly bleached remnant dose. Previous studies have avoided this issue by deliberately sampling modern or known-age sediment. Such studies have indicated that bleaching is better in coarse sand-sized grains compared to finer grains (Olley et al., 1998; Truelsen and Wallinga, 2003), and may be dependent on depositional context (Murray et al., 1995; Schielein and Lomax, 2013) and transport distance (Stokes et al., 2001; Jain et al., 2004; and references therein).

Nevertheless, the inherent variability from sample to sample makes definitive conclusions hard to come by. The main problem arises in distinguishing signal from noise: how much of the sample-to-sample variation in bleaching is due to physical processes, as opposed to random statistical fluctuations? Studies focusing on modern or known-age deposits seldom have enough samples for confident conclusions to be drawn, and no study has quantified the variation between adjacent samples. Moreover, the review of Jain et al. (2004) showed a discrepancy in residual doses of modern fluvial samples compared to young known-age samples, with modern samples yielding larger residual doses. They argued that modern deposits may yet be remobilised, so their transport history is not representative of deposits preserved in the stratigraphic record.

Here we focus not on modern samples but on samples of various ages that have already been used for age estimation. This approach allows for more samples to be included, and avoids the bad-modern-analogue issue, but presents the additional problem of separating out the burial dose from the remnant dose. For this purpose we have designed an age model specifically for these young, partially bleached D_e distributions. We define the degree of bleaching by the proportion of grains that were well bleached upon deposition, rather than by the remnant dose. We apply the model to a suite of 46 samples from embanked floodplains of the lower Rhine in the Netherlands, and correlate the outcome with geomorphic data for each sample.

2 Methods

2.1 Samples and measurements

We use a data set of OSL measurements on a suite of 46 samples from embanked floodplain deposits formed during the past 700 years. Different parts of the data set have been presented by Hobo et al. (2010, 2014) and Wallinga et al. (2010). Samples come from four different sites, all located in the Rhine Delta in the Netherlands (Fig. 1). At each of the sites, several cores (diameter 14–19 cm) were taken in a cross section perpendicular to the river course (see Hobo et al., 2010, for examples). Samples were extracted from the cores in subdued orange light and prepared using methods described by Wallinga et al. (2010). For each of the sample sites, cross sections were constructed based on the borehole database of Utrecht University (Berendsen and Stouthamer, 2002) and additional hand corings. The cross sections were interpreted to identify morphogenetic units (see also Hobo et al., 2014).

Figure 1. Map showing the sample sites. The sites Brummen and Zwolle are along the river IJssel, whereas the other two sites (Neerijnen and OB1-3) are along the river Waal. Both are branches of the river Rhine. OB1-3 refers to two cores from the Hiensche Uiterwaarden and one core from the Gouverneursche polder.

For all samples, radionuclide concentrations were determined with high-resolution gamma-ray spectroscopy, from which dose rates were estimated using standard conversion factors. Sand-sized quartz grains were extracted for single-aliquot regeneration OSL measurements (Murray and Wintle, 2003) for equivalent-dose measurements. Details of the procedure are described by Hobo et al. (2010) and Wallinga et al. (2010). The grain-size fractions varied between samples (180–212 μm, 180–250 μm or 90–180 μm), due to differences in texture of the sampled deposit. The measurement protocols were similar for all samples. The dose response was defined using a single regenerative dose (Wallinga et al., 2010), net OSL signals were defined using the early background subtraction (Cunningham and Wallinga, 2010) and low preheat temperatures were selected to avoid thermal transfer (e.g. Truelsen and Wallinga, 2003).

We identified nine variables that could influence the bleaching of the sample either directly or by proxy. The choice of variables is based on our judgement of possible relevance and data availability. With regard to sample position, we considered the average river water level at the site (recorded in 2001), the height of the present surface at the sample location, the depth of the sample below the present surface, and the depth of the sample relative to the 2001 average water level. With regards to the sample nature, we considered the depositional environment (ordinal classes of distal overbank, overbank, proximal overbank, channel); the sediment texture (clay, loam, silt, sand, coarse sand/gravel); the dose rate; the D_e; and the OSL age (although D_e and OSL age are derived from the model, they can also be considered as predictor variables; the dose rate may indirectly affect the bleaching statistic). Table S1 (Supplement) provides an overview of all variables that are considered.

2.2 Statistical rationale

We seek to define a poor-bleaching score based on the measured D_e distribution, which can then be used to compare bleaching between samples. Previous attempts have applied a statistical model directly to the D_e distribution to define a summary statistic (e.g. the F statistic; Spencer et al., 2003 – skewness and kurtosis applied to single-grain (SG) distributions; Bailey and Arnold, 2006). This type of approach may be valid if the observed D_e distribution is a function of the burial dose and remnant dose. For multi-grain (MG) aliquots, the OSL signal comes from many grains; the D_e for an aliquot is the average of those grains, weighted by their OSL sensitivity (e.g. Wallinga, 2002b; Duller, 2008; Cunningham et al., 2011). Therefore, for MG data sets, the D_e distribution is a function of the burial and remnant doses, and also the aliquot size and the single-grain OSL sensitivity distribution.

Aliquot size and SG sensitivity may vary between samples, so for a statistic to be useful, it must be independent of these factors for the range of samples considered. A model defined directly on the D_e distribution of young samples is also likely to be sensitive to the burial dose, as the measurement precision decreases with decreasing D_e. Our data set contains many samples, measured over several years on different OSL readers. While the SG sensitivity distributions are likely to be similar (as all samples are from Rhine deposits), the aliquot size varies both between and within samples: measurements used either 2 or 3 mm mask size, with grain sizes of 180–212, 180–250 or 90–180 µm. A statistic defined from the MG aliquot D_e distribution (such as the burial dose, overdispersion, degree of bleaching) may not have any real-world meaning, because the data are affected by the confounding variables of aliquot size and SG sensitivity. The meaningful parameters operate at the single-grain level, so the approach we take here is to estimate what combination of single-grain parameters would lead to the measured MG D_e distribution. There are two parts to the procedure. First, we define how the MG D_e distribution is derived from single-grain parameters. Second, we use Bayesian computational statistics to estimate the value each parameter must take to reproduce the observed MG D_e distribution.

2.3 From single-grain parameters to the multi-grain-aliquot distribution

2.3.1 The single-grain sensitivity distribution

The magnitude of the OSL signal induced by a given radiation dose varies from grain to grain. The sensitivity distribution also varies between samples (Duller et al., 2000). Quantifying the SG sensitivity is important for dating partially bleached samples, because it governs the extent of averaging across multi-grain aliquots (Cunningham et al., 2011). We therefore need to define the SG sensitivity distribution in order to simulate an MG D_e distribution. While this can be done using a single-grain measurement system, there are practical difficulties: some grain holes may be empty and some may contain more than one grain, and with many sensitivity values clustered around zero, it is difficult to distinguish signal from noise.

Here we use computational Bayesian statistics to estimate the SG sensitivity distribution from the MG sensitivity data. The first step is to parameterise the SG sensitivity, for which we use the gamma distribution. The gamma distribution can be formatted with two parameters: a shape parameter a and a scale parameter b. By altering these parameters, the gamma distribution can comfortably fit a range of measured SG sensitivity distributions (Fig. 2a). Moreover, when a MG aliquot is simulated from SG sensitivity data, the MG sensitivity distribution can also be fitted with a gamma distribution (Fig. 2b). For measured data, we already know the MG sensitivity distribution (from the regenerative-dose signal) and the approximate number of grains in the aliquot (from the grain size and mask size); we can therefore estimate the parameters a and b of the SG sensitivity distribution using a computational Bayesian procedure similar to that described below.

2.3.2 Modelling the D_e distribution

The single-grain parameters are as follows:

a. SG sensitivity, drawn from the gamma distribution with parameters:
   - a describes the shape (i.e. skewness);
   - b describes the scale.

b. The burial dose, drawn from a normal distribution with parameters:
   - γ: mean, in Gy;
   - σ_b^{SG}: relative standard deviation of the burial dose (fixed as 0.20 for this work).

c. The remnant dose, drawn from the positive part of a normal distribution with mean of 0 and:
   - σ: standard deviation;

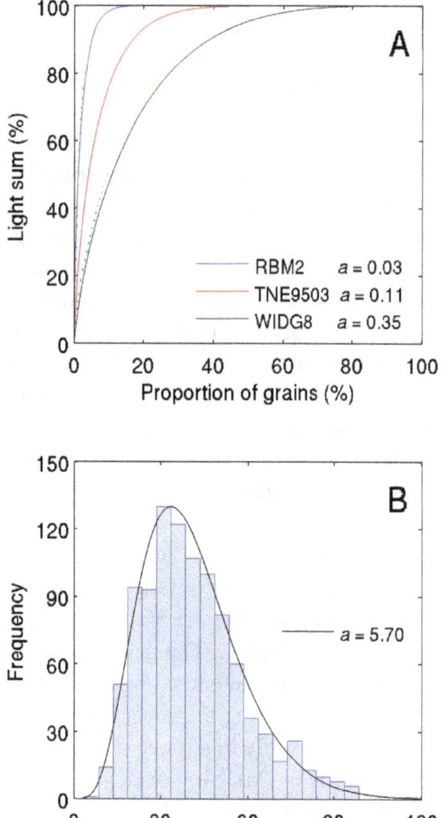

Figure 2. **(a)** Three SG sensitivity distributions presented by Duller et al. (2000) from SG measurements (dotted lines). Each has been approximated using the gamma distribution, with sample name and shape parameter a indicated. **(b)** Simple stochastic simulation of a multi-grain-aliquot sensitivity distribution. The simulation uses the measured SG sensitivity data set RBM2 (from Duller et al., 2000), with parameters $n_g = 200$ and $n_a = 1000$. The MG sensitivity distribution can also be fitted using the gamma distribution, with $a = 5.70$. The shape of the gamma distribution is indicated in the figure, with the y scale normalised to the peak of the histogram.

- p: proportion of well-bleached grains.

d. Additional parameters:

- n_g: number of grains in each aliquot;
- n_a: number of aliquots.

The simulated natural OSL signal from n_a aliquots is the sum of the signal from n_g grains, with Poisson noise added. Each grain is assigned a sensitivity value (per Gy) drawn from the gamma distribution with parameters a and b, and an indicative dose. The indicative dose combines the burial dose, drawn from a normal distribution, and a remnant dose, drawn from a half-normal distribution. The number of grains in each aliquot that have a remnant dose is drawn from the binomial distribution with parameters n_g and $1 - p$. The D_e

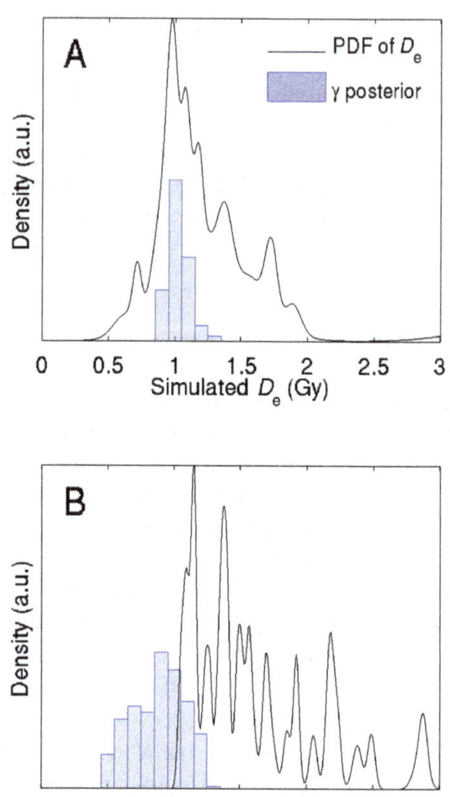

Figure 3. Posterior distribution of the burial dose γ for simulated data of aliquots of **(a)** 80 grains and **(b)** 300 grains. The "given" burial dose is 1 Gy; other parameters are specified in Table 1. The simulated D_e distributions are visualised with probability density functions (PDFs).

is determined by constructing a dose-response curve in the same way as measured data, i.e. one regenerative point of 3 Gy, sensitivity-corrected (although no sensitivity change is added), and subject to the same rejection criteria. Where different aliquot sizes are used in the measured data, these are replicated in the simulation. The number of grains per aliquot (n_g) is approximated using the known mask size and grain size, assuming spherical grains and a 0.7 packing density (i.e. quartz grains cover 70 % of the mask surface).

2.4 Computational Bayesian solution

With the D_e distribution simulated by single-grain parameters, we seek to identify which values the parameters must take to result in the best match between the simulated MG distribution and the measured MG distribution. In Bayesian terms, we seek the posterior distribution, which measures how plausible we consider each possible value of the parameters after we have observed the data. For complex models such as this, the posterior density cannot be calculated directly. Instead, inferences are based on random sampling of the posterior distribution, which requires intensive computation.

In our model, the posterior is sampled using a Markov chain Monte Carlo (MCMC) process. Single-grain parameters for four Markov chains are drawn from a starting distribution, and these parameters are then corrected to better approximate the target posterior. The approximate distributions are improved at each step using the Metropolis algorithm. When the simulation has run long enough, each step can be considered a random draw from the target distribution. The length of the sequence is determined by the convergence of the Markov chains; this is monitored by comparing the within-chain and between-chain variance, following the procedure of Gelman et al. (2004). The first half of each Markov chain is discarded to ensure that the choice of starting values does not influence the result.

2.4.1 Priors

Five parameters are determined in the computational processing: γ, σ, p, a and b. Each of these is assigned a prior distribution, which represents our knowledge of these parameters before any measurements are undertaken. The priors could in future be determined from previous measurement data or, in the case of γ, from the stratigraphic order of the samples. For a and b, the priors are given by the posteriors obtained from the SG sensitivity model. For γ we use a uniform-positive prior, as we have no information on the burial dose and do not want to restrict it. The situation is different when it comes to p and σ. There is an unavoidable conflict when estimating these two parameters: as the residual dose gets smaller, at some point the D_e values cannot be distinguished from well-bleached D_e values. This problem, not unique to our model, could erroneously assign a low p to a well-bleached sample, or high p to a poorly bleached sample (if σ is very small). To prevent this error, we specify arbitrary cut-offs in the priors: the lower limit for σ is 0.25 Gy, and for p it is 0.05. As such, any sample that is poorly bleached to a very small extent ($\sigma < 0.25$ Gy) will be inferred to be well bleached (high p). Very few samples are affected by the low-cut priors. In future, it will be useful to find an objective value for the low cut-off, probably varying with measurement precision.

2.4.2 Density evaluation

Parameters with positive values (γ, σ, a, b) are estimated on the log scale; p must lie between 0 and 1, and is therefore estimated on the logit scale (logit $p = \log(p/(1-p))$; this transforms the unit interval to the real number line). The simulated MG D_e distribution is compared to the measured distribution using the models of Galbraith et al. (1999) as summary statistics. This provides four summary statistics to compare with the measured data (three from the three-component minimum-age model (MAM3), and one from the central-age model (CAM)). The likelihood term is defined by projecting these values onto the bootstrap likelihood distribution for the measured data (see Cunningham and Wallinga, 2012).

The model is run in two phases. The first is a short run, giving an approximate range of the parameter space. The output of the first run is summarised by a multivariate normal distribution, which is used to define the starting distribution and jumping distributions for phase two. The second phase is run until convergence.

Table 1. Results of the simulation recovery: "recovered" values are defined by the mean and standard deviation of the posterior distribution. $n_a = 40$.

	Given	$n_g = 80$	$n_g = 300$
γ	1	1.03 ± 0.08	0.88 ± 0.20
σ	2.5	2.81 ± 1.66	2.96 ± 1.32
p	0.7	0.77 ± 0.13	0.59 ± 0.21
a	0.03	0.042 ± 0.011	0.025 ± 0.008
b	600	604 ± 196	539 ± 188

2.5 Model validation

Here we perform a simulation-recovery test to check that the model is performing as expected. Single-grain parameters are chosen, and then used to simulate D_e data for two different aliquot sizes (80 and 300 grains). Each data set is used as input for the model, and the SG parameters are then reconstructed. The results are given in Table 1, and plotted in Fig. 3 for the burial dose γ.

For both aliquot sizes, the SG parameters can be reconstructed (Table 1). Reconstruction of the 1 Gy burial dose is reasonably precise (8 %) for the 80-grain aliquots, and very close to the bootstrapped MAM3 estimate of the burial dose on the MG aliquot data set (1.03 ± 0.08 Gy, using σ_b of 0.16. σ_b is used in the MAM3 to allow for the expected dispersion in D_e from well-bleached samples, and it changes with aliquot size). For the 300-grain aliquots, the estimate of the burial dose is imprecise but accurate, and lies mostly outside the range of the MG aliquot D_e distribution. For multi-grain aliquots, it is quite possible that none of the aliquots are indicating the burial dose, if at least one grain contributing to the OSL signal on each aliquot is poorly bleached. The model is able to explore this possibility by making use of the MG sensitivity distribution and aliquots size. In contrast, the bootstrap MAM3 applied to the MG data assumes some "well-bleached" aliquots exist, and therefore gives an overestimated burial dose of 1.15 ± 0.03 Gy for this data set ($\sigma_b = 0.08$).

As a further step, it would be interesting to see how the age model applied to multi-grain aliquot data compares to single-grain data from the same sample. However, this comparison is not as simple as it sounds. Our model uses multi-grain aliquot data to estimate the assumed parameters of the SG D_e

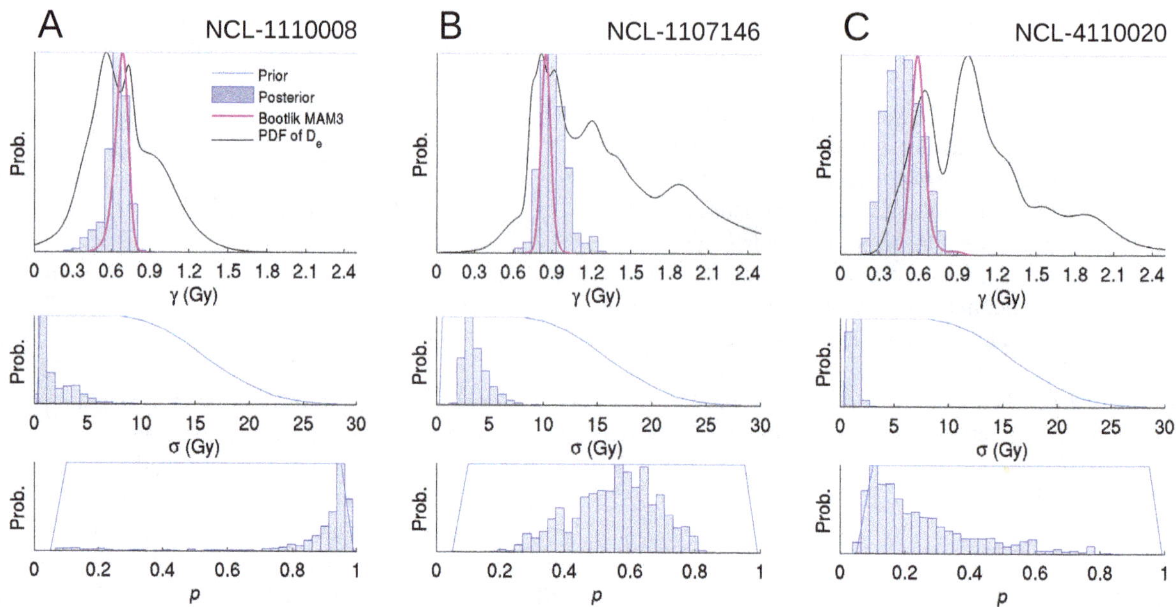

Figure 4. Example model output for three samples that are **(a)** relatively well bleached, **(b)** moderately bleached and **(c)** poorly bleached. For each sample the posterior draws are indicated by the blue histograms, showing the burial dose γ (along with the bootstrap likelihoods of the MAM3 (bootlik MAM3) and a probability density function of the D_e distribution), the residual dose σ, and the proportion of well-bleached grains p. Two sensitivity parameters, a and b, are also estimated in the model, but not shown here. All data sets are normalised to their maximum value.

distribution; it does not reconstruct the SG distribution itself. Testing the model against SG data for a real sample would not distinguish between the performance of the model and validity of the assumptions about SG parameters. The way around this would be to construct an artificial sample with a known dose distribution, like Roberts et al. (2000) and Sivia et al. (2004), but such an elaborate approach is outside the scope of this paper. Also, the mode of optical stimulation in single-grain measurement systems (green laser) differs from that used for MG aliquots for our study (blue LEDs). This prevents direct comparison between SG and MG, as component separation is wavelength-dependent (e.g. Singarayer and Bailey, 2003).

3 Results

The reconstructed SG sensitivity distribution is similar for all samples measured here, not surprising as they are all from recent Rhine deposits. The shape parameter a has a mean of 0.008 and standard deviation 0.003, indicating a highly skewed sensitivity distribution (more so than all of the example distributions in Fig. 2). The averaging effect on multi-grain aliquots is therefore very weak. The scale parameter b has a mean of 400 and standard deviation of 220. The estimates of p are evenly spread between 0.2 and 0.95 (not shown). The uncertainty on p is typically large, except for those values close to 1.

The distribution of σ is positively skewed; the mean is 2 Gy, and most values are below 8 Gy. However, high values of σ have very poor precision, coming from samples with high p in which σ has little influence on the D_e distribution. The susceptibility of σ to outliers and to p makes it unsuitable as an indicator of bleaching. The degree of bleaching is best defined by p, the proportion of well-bleached grains.

Three examples of the model output are shown in Fig. 4, each indicating a different degree of bleaching. For each sample, the histograms indicate the posterior density of the burial dose γ and the two bleaching parameters σ and p. The two sensitivity parameters a and b are of lesser interest for this study, and are not shown. In Fig. 4a, the sample is inferred to be well bleached, and the burial-dose posterior is similar to the MAM3 bootstrap likelihood. This sample is affected by the problematic low-σ/high-p distinction alluded to earlier; the posterior σ is being influenced by the low-cut prior, causing a low tail in the burial-dose posterior. Figure 4b shows a moderately bleached sample, with mean $p = 0.56$. The posterior σ is well clear of the low-cut prior, and the burial dose is less precise but potentially more accurate than the MAM3 burial-dose estimate. A poorly bleached sample is shown in Fig. 4c (mean $p = 0.25$, but affected by the prior). The burial-dose estimate is much less precise than the MAM3 estimate, but is permitted to be smaller than any of the measured MG D_e.

When the burial dose is very close to zero, the posterior distribution is shaped like an exponential decay (not shown here). Such distributions are not well described by the mean and standard deviation, and thus may need to be summarised differently. We will not dwell on the issue here, but will leave it for future consideration.

The mean (and standard deviation) of the p posterior provides a useful summary statistic for between-sample comparison. In Fig. 5, p is plotted against the potential geomorphic variables (Fig. 5a–g), the other model-derived statistics (Fig. 5h–k), and finally the dose rate and model-derived age (Fig. 5l, m). A crude table of correlations is provided in the Supplement (Table S2), but is of limited use because it treats the ordinal variables of "depositional environment" and "texture" as if they are interval types. Nevertheless, there are some correlations among the predictor variables. For example, deeper samples are older and coarser (being channel deposits), and thus they have lower dose rates than the silty overbank samples. There are also significant correlations between the sensitivity parameters a and b and several predictor variables; these are probably due to inadequate aliquot-size estimates, as discussed in Sect. 4.

While there are no clear relationships between p and predictor variables, some of the plots appear to indicate structure beyond that expected through chance. Figure 5g shows that most of the best-bleached samples were located close to the modern mean water level. Figure 5k and m show a trend in p for the relatively young samples. In Fig. 6, the burial-dose estimate is compared directly to the original MAM3. The weighted mean ratio of γ to MAM3 minimum age is 0.97 ± 0.025. Samples inferred to be relatively well bleached return very similar burial doses in both models, with comparable precision. For poorly bleached samples, our model provides a smaller and less precise burial-dose estimate.

4 Discussion

4.1 Influences on bleaching

There appear to be significant relationships between the sensitivity parameters a and b, and several predictor variables: both a and b are correlated with sample depth/elevation (Table S1). These relationships are difficult to explain in geomorphic terms, but may be a manifestation of subtle differences in grain size. Most measurements were carried out on grain-size range of 180–250 μm, with the aliquot size n_g estimated from the grain size and mask size. This grain-size range still allows differences in the grain-size distribution of the natural sediment to be reflected on the disc. For fine sediments, the selected grain-size range will be at the upper tail of the grain-size distribution, resulting in a prepared fraction with many grains at the lower end of the sieved fraction (i.e. 180 μm or just larger). In contrast, for coarse sediments the prepared fraction will be dominated by grains at the upper end of the sieved fraction (i.e. 250 μm and somewhat smaller). The aliquots prepared from overbank sediment will therefore contain more grains than assumed in the model, and aliquots prepared from channel sediments will contain fewer. These biases lead to an error in the model's estimate of the SG sensitivity. The error would also feed through into the estimate of bleaching parameters. We should therefore ignore the correlations involving the sensitivity parameters for this data set, and be cautious about any relationship between bleaching parameters and depositional environment or depth, as both are correlated with sediment texture.

The large degree of uncertainty in our model results prevents convincing conclusions from being drawn. This uncertainty is again down to aliquot size. Firstly, the aliquot sizes used were often too large. Second, our post hoc estimates of the aliquot size were not sufficiently accurate (as noted by Heer et al., 2012). These issues affected model efficiency and outcome by amplifying the difficulty of distinguishing high p and low σ. By contrast, recent measurements by Cunningham et al. (2015) included a grain-counting step, and allowed more emphatic conclusions to be drawn. Here, we do not feel confident in claiming either the existence or absence of geomorphic controls on the basis of our results.

With these caveats established, it is worth considering two structures in the modelled data that might, possibly, have geomorphic significance. These concern the relationship between p and mean water level, and between p and model age, which are enlarged for clarity in Fig. 7. Figure 7a shows a cluster of relatively well-bleached samples coming from sediment deposited close to, or just above, the modern mean water level. This cluster might be telling something about the comparative strengths of bleaching during transport and deposition. The attenuation of light (especially UV/blue) underwater is well established (Berger, 1990), and if light intensity at deposition was the main control on bleaching, we might expect shallower sediments to be better bleached (Wallinga, 2002a). The clustering of high p values around the mean water level may therefore reflect a period of bleaching that occurs at deposition. However, close examination of the samples that are best bleached (NCL-111004, -5, -7, -8; NCL-2107157, -59, -61; NCL-4110018) shows that some of these are sandy channel deposits, whereas others are sand beds within silty overbank deposits, or silty overbank deposits with sand admixtures. All these samples are indeed within a metre from the transition of channel to overbank deposits. For the samples classified as channel deposits, deposition likely occurred on top of point bars, potentially with swash backwash operating and sub-aerial exposure likely (analogous to coastal beaches, which produce well-bleached quartz; e.g. Ballarini et al., 2003). For the well-bleached samples from overbank deposits, we hypothesise that sand grains may have experienced aeolian reworking prior to final deposition and burial. Such aeolian reworking of sandy flood deposits has been documented following sand deposition on overbanks during high-discharge events of the Waal and Lek (Isarin et al., 1995; illustrated in Fig. 8).

There is also structure in Fig. 7b, which shows p in relation to modelled age. Highest p values, indicative of best-

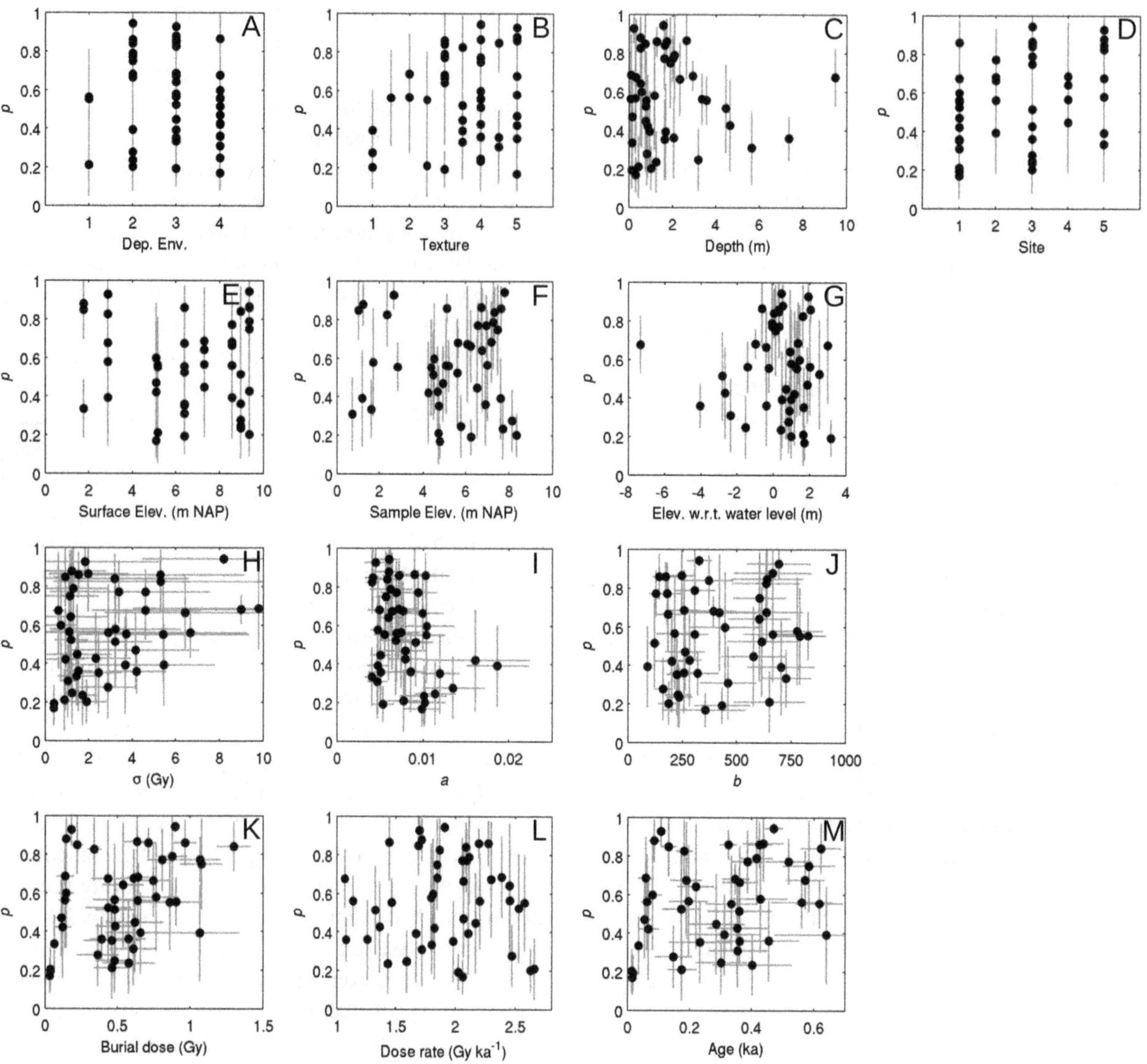

Figure 5. The bleaching statistic p (proportion of grains well bleached) plotted against possible predictor variables **(a–g)**, and other model-derived statistics and the dose rate **(h–m)**. **(a)** Depositional environment on a scale of distal overbank (1) to channel (4). **(b)** Sediment texture is classed from clay (1) to coarse sand (5). Site classification in **(d)** is nominal. NAP is Dutch ordnance datum ($\approx$ mean sea level in Amsterdam). For classification details see Table S1.

bleached deposits, are obtained for samples with model ages around $\sim$0.10 ka. The youngest samples appear to be less well bleached (i.e. low p), and p varies a lot for the samples older than $\sim$0.10 ka. The same trends are observed when burial dose (Fig. 6k) or MAM3 age (not shown) is used instead of modelled age.

There are three possible reasons for the structure. It could be an artefact of the model through the high-p/low-σ problem, but examination of the model output for these samples does not confirm this. Alternatively, it might be a sampling effect caused by coring from the floodplain down towards the high-p cluster at mean water level, but most of the high-p samples are older. A more intriguing explanation involves changes in river management over the last few hundred years. Major changes occurred with dike construction from AD 1000–1300, and river normalisation and groyne emplacement after AD 1850 (Middelkoop, 1997; Hesselink, 2002). Dike construction forced river bends to migrate downstream rather than laterally (see also Hobo et al., 2014), causing the river to rework some of its recent deposits. The lower residual dose in these deposits means that less light exposure would be required for an acceptable degree of bleaching to take place. During this period, sand bars were exposed in the river during low flow, enhancing bleaching conditions for river-transported sediments. Bleaching conditions may have been reduced following the construction of groynes (after AD 1850), for a number of reasons. Firstly, the groynes prevented bend migration and thus reworking of recent river sed-

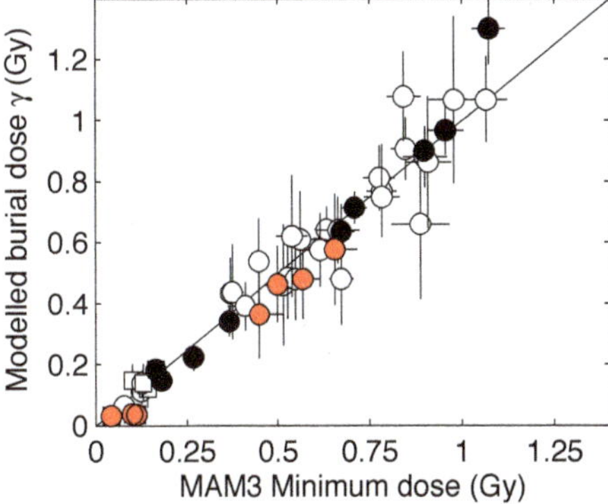

Figure 6. Comparison of burial-dose estimates. The modelled dose is defined by the mean and standard deviation of the posterior, and is compared to the original MAM3 minimum dose (using σ_b of 12 %; Galbraith et al., 1999). Squares indicate where the unlogged MAM3 was used (Arnold et al., 2009). The best-bleached samples are filled black ($p > 0.8$), and the worst-bleached samples in red ($p < 0.3$).

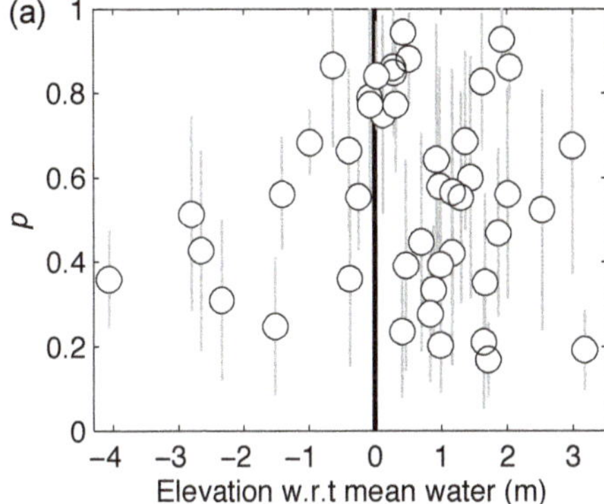

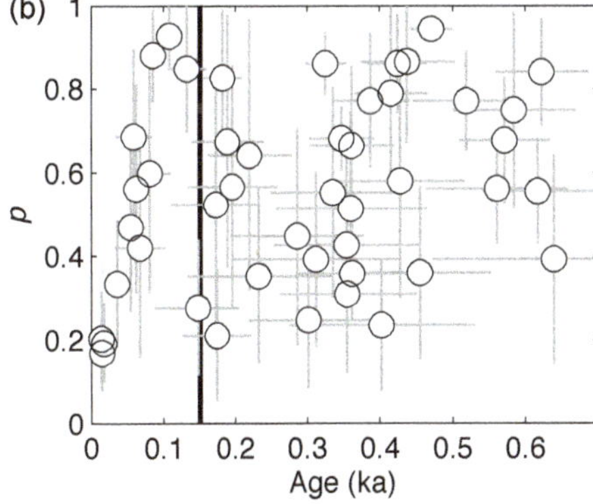

Figure 7. Two possible structures in the proportion of well-bleached grains p versus predictor variables of Fig. 5, enlarged here for clarity. **(a)** Plotted against the sample elevation with respect to mean water level at each sampling site. A cluster of well-bleached samples appears at about the (modern) mean water level. **(b)** Plotted against model-derived age, showing a trend for the youngest samples (age < 0.10 ka). The onset of groyne construction is indicated at 0.15 ka.

Figure 8. A sand bar deposited close to the river Waal during high discharge (photo by Gilbert Maas, Alterra). Due to the absence of vegetation, such deposits may be reworked through aeolian processes, which may enhance bleaching for deposits formed above the mean water level.

iments. Secondly, sand bars within the channel disappeared, perhaps reducing light exposure of channel deposits during low flow. Thirdly, the groynes caused deepening of the channel through bottom scour, enhancing the reworking of the underlying, high-residual-dose Pleistocene deposits.

4.2 Modelling the burial dose

The requirements of this project led us to develop a specific "age model" for partially bleached, multi-grain-aliquot data. It uses Bayesian computational methods to estimate the parameters of the single-grain dose distribution, without the need for any single-grain measurements. Along the way, the parameters of the single-grain sensitivity distribution are estimated from multi-grain aliquot sensitivity data. Our approach has significant advantages over existing models:

- The interaction of aliquot size and SG sensitivity is incorporated, meaning that prior quantification of the averaging effect is not necessary.
- It includes uncertainty deriving from the number of aliquots consistent with the burial dose.
- It should provide an unbiased estimate of the burial dose, even when no aliquots are "well bleached". Poorly bleached samples give a very imprecise, but still accurate, estimate of the burial dose.
- The degree of bleaching is quantified, and is potentially independent of the SG sensitivity, aliquot size and burial dose.
- Different data sets from the same sample (i.e. different aliquot sizes) can be combined to produce a single estimate of the burial dose.

Of course, the validity of the outcome rests on a number of assumptions. The parameterisation of the SG dose and sensitivity distributions must be appropriate and, crucially, the estimate of aliquot size should be reasonable. This paper uses archive data, so aliquot size was estimated only roughly. When applied in future, careful grain counting should take place; this could be performed manually, or with a digital camera plus image-recognition software.

Compared to familiar and well-used age models in OSL dating (e.g. CAM and MAM3), this model is a different beast altogether. It requires more data to be input per sample, and careful consideration and specification of model parameters and priors. It includes the MAM3 (Galbraith et al., 1999) and bootstrap likelihoods (Cunningham and Wallinga, 2012) as a small part of it, and requires ~ 1 h of computer time to run per sample, provided no errors arise. We provide the code in the Supplement in order to spur further development in this field; we do not present a refined model for general application.

The model could be immediately improved by treating σ_b^{SG} as an unknown parameter. At present, the model assumes that scatter in the single-grain burial-dose population is exactly 20 %. If it becomes possible to create a sample-specific estimate of σ_b^{SG} (e.g. through radiation transport modelling; Cunningham et al., 2012; Guerin et al., 2012), it could be incorporated as a prior. The posterior σ_b^{SG} would then be estimated along with γ, σ and p. A further step would be to incorporate stratigraphic information on sample order and/or age (e.g. Rhodes et al., 2003; Cunningham and Wallinga, 2012), although this would significantly increase computational time.

5 Conclusions

There are a particular set of challenges in estimating the degree of bleaching for unknown-age OSL samples, and these problems relate closely to the burial-dose calculation. We have begun to refashion the burial-dose calculation by using Bayesian computational statistics to reduce the D_e distribution into meaningful statistics. There are many novel aspects to our approach, such as parameterising the OSL sensitivity distribution and inferring single-grain statistics from small-aliquot measurements. We found that a good aliquot-size estimate is particularly important for the model, and that our poor knowledge of aliquot size hampered our application of the model to archive data.

Nevertheless, the results do show some interesting features that may point to geomorphic controls on sediment bleaching. We found a concentration of well-bleached samples around the modern mean water level, indicating to us that sediment receives a "kick" of bleaching upon deposition, through local reworking, in addition to the bleaching that occurs during transport. We also speculate on whether changes in the degree of bleaching over time could relate to river management changes, especially the construction of groynes in the lower Rhine around AD 1850.

Despite the limitations of this study, it seems clear that processes of sediment provenance, transport and deposition can influence the measured OSL signal. The challenge lies in extracting meaningful information from the OSL data. The computational approach explored in this study has real potential, and we hope aspects of our model will be taken forward.

Acknowledgements. This work was partially funded through a Technology Foundation (STW) VIDI grant (DSF.7553). We would like to thank Geoff Duller (Aberystwyth University) for making the data set on single-grain sensitivity distributions available. Zhixiong Shen and an anonymous reviewer are kindly thanked for their constructive comments. Matlab code is available in the supplement files.

Edited by: A. Lang

References

Arnold, L. J., Roberts, R. G., Galbraith, R. F., and DeLong, S. B.: A revised burial dose estimation procedure for optical dating of youngand modern-age sediments, Quatern. Geochronol., 4, 306–325, doi:10.1016/j.quageo.2009.02.017, 2009.

Bailey, R. M. and Arnold, L. J.: Statistical modelling of single grain quartz De distributions and an assessment of procedures for estimating burial dose, Quaternary Sci. Rev., 25, 2475–2502, 2006.

Ballarini, M., Wallinga, J., Murray, A. S., Van Heteren, S., Oost, A. P., Bos, A. J. J., and Van Eijk, C. W. E.: Optical dating of young coastal dunes on a decadal time scale, Quaternary Sci. Rev., 22, 1011–1017, 2003.

Berendsen, H. J. A. and Stouthamer, E.: Palaeogeographic development of the Rhine-Meuse delta, The Netherlands, Van Gorcum, Assen, p. 268, 2002

Berger, G. W.: Effectiveness of natural zeroing of the thermoluminescence in sediments, J. Geophys. Res.-Solid Ea., 95, 12375–12397, 1990.

Cunningham, A. C. and Wallinga, J.: Selection of integration time-intervals for quartz OSL decay curves, Quaterna. Geochronol., 5, 657–666, 2010.

Cunningham, A. C. and Wallinga, J.: Realizing the potential of fluvial archives using robust OSL chronologies, Quatern. Geochronol., 12, 98–106, 2012.

Cunningham, A. C., Wallinga, J., and Minderhoud, P. S. J.: Expectations of scatter in equivalent-dose distributions when using multi-grain aliquots for OSL dating, Geochronometria, 38, 424–431, 2011.

Cunningham, A. C., DeVries, D. J., and Schaart, D. R.: Experimental and computational simulation of beta-dose heterogeneity in sediment, Radiat. Measure., 47, 1060–1067, 2012.

Cunningham, A. C., Evans, M., and Knight, J.: Quantifying bleaching for zero-age fluvial sediment: a Bayesian approach, Radiat. Measure., in press, 2015.

Duller, G. A. T.: Single-grain optical dating of Quaternary sediments: why aliquot size matters in luminescence dating, Boreas, 37, 589–612, 2008.

Duller, G. A. T., Botter-Jensen, L., and Murray, A. S.: Optical dating of single sand-sized grains of quartz: sources of variability, Radiat. Measure., 32, 453–457, 2000.

Galbraith, R. F.: On plotting OSL equivalent doses, Ancient TL, 28, 1–10, 2010.

Galbraith, R. F., Roberts, R. G., Laslett, G. M., Yoshida, H., and Olley, J. M.: Optical dating of single and multiple grains of quartz from Jinmium rock shelter, northern Australia: Part I. Experimental design and statistical models, Archaeometry, 41, 339–364, 1999.

Gelman, A., Carlin, J. B., Stern, H. S., and Rubin, D. B.: Bayesian data analysis, Chapman & Hall/CRC, Boca Raton, Florida, 2004.

Guerin, G., Mercier, N., Nathan, R., Adamiec, G., and Lefrais, Y.: On the use of the infinite matrix assumption and associated concepts: A critical review, Radiat. Measure., 47, 778–785, 2012.

Heer, A. J., Adamiec, G., and Moska, P.: How many grains are there on a single aliquot?, Ancient TL, 30, 9–16, 2012.

Hesselink, A. W.: History makes a river: Morphological changes and human interference in the river Rhine, the Netherlands, Neth. Geogr. Stud., 292, 177, 2002.

Hobo, N., Makaske, B., Middelkoop, H., and Wallinga, J.: Reconstruction of floodplain sedimentation rates: a combination of methods to optimize estimates, Earth Surf. Proc. Land., 35, 1499–1515, 2010.

Hobo, N., Makaske, B., Middelkoop, H., and Wallinga, J.: Reconstruction of eroded and deposited sediment volumes of the embanked river Waal, the Netherlands, for the period AD 1631–present, Earth Surf. Proc. Land., 39, 1301–1318, 2014.

Isarin, R. F. B., Berendsen, H. J. A., and Schoor, M. M.: De morfodynamiek van de rivierduinen langs de Waal en de Lek, in: Publications and reports of the project 'Ecological Rehabilitation of the Rivers Rhine and Meuse', RIZA report 49, RIZA – Rijksinstituut voor Integraal Zoetwaterbeheer en Afvalwaterbehandeling, Lelystad, 1995.

Jain, M., Murray, A. S., and Botter-Jensen, L.: Optically stimulated luminescence dating: how significant is incomplete light exposure in fluvial environments?, Quaternaire, 15, 143–157, 2004.

Middelkoop, H.: Embanked floodplains in the Netherlands; Geomorphological evolution over various time scales, Netherlands Geographical Studies 224, Boca Raton, Florida, p. 341, 1997.

Murari, M. K., Achyuthan, H., and Singhvi, A. K.: Luminescence studies on the sediments laid down by the December 2004 tsunami event: Prospects for the dating of palaeo tsunamis and for the estimation of sediment fluxes, Current Sci., 92, 367–371, 2007.

Murray, A. S. and Wintle, A. G.: The single aliquot regenerative dose protocol: potential for improvements in reliability, Radiat. Measure., 37, 377–381, 2003.

Murray, A. S., Olley, J. M., and Caitcheon, G. G.: Measurement of equivalent doses in quartz from contemporary water-lain sediments using optically stimulated luminescence, Quaternary Sci. Rev., 14, 365–371, 1995.

Olley, J., Caitcheon, G., and Murray, A.: The distribution of apparent dose as determined by optically stimulated luminescence in small aliquots of fluvial quartz: implications for dating young sediments, Quaternary Sci. Rev., 17, 1033–1040, 1998.

Reimann, T., Notenboom, P. D., De Schipper, M. A., and Wallinga, J.: Testing for sufficient signal resetting during sediment transport using a polymineral multiple-signal luminescence approach, Quatern. Geochronol., 25, 26–36., doi:10.1016/j.quageo.2014.09.002, 2015.

Rhodes, E. J., Bronk Ramsey, C., Outram, Z., Batt, C., Willis, L., Dockrill, S., and Bond, J.: Bayesian methods applied to the interpretation of multiple OSL dates: high precision sediment ages from Old Scatness Broch excavations, Shetland Isles, Quaternary Sci. Rev., 22, 1231–1244, doi:10.1016/S0277-3791(03)00046-5, 2003

Roberts, R. G., Galbraith, R. F., Yoshida, H., Laslett, G. M., and Olley, J. M.: Distinguishing dose populations in sediment mixtures: a test of single-grain optical dating procedures using mixtures of laboratory-dosed quartz, Radiat. Measure., 32, 459–465, 2000.

Schielein, P. and Lomax, J.: The effect of fluvial environments on sediment bleaching and Holocene luminescence ages – A case study from the German Alpine Foreland, Geochronometria, 40, 283–293, doi:10.2478/s13386-013-0120-y, 2013.

Singarayer, J. S. and Bailey, R. M.: Further investigations of the quartz optically stimulated luminescence components using linear modulation, Radiat. Measure., 37, 451–458, doi:10.1016/S1350-4487(03)00062-3, 2003.

Sivia, D. S., Burbidge, C., Roberts, R. G., and Bailey, R. M.: A Bayesian approach to the evaluation of equivalent doses in sediment mixtures for luminescence dating, Am. Inst. Phys. Conf. Proc., 735, 305–311, 2004.

Spencer, J., Sanderson, D. C. W., Deckers, K., and Sommerville, A. A.: Assessing mixed dose distributions in young sediments identified using small aliquots and a simple two-step SAR procedure: the F-statistic as a diagnostic tool, Radiat. Measure., 37, 425–431, 2003.

Stokes, S., Bray, H. E., and Blum, M. D.: Optical resetting in large drainage basins: tests of zeroing assumptions using single-aliquot procedures, Quaternary Sci. Rev., 20, 879–885, 2001.

Truelsen, J. L. and Wallinga, J.: Zeroing of the OSL signal as a function of grain size: investigating bleaching and thermal transfer for a young fluvial sample, Geochronometria, 22, 1–8, 2003.

Wallinga, J.: Optically stimulated luminescence dating of fluvial deposits: a review, Boreas, 31, 303–322, 2002a.

Wallinga, J.: Detection of OSL age overestimation using single-aliquot techniques, Geochronometria, 21, 17–26, 2002b.

Wallinga, J., Hobo, N., Cunningham, A. C., Versendaal, A. J., Makaske, B., and Middelkoop, H.: Sedimentation rates on embanked floodplains determined through quartz optical dating, Quatern. Geochronol., 5, 170–175, 2010.

4

A linear inversion method to infer exhumation rates in space and time from thermochronometric data

M. Fox[1], F. Herman[1,*], S. D. Willett[1], and D. A. May[1]

[1]Institute of Geology, Swiss Federal Institute of Technology, ETH Zürich, Switzerland
[*]now at: Institute of Earth Science, University of Lausanne, Switzerland

Correspondence to: M. Fox (matthew.fox@erdw.ethz.ch)

Abstract. We present a formal inverse procedure to extract exhumation rates from spatially distributed low temperature thermochronometric data. Our method is based on a Gaussian linear inversion approach in which we define a linear problem relating exhumation rate to thermochronometric age with rates being parameterized as variable in both space and time. The basis of our linear forward model is the fact that the depth to the "closure isotherm" can be described as the integral of exhumation rate, $\dot{e}$, from the cooling age to the present day. For each age, a one-dimensional thermal model is used to calculate a characteristic closure temperature, and is combined with a spectral method to estimate the conductive effects of topography on the underlying isotherms. This approximation to the four-dimensional thermal problem allows us to calculate closure depths for data sets that span large spatial regions. By discretizing the integral expressions into time intervals we express the problem as a single linear system of equations. In addition, we assume that exhumation rates vary smoothly in space, and so can be described through a spatial correlation function. Therefore, exhumation rate history is discretized over a set of time intervals, but is spatially correlated over each time interval. We use an a priori estimate of the model parameters in order to invert this linear system and obtain the maximum likelihood solution for the exhumation rate history. An estimate of the resolving power of the data is also obtained by computing the a posteriori variance of the parameters and by analyzing the resolution matrix. The method is applicable when data from multiple thermochronometers and elevations/depths are available. However, it is not applicable when there has been burial and reheating. We illustrate our inversion procedure using examples from the literature.

1 Introduction

Since the initial work of Clark and Jäger (1969) and Wagner and Reimer (1972), the potential for in situ thermochronometry to quantify exhumation rates through time has been widely recognized. According to the definition of closure temperature (Dodson, 1973), a thermochronometric age represents the time elapsed since a rock cooled through a specific temperature (T_c). Provided with an estimate of the thermal structure of the crust, a closure temperature can be related to a depth in Earth and a thermochronometric age used to calculate time-averaged exhumation rate (e.g., Clark and Jäger, 1969; England and Richardson, 1977). Therefore, a variety of thermochronometers and modeling techniques have gained a role in a range of geological research: from tectonic processes that build topography (Harrison et al., 1992; Batt and Braun, 1999; House et al., 1998; McQuarrie et al., 2005; Sutherland et al., 2009) to the erosional processes that destroys it (Schildgen et al., 2010; Shuster et al., 2011; Glotzbach et al., 2011). Furthermore, measurements of exhumation rate have fueled debate regarding how tectonics, climate and erosional processes interact to shape Earth's surface (Molnar and England, 1990; Willett and Brandon, 2002; Dadson et al., 2003; Valla et al., 2012; Glotzbach et al., 2011).

A variety of methods have been proposed to convert thermochronometric ages to exhumation rates. For example, a common approach is to collect and date a suite of samples distributed in elevation. If the samples exhumed together at a constant rate through a steady temperature field, the ages will show a linear relationship with elevation. The slope of this age–elevation relationship (AER) is equal to the exhumation rate and changes in slope can be interpreted as changes in exhumation rate through time (e.g., Wagner and Reimer, 1972; Wagner et al., 1979; Fitzgerald and Gleadow, 1988; Fitzgerald et al., 1995).

Thermal models have also been used to investigate exhumation rates. In this case, ages are calculated by tracking material paths through space and time and using the resulting temperature–time ($T-t$) paths. Various analytical solutions have been presented to account for heat advection in response to erosion (e.g., Zeitler, 1985; Mancktelow and Grasemann, 1997; Brandon et al., 1998; Moore and England, 2001; Willett and Brandon, 2013) as well as the perturbation of the thermal field by topography (e.g., Lees, 1910; Birch, 1950; Stüwe et al., 1994; Mancktelow and Grasemann, 1997; Braun, 2002). Otherwise, numerical thermo-kinematic models have been used, in which spatial variations in rock uplift due to differential tectonic movements may be specified by structural or tectonic models combined with a surface erosion model (e.g., Grasemann and Mancktelow, 1993; Batt and Braun, 1997; Harrison et al., 1997; Batt and Braun, 1999; Stüwe and Hintermuller, 2000; Ehlers et al., 2003).

Measured and predicted ages can be compared through formal inverse methods. For example, two- and three-dimensional thermal models have been combined with nonlinear inversion schemes, which minimize the difference between model predicted and measured ages. Using such an approach, the optimum exhumation rates may be inferred when exhumation rates, tectonics or topography vary as a function of time (e.g., Glotzbach et al., 2011; Campani et al., 2010; Herman et al., 2010; Braun et al., 2012). With these methods it is often necessary to solve the two- or three-dimensional thermo-kinematic model several thousand times, which can become computationally expensive.

An alternative to these computationally expensive three-dimensional thermo-kinematic inverse models is to simply map spatial patterns in exhumation by converting isolated ages into time-averaged exhumation rates (e.g., Van den Haute, 1984; Omar et al., 1989; Brandon et al., 1998; Berger et al., 2008; Thomson et al., 2010). Vernon et al. (2008) computed the difference in elevation of isoage surfaces, which are interpolated from the data distributed in three dimensions, to estimate variations in exhumation rate. Alternatively, in a more sophisticated analysis Sutherland et al. (2009) implemented a nonlinear weighted least squares regression scheme for closure depth as a function of sample age. This approach enabled the authors to obtain a regional exhumation history, variable in space and time, for an application to ages from Fiordland, New Zealand.

In addition to inferring exhumation rates, complex cooling histories can be inferred from thermochronometric data. For example, the thermally controlled annealing of fission tracks provides constraints on $T-t$ paths between ~110°C to ~60°C for apatite (e.g., Green et al., 1985; Gleadow and Duddy, 1981; Green et al., 1989), and nonlinear inverse methods have been used to infer complex cooling and reheating histories (Gallagher, 1995, 2012; Willett, 1997; Ketcham et al., 2007). Gallagher et al. (2005) simultaneously exploited the variable $T-t$ information preserved in track length distributions from samples composing an AER. Due to the inclusion of the additional data from different elevations with different $T-t$ sensitivity, the resolution of the inferred $T-t$ history is greatly improved (Gallagher et al., 2005). Stephenson et al. (2006) investigated spatial and temporal variations in cooling rate using partition modeling to determine discrete spatial units with different thermal histories. However, the conversion of cooling rate to exhumation rate is not straightforward and requires a thermal model.

In this paper, we present a new method for estimating exhumation rates that are variable in both space and time. The key to our approach is to combine a general space–time function for exhumation rate with a four-dimensional model of the temperature field. We simplify the kinematic problem by assuming that rock material follows a vertical path. The four-dimensional thermal field is approximated by a transient, one-dimensional thermal model, calculated independently at each data location. A temperature perturbation due to the conductive effects of the local surface topography is calculated. Although, with this formulation, each datum is treated independently, we link solutions by requiring that the exhumation rate function, and thus the temperature field, be spatially correlated and vary smoothly in space. This formulation permits even large data sets, consisting of ages from multiple thermochronometric systems distributed in both space and elevation, to be treated simultaneously and efficiently.

2 Exhumation rates derived from thermochronometric ages co-located in space

In this section, we derive the relationship between thermochronometric ages and exhumation rates for the case where multiple ages have been obtained from samples distributed in elevation but effectively co-located in the horizontal plane. This is the situation in which an AER is typically constructed. Note that due to the multiple variables mentioned in this section and the next, see Table 1 for a list of variables and where they are defined in the text.

2.1 Forward and inverse formulations

For an isolated thermochronometric age, the distance between the sample elevation and the elevation of the corresponding closure isotherm, z_c, can be described as the integral of exhumation rate, $\dot{e}$, from the present day to the thermochronometric age,

$$\int_0^{\tau} \dot{e}\,\mathrm{d}t = z_c, \quad (1)$$

where τ is the thermochronometric age. To obtain a numerical solution to this problem, we discretize the integral in Eq. (1),

$$\sum_{j=1}^{M-1} \dot{e}_j \Delta t + \dot{e}_M R = z_c, \quad (2)$$

where $\dot{e}_j$ corresponds to the exhumation rate within a time interval of length Δt. M is the number of time intervals and R is the remainder of the division of τ by $M-1$, so that

$$R = \tau - (M-1)\Delta t. \quad (3)$$

In the case where two ages, τ_1 and τ_2 ($\tau_1 < \tau_2$), are obtained from samples that are effectively co-located in space, but from different elevations, they will integrate over different time intervals of the same exhumation history. Therefore, Eq. (2) applies to each age, yielding two independent equations expressed in matrix form as

$$\begin{bmatrix} \Delta t & \cdots & \Delta t & R_1 & 0 & 0 \\ \Delta t & \cdots & \cdots & \cdots & \Delta t & R_2 \end{bmatrix} \begin{bmatrix} \dot{e}_1 \\ \vdots \\ \vdots \\ \dot{e}_{M_1} \\ \vdots \\ \dot{e}_{M_{\max}} \end{bmatrix} = \begin{bmatrix} z_{c1} \\ z_{c2} \end{bmatrix}, \quad (4)$$

where $M_{\max}$ is the number of time intervals sampled by the oldest age, τ_2, and $\dot{e}_{M_{\max}}$ is the exhumation rate during this time interval.

For the general case of N samples, we express Eq. (4) as

$$\mathbf{A}\dot{\mathbf{e}} = \mathbf{z}_c, \quad (5)$$

where $\mathbf{A}$ is the forward model matrix[1] and has dimensions $N \times M_{\max}$, $\dot{\mathbf{e}}$ is a vector containing the exhumation rates over each time interval and has length $M_{\max}$, and $\mathbf{z}_c$ is a vector of length N containing the set of differences between sample elevation and elevation of the closure isotherm. Recognizing that z_c is an observation with some uncertainty, ϵ, Eqn. 5 is only an exact equation by including this uncertainty explicitly:

$$\mathbf{A}\dot{\mathbf{e}} + \epsilon = \mathbf{z}_c. \quad (6)$$

The unknown quantity in this matrix equation is $\dot{\mathbf{e}}$ and its estimation requires solving the linear inverse problem. In the case where $M_{\max}$ is greater than N, the problem is non-unique. Even if N is greater than $M_{\max}$, the calculations of Eq. (6) will rarely be independent, so we must use formal inverse methods for underdetermined or mixed-determined problems. We use a method of linear Gaussian inversion in which both the unknowns (i.e., model parameters) and data are described as stochastic, or random, variables that have a Gaussian probability density characterized by a mean and a variance. Using this approach requires some a priori knowledge of the model parameters in order to construct a covariance matrix for the model parameters (Backus and Gilbert, 1968, 1970; Jackson, 1972; Tarantola and Valette, 1982).

In our case, the stochastic variables are the exhumation rate and the errors on the closure depth. Therefore, we define an a priori exhumation rate model that is independent of the data and consists of an estimate of the mean exhumation rate, $\dot{e}_{\mathrm{pr}}$, and an estimate of the variance about this mean, σ^2_{pr}. This variance is used to construct the model parameter covariance matrix, $\mathbf{C}$, which has dimensions $M_{\max} \times M_{\max}$ and diagonal

[1] Henceforth, bold lower case variables will denote vectors and bold upper case variables will denote matrices.

Table 1. Variables and location of definition in the text.

Variable	Definition	Defined
z_c	Closure depth	Eq. (1)
$\dot{e}$	Exhumation rate	Eq. (1)
τ	Thermochronometric age	Eq. (1)
ΔT	Time interval length	Eq. (2)
M	Number of discrete time intervals	Eq. (2)
R	Remainder of $\tau/\Delta T$	Eq. (3)
$M_{\max}$	Maximum number of discrete time intervals	Eq. (4)
N	Number of ages	Eq. (5)
$\mathbf{A}$	Forward model matrix	Eq. (5)
ϵ	Uncertainty on closure depth	Eq. (6)
σ^2_{pr}	a priori variance	Eq. (7)
$\mathbf{C}$	Model parameter covariance matrix	Eq. (7)
$\mathbf{C}_\epsilon$	Data covariance matrix	Eq. (8)
$\dot{\mathbf{e}}_{\mathrm{po}}$	a posteriori exhumation rate	Eq. (9)
$\dot{\mathbf{e}}_{\mathrm{pr}}$	a priori exhumation rate	Eq. (9)
$\mathbf{C}_{\mathrm{po}}$	a posteriori covariance matrix	Eq. (8)
$\mathbf{R}$	Model resolution matrix	Eq. (12)
T	Temperature	Eq. (14)
κ	Thermal diffusivity	Eq. (14)
T_{m}	Mean temperature with depth	Eq. (15)
T_{d}	Temperature perturbation due to topography	Eq. (15)
q_l	Basal heat flux	Eq. (17)
k_l	Thermal conductivity	Eq. (17)
t^*	Onset of exhumation	Eq. (17)
T_{c}	Closure temperature	Sect. 2.2
z_{m}	Closure depth with respect to z=0	Sect. 2.2
h	Sample elevation	Sect. 2.2
ψ^2	Misfit criterion	Eq. (19)
$\rho(u)$	Spatial correlation	Eq. (21)
u	Separation distance	Eq. (21)
ϕ	Length scale parameter	Eq. (21)
$h(x,y)$	Earth surface topography	Sec. 3.3
$p(x,y)$	Isotherm topography with respect to z_{m}	Sect. 3.3
γ_a	Atmospheric lapse rate	Sect. 3.3
γ	General geothermal gradient as a function of depth	Sect. 3.3
ζ	Continuation factor	Eq. (24)

elements equal to the a priori variance $\sigma_{\rm pr}^2$:

$$C_{ij} = \sigma_{\rm pr}^2 \delta_{ij}, \quad i, j \in [1, M_{\rm max}], \tag{7}$$

where δ_{ij} is the Kronecker delta. Off-diagonal elements of **C** are equal to zero as we assume exhumation rates are not correlated in time. We then use the estimate of data errors, $\boldsymbol{\epsilon}$, to describe the data covariance matrix, $\mathbf{C}_\epsilon$, with dimensions $N \times N$. In our problem, the errors in the system come from analytical errors in the ages, ε_i, and from uncertainty in how this relates to a closure depth. If the errors are uncorrelated, the data covariance matrix is diagonal, with components

$$(C_\epsilon)_{ij} = \begin{cases} \dot{e}_{\rm pr}\varepsilon_i & \text{if } i = j \\ 0 & \text{if } i \neq j \end{cases} \tag{8}$$

Given this formulation of the model parameter and data covariance matrices and the a priori estimate of the model parameters, the underdetermined problem has a maximum likelihood estimate of the model parameters (Franklin, 1970; Tarantola, 2005):

$$\dot{\mathbf{e}}_{\rm po} = \dot{\mathbf{e}}_{\rm pr} + \mathbf{C}\mathbf{A}^T \left(\mathbf{A}\mathbf{C}\mathbf{A}^T + \mathbf{C}_\epsilon\right)^{-1} (\hat{\mathbf{z}}_{\rm c} - \mathbf{A}\dot{\mathbf{e}}_{\rm pr}), \tag{9}$$

where $\dot{\mathbf{e}}_{\rm pr}$ is a vector of length N containing the $\dot{e}_{\rm pr}$. The model parameter estimate, $\dot{\mathbf{e}}_{\rm po}$, is the maximum likelihood estimate or the a posteriori exhumation rate. It is also treated as a Gaussian variable with a covariance matrix, $\mathbf{C}_{\rm po}$, which provides a measure of the uncertainty on the parameter estimate with respect to the a priori variance and is calculated as

$$\mathbf{C}_{\rm po} = \mathbf{C} - \mathbf{H}\mathbf{A}\mathbf{C}, \tag{10}$$

where the diagonal elements of $\mathbf{C}_{\rm po}$ show the a posteriori variance, $\sigma_{\rm po}^2$, and **H** is the inverse operator from Eq. 9:

$$\mathbf{H} = \mathbf{C}\mathbf{A}^T \left(\mathbf{A}\mathbf{C}\mathbf{A}^T + \mathbf{C}_\epsilon\right)^{-1}. \tag{11}$$

One advantage of linear inverse methods is that one can estimate the resolution of the inferred model parameters. The model resolution matrix, R

$$\mathbf{R} = \mathbf{H}\mathbf{A}, \tag{12}$$

provides a measure of this resolution. This resolution matrix gives information on how well the estimate of the model parameters correspond to the true parameters, and is a function of the age distribution and age uncertainty. **R** relates the difference between the a priori and a posteriori exhumation rates to the difference between the a priori and the actual, or true model parameters, $\dot{\mathbf{e}}_{\rm true}$, through

$$\dot{\mathbf{e}}_{\rm po} - \dot{\mathbf{e}}_{\rm pr} = \mathbf{R}(\dot{\mathbf{e}}_{\rm true} - \dot{\mathbf{e}}_{\rm pr}). \tag{13}$$

If the resolution is perfect, $\dot{\mathbf{e}}_{\rm po} - \dot{\mathbf{e}}_{\rm pr}$ is equal to $\dot{\mathbf{e}}_{\rm true} - \dot{\mathbf{e}}_{\rm pr}$. Perfect resolution is obtained when the resolution matrix is equal to the identity matrix. In this instance, the exhumation rates inferred for a specified time interval are resolved independently of exhumation rates in other time intervals. The further **R** is from the identity matrix, the less resolved the model parameters are.

2.2 Computation of closure depth

Deriving an exhumation rate from thermochronological data requires information on the depth to the closure temperature. This can be achieved by solving the heat transfer equation:

$$\frac{\partial T}{\partial t} - \dot{e}\frac{\partial T}{\partial z} = \kappa \nabla^2 T, \tag{14}$$

where $T(x,y,z,t)$ is the temperature, t is time, $\dot{e}$ is the exhumation, and κ is the thermal diffusivity. Heat production is not explicitly included. However, as we calibrate model heat flow to measured surface heat flow, the contribution of heat production to near-surface geothermal gradients is accounted for. The surface boundary condition is the fixed temperature of the Earth's surface and there is a basal condition reflecting the deep Earth heat flux.

For the thermochronometry problem the two effects we need to consider are perturbation of the isotherms by surface topography and vertical advection of heat by surface erosion and the upward motion of rock. We deal with each of these separately by decomposing the thermal field into two components (Turcotte and Schubert, 1982): an average temperature, $T_{\rm m}$, which varies with depth and time, and a perturbation away from this mean induced by the topography, T_d,

$$T(x,y,z,t) = T_{\rm m}(z,t) + T_{\rm d}(x,y,z,t). \tag{15}$$

The one-dimensional problem to solve for the mean temperature with depth is described by the simpler form of the heat equation:

$$\frac{\partial T_{\rm m}}{\partial t} - \dot{e}\frac{\partial T_{\rm m}}{\partial z} = \kappa\frac{\partial^2 T_{\rm m}}{\partial z^2}, \tag{16}$$

which we solve using a finite difference method in space. The solution in time is obtained using a Crank–Nicolson time integration. We use an integration time step that provides a Courant–Friedrichs–Lewy number of ~1 to ensure accuracy. The upper surface for the thermal model, $z = 0$, is chosen as the average elevation, with respect to sea level, of the area covered by the data and the temperature at this elevation is T_0. The boundary conditions are thus,

$$\begin{cases} \left.\dfrac{\partial T_{\rm m}}{\partial z}\right|_{z=l} = \dfrac{q_l}{k_l} \\ \\ \left.T_{\rm m}\right|_{z=0} = T_0, \end{cases} \tag{17}$$

where l is the depth to the base of the model, q_l is the heat flow into the base of the model, and k_l is the thermal conductivity. The initial condition for the problem (i.e., at time t^*) is a constant geothermal gradient, and we run the model from some assumed time of initiation of exhumation until the present day.

The next step consists of deriving the closure temperature and its corresponding closure depth from the thermal model. The closure temperature depends on the cooling rate at the time of closure (Dodson, 1973). We use kinetic parameters for the Dodson equation from the literature: helium diffusion in apatite (Farley, 2000), fission track annealing in apatite (Ketcham et al., 1999) and helium diffusion in zircon (Reiners et al., 2004). For an exhuming rock, the appropriate cooling rate is obtained from a transient geotherm through the material derivative:

$$\frac{DT_{\mathrm{m}}}{Dt} = \frac{\partial T_{\mathrm{m}}}{\partial t} + \dot{e}\frac{\partial T_{\mathrm{m}}}{\partial z}. \tag{18}$$

This cooling rate is then used to predict the closure temperature. With a transient geotherm $T_{\mathrm{m}}(t,z)$, cooling rate is a function of depth and time, but for any given time, there is a single depth where the temperature and the closure temperature are equal (Fig. 1). The distance traveled by a rock from this depth to the surface is the closure depth and is equal to $z_{\mathrm{m}} + h$, where h is the elevation of the sample with respect to the mean elevation.

In order to calculate $T_{\mathrm{m}}(t,z)$, and thus cooling rate and closure temperature, we require the exhumation rate. This is unknown and is the quantity that we are trying to obtain. This defines a nonlinear problem in $\dot{e}$, which we solve by direct iteration of Eq. (6). Fortunately, for most exhumation rates, this nonlinearity is weak and necessitates only a few iterations. We initialize the problem by using the a priori value for the exhumation rate, $\dot{e}_{\mathrm{pr}}$.

2.3 Age elevation profile

The inversion scheme outlined above is now illustrated with a vertical age profile, where the AER can be easily interpreted. We use published apatite fission track (AFT) data from samples collected within the footwall of the Denali Fault, taken from Fitzgerald et al. (1995) (Fig. 2). These data show a break in slope of the AER, which can be interpreted as a change in exhumation rate. Our aims are to assess how well we can resolve this apparent change in exhumation rate and to determine the dependencies of our inferred exhumation rates on (1) the time interval length, Δt; (2) the thermal parameters; and (3) the a priori model for the exhumation rate. To assess the quality of fit of the predicted to the observed ages for the different models, we report a misfit criterion, ψ^2, defined as

$$\psi^2(\dot{\mathbf{e}}) = \sqrt{\frac{1}{N}\sum_{i=1}^{N}\left(\frac{\tau_{pred,i} - \tau_i}{\epsilon_i}\right)^2}, \tag{19}$$

where predicted ages, τ_{pred}, are calculated from the $T-t$ paths using Dodson's expression, as described above. In the examples below we present the misfit values for the ages using both the a priori and a posteriori model parameters. These are designated ψ^2_{pr} and ψ^2_{po}, respectively. The difference between these numbers provides a quantitative measure of the model fit relative to the fit provided by a model with a constant exhumation rate.

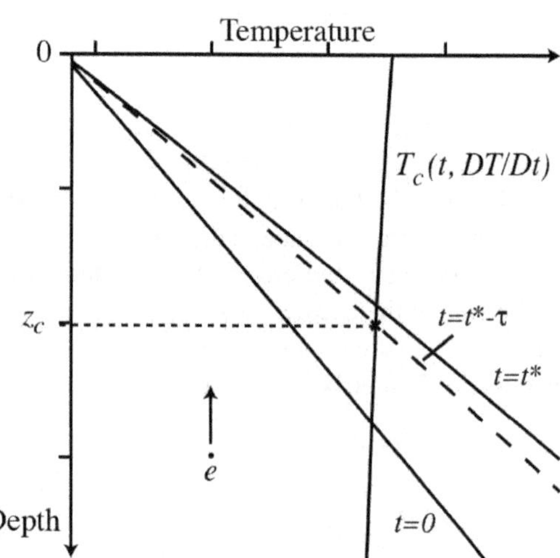

Figure 1. The evolution of crustal temperature through time in the presence of erosion rate at a rate of $\dot{e}$. Closure temperature, T_{c}, is shown at time $t^* - \tau$ (Dodson, 1973). The initial geotherm ($t = 0$) has a constant gradient. A cooling age of τ records closure at a time of, $t = t^* - \tau$, and a depth of, z_{c}.

We first describe results of a reference inverse model and then develop subsequent models by systematically varying parameterizations and a priori assumptions.

2.3.1 Reference inverse model

We start this reference model at a time equal to 25 Ma and discretize the exhumation history into time intervals, Δt, of 2.5 Myr. All elevations and depths are calculated with respect to a mean elevation of 4020 m. The thermal boundary condition is also fixed at this elevation, which is held at a temperature of $-12\,^{\circ}$C. We assume an a priori exhumation rate of 0.5 ± 0.15 km Myr^{-1}. The initial geothermal gradient is $24\,^{\circ}$C km^{-1}. With a constant exhumation rate of 0.5 km Myr^{-1}, the present day surface geothermal gradient would be $38.9\,^{\circ}$C km^{-1}. Values for other parameters used in the reference model are shown in Table 2.

Solving Eq. (9) for the age data shown in Fig. 2 gives the exhumation estimate shown in Fig. 3. The solid line gives the inferred exhumation rate and the grey shaded region gives this rate plus or minus one standard deviation, which we obtain as the square root of the diagonal terms of the a posteriori covariance matrix, Eq. (10). The a posteriori data misfit, ψ^2_{po}, is smaller than the a priori misfit, ψ^2_{pr}, reflecting the improved data fit.

The inversion results reveal that prior to the closure of the oldest age (17 Ma) we have only the a priori information, and the data have no influence on the a posteriori exhumation

Table 2. Parameters used to determine reference inverse model for the Denali Fault in Sect. 2.3.1.

Parameter	Value	Units
h_m	4020	m.a.s.l
T_0	−12	°C
q_l	76	$\mathrm{mW\,m^{-2}}$
κ	30	$\mathrm{km^2\,Myr^{-1}}$
k_l	2.6	$\mathrm{Wm^{-1}\,K^{-1}}$
l	80	km
t^*	25	Ma
Δt	2.5	Myr
$\bar{e}_{pr}$	0.5	$\mathrm{km\,Myr^{-1}}$
σ_{pr}	0.15	$\mathrm{km\,Myr^{-1}}$

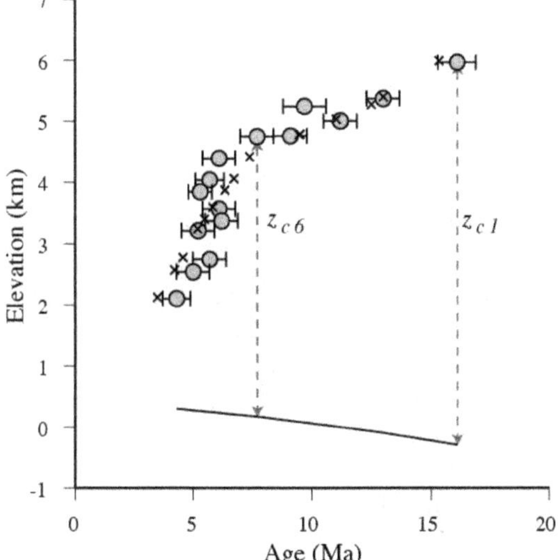

Figure 2. Apatite fission track ages from Mount Denali (Fitzgerald et al., 1995) plotted as a function of elevation. There is a break in slope at 6 Ma, interpreted as a response to an increase in exhumation rate. The lower curve shows the evolution of the closure depth through time in response to changes in the geotherm, as discussed in the text. Also plotted are the predicted ages (crosses) using the exhumation rate history shown in Fig. 3b.

rate. For these time intervals, the a posteriori variance is equal to the a priori variance, indicating that the data have no influence on the a posteriori rate. Subsequently, the cooling ages resolve a slow exhumation rate (with reduced variance) of about ~0.2 $\mathrm{km\,Myr^{-1}}$ from 15–7.5 Ma and a high rate of 0.6 $\mathrm{km\,Myr^{-1}}$ from 7 Ma to the present day.

During the iterative process the computed closure temperatures and closure depths evolve subject to the exhumation rate. As shown in Fig. 4, the computed closure temperatures and closure depths with respect to the mean elevation remain constant after three iterations.

It is also important to establish the resolution of the inferred exhumation rate history. To this end, we calculate and show the resolution matrix **R**, Eq. (12), in Fig. 5a. It shows

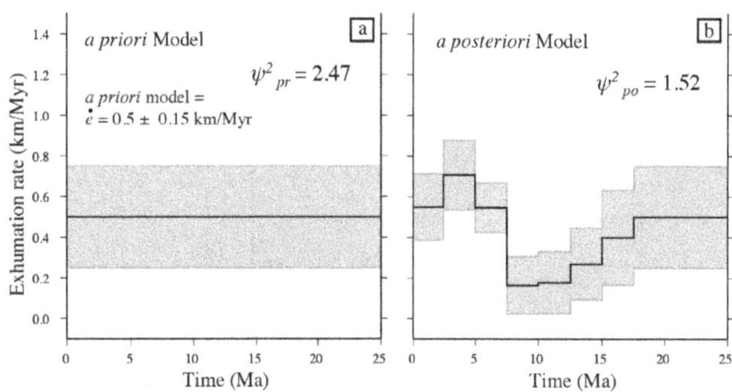

Figure 3. **(a)** The a priori exhumation rate and **(b)** the a posteriori exhumation rate, which is variable in time. Grey envelopes indicate one standard deviation in the a priori or a posteriori exhumation rate. Also given are a norm of the misfit values to the ages using the a priori model and the a posteriori model, ψ^2_{pr} and ψ^2_{po}, respectively.

that parameters relating to the early stages of the model (25–22.5 to 20–17.5 Ma) are unresolved, indicated by values of zero in the upper left-hand block of the matrix. This is expected as there are no age data within these time intervals. The exhumation rate from 17.5 Ma to the present is partially-to-fully resolved by the data, reflected by the range of the measured ages. The highest value of resolution is the diagonal element of the matrix corresponding to the time interval of 7.5–5 Ma. This corresponds to the high gradient segment of the AER, so that there are several ages within this interval constraining its exhumation rate. This is also reflected in the near zero values of the off-diagonal components of this row of the matrix. The most recent phase of exhumation, from 5 to 0 Ma, includes two time intervals (last two rows of the resolution matrix). No ages fall into this time interval, and the exhumation rate for these two intervals is not well-resolved, as indicated by the near equal values of the four lower-right components of the resolution matrix. This shows that the exhumation rate between 5 Ma and the present cannot be resolved into two independent time intervals.

A further measure of parameter resolution is provided by assessing the covariance between model parameters. The full posterior covariance matrix provides that information. We scale covariance by the diagonal entries of the posterior covariance matrix to assess the correlation, thereby providing a correlation parameter that varies from −1 to 1. We convert covariance between two parameters, ξ and β, denoted $C_{\xi\beta}$, to correlation between these parameters $\hat{C}_{\xi\beta}$ (Tarantola, 2005):

$$\hat{C}_{\xi\beta} = \frac{C_{\xi\beta}}{\sqrt{C_{\xi\xi}}\sqrt{C_{\beta\beta}}}. \tag{20}$$

The complete correlation matrix is shown in Fig. 5b. Correlation values along the diagonal part of the correlation matrix (Fig. 5b) are equal to 1. Off-diagonal elements are dominated by negative values, indicating that the uncertainties are anti-correlated, which implies that exhumation rate parameters have not been independently resolved by the data.

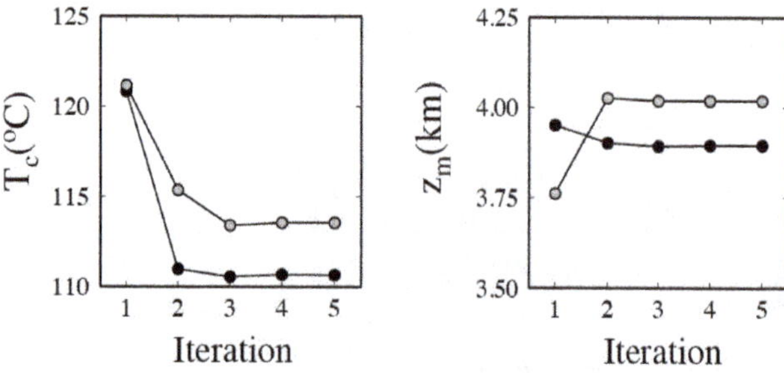

Figure 4. Evolution of closure temperature and closure depth during the iterative process. The black circles show a sample collected from high in the vertical profile with an age of 9.2 Ma. The grey circles show a sample from lower elevations with an age of 5.2 Ma. Both ages show that after three iterations closure temperature and closure depths remain constant during the iteration process.

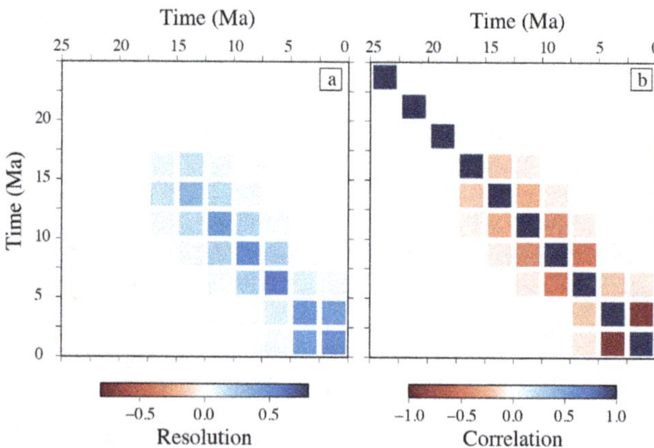

Figure 5. The model resolution matrix **(a)** and the correlation matrix **(b)** for the inversion shown in Fig. 3. Each element in these matrices relates exhumation rate during one time interval to exhumation rate during a different time interval. Large positive values of the diagonals of the resolution matrix indicate well-resolved parameters.

Anti-correlation is also expected: if two time intervals are available to exhume a rock from a given depth, by increasing the exhumation rate during one interval, the rate during the other interval is required to decrease. In the early stages of the model (25–22.5 to 20–17.5 Ma), the correlation matrix suggests that the model parameters are perfectly resolved. However, this is because the a posteriori covariance matrix is equal to the a priori covariance matrix, which is diagonal. In the time intervals younger than 17.5 Ma, the relative importance of negative off-diagonal elements is lowest during the 7.5–5 Ma time interval as this parameter has the highest resolution. Conversely, the off-diagonal elements during more recent exhumation, from 5 to 0 Ma, have values close to minus one. These large negative values demonstrate that the most recent phase of exhumation is unresolved.

2.3.2 Effect of the time interval length

The inferred exhumation rates vary with the selection of the time interval length, Δt. Figures 6d and 6g show inversion results where the time interval length has been changed relative to the reference model. In the case of shorter time intervals (Fig. 6d), the a posteriori standard deviation of the exhumation rate is generally larger than that of the reference model in Figure 6a. The exhumation history is also more poorly resolved. In spite of the fact that there is more generality to the exhumation rate history, the data misfit is increased and ψ^2_{po} is slightly higher compared to Fig. 6d. This is because, by having shorter, less well resolved time intervals, deviations from the a priori value are reduced and changes in exhumation rate are smoothed. Finally, the timing of the change from slow exhumation to fast exhumation is approximately 7 Ma. This age of increased exhumation rate is very similar to age of the break in slope of the AER (see Fig. 2).

With longer time interval length the a posteriori standard deviation of the exhumation rate is reduced. This is because there are fewer model parameters and, therefore, fewer possible exhumation rate histories that could fit the data. However, ψ^2_{po} is larger than for the reference model. This is due to the fact that the long time intervals changes in exhumation rate may not be possible at the appropriate time. For example, the time interval 10–5 Ma spans the break on slope (~ 7 Ma) and therefore the data fit is reduced.

The diagonal elements of the resolution matrices provide information about how well resolved exhumation rates are through time. For the reference model, a maximum resolution value is found for the time interval of 7.5–5 Ma (and is ~ 0.6). With shorter time intervals, exhumation rates are not resolved independently of one another and the maximum value is only 0.3. Conversely, with long time intervals the maximum resolution value is 0.97 for the time interval 5–0 Ma, indicating that this parameter is almost perfectly resolved.

This example highlights the trade-off between model complexity and variance reduction or parameter resolution. A greater number of time intervals permits greater variability and thus the precise timing of changes in exhumation rate to be identified. However, variance reduction and parameter resolution are decreased, indicating that model parameters are not resolved independently.

2.3.3 Effect of the thermal model

The assumptions made for the thermal model calculations have important implications for the estimated exhumation rates. In the case of a single age, the exhumation rate depends directly on the depth to the closure isotherm. Any error in the thermal model translates directly into an error of the exhumation rate estimate. Multiple ages provide a redundancy that partially overcomes this effect, but sensitivity to the thermal model remains. This is highlighted by the inversion results shown in Fig. 6h and 6i. In these models, the thermal parameters were selected such that the initial geothermal gradients are 18.6 °C km^{-1} and 29.3 °C km^{-1}, respectively, compared to the reference model, which has a gradient of 24 °C km^{-1}. In each case, we observe that when the exhumation rate is constrained by the vertical separation between two ages, the a posteriori exhumation rate estimate changes little. However, exhumation rates during the most recent stages vary widely, depending on the thermal model. For example, the inferred exhumation rate during the final time interval is 0.4 km Myr^{-1} for the case in which the surface geothermal gradient is relatively high (Fig. 6i), compared to 0.8 km Myr^{-1} for the case with a low gradient (Fig. 6h).

2.3.4 Effect of changes in the a priori exhumation rate

The a priori exhumation rate includes the expected value (mean) of exhumation rate and its variance. First, we investigate the effect of changing the a priori variance, which acts as a penalty parameter. A lower variance forces the inverse solution to attain a value close to the a priori value of exhumation rate. Conversely, a higher a priori variance permits more variation in the inverse solution, resulting in a better fit to the data. This is shown in Fig. 6b and c where the a priori variance is decreased and increased compared to the reference inverse model (Fig. 6a), respectively. With a lower a priori variance, a smoother solution is obtained as model parameters are penalized more strongly for deviations away from the a priori value. On the other hand, a large a priori variance (Fig. 6c) produces larger variations in the a posteriori model parameters. In addition, the a priori variance has an important effect on the a posteriori variance. Therefore, in order to interpret the a posteriori variance as parameter uncertainty it is important to compare it to the a priori variance.

Second, we explore the effect of the a priori mean exhumation rate on the parameter estimation. The cases in which the a priori exhumation rate is lower and higher are shown in Fig. 6e and f, respectively. In these models, the time intervals with no age constraints (17–25 Ma) simply have the a priori value of exhumation rate. However, once the ages constrain the exhumation rate, the a priori value has little influence on the a posteriori exhumation rate. It is also worth noting that we use the a priori exhumation rate to calculate the evolution of the thermal model, which explains the observed differences in exhumation rates in Fig. 6a, e and f.

3 Inversion of spatially distributed data

Here we extend the previous analysis to account for thermochronometric data that are spatially distributed and therefore have the potential to resolve spatial variability in exhumation rate. We illustrate the methodology with a case study using topography and published data from the Dabie Shan, China.

3.1 Correlation in Space

We include spatial variability in exhumation rate by describing exhumation rate as a spatial stochastic process. As a spatial stochastic process, a variable is described by not only a mean value and a variance but also by a spatial covariance. We thus define the covariance of exhumation rate, $\mathbf{C}$, as a spatial process with a spatial correlation function. The correlation function describes how exhumation rates vary in space. We define this as a function of the separation distance, u:

$$\rho(u) = \exp(-(u/\phi)^k), \quad (21)$$

where ϕ is a length scale parameter, and the function is a Gaussian function when $k = 2$ and an exponential function when $k = 1$.

3.2 Inverse Problem

The inversion process is constructed in the same manner as outlined for the co-located samples. However, we now define an independent exhumation history for each sample but require that exhumation rate varies smoothly in space through the spatial correlation function. Hence, exhumation rate, $\dot{\mathbf{e}}$, has length $M_{\max}N$ and the forward model operator, $\mathbf{A}$, has size $N \times M_{\max}N$. The covariance matrix now contains the spatial correlation structure between the exhumation rates defined in Eq. (21) (Tarantola and Nercessian, 1984; Willett, 1990). For a single time interval, a block of the covariance matrix is constructed using the separation distance between the ith and jth data (u) and the correlation function, $\rho(u)$:

$$\mathbf{C}_{ij} = \sigma_{\mathrm{pr}}^2 \delta_{ij} \rho(u). \quad (22)$$

It is assumed that exhumation rate is not correlated in time, so there is an independent matrix of form Eq. (22) for each time interval. These can be combined into a global matrix, setting cross time interval terms to zero.

Aside from this change in definition of the covariance matrix, the inversion process and the definition of the inverse operator, Eq. 9, remain unchanged. The a posteriori covariance matrix contains information about how inferred rates covary in time and space. The resolution matrix contains information about which parameters are determined independently of one another in time and space.

We have introduced a new model parameter, the characteristic length scale, ϕ. Its value can either be prescribed in a way to give a desired spatial smoothing, or it can be estimated through geostatistical methods (e.g., Matheron, 1963). The value of ϕ provides a means to trade-off model smoothness against variance reduction. A smaller ϕ reduces the number and weight of age data used to infer exhumation rates at any given location. A larger ϕ provides more spatial smoothing.

Finally, for ease of visualization and interpretation, we evaluate exhumation rate at arbitrary points defined on a regular grid. The exhumation rate at an arbitrary location is a

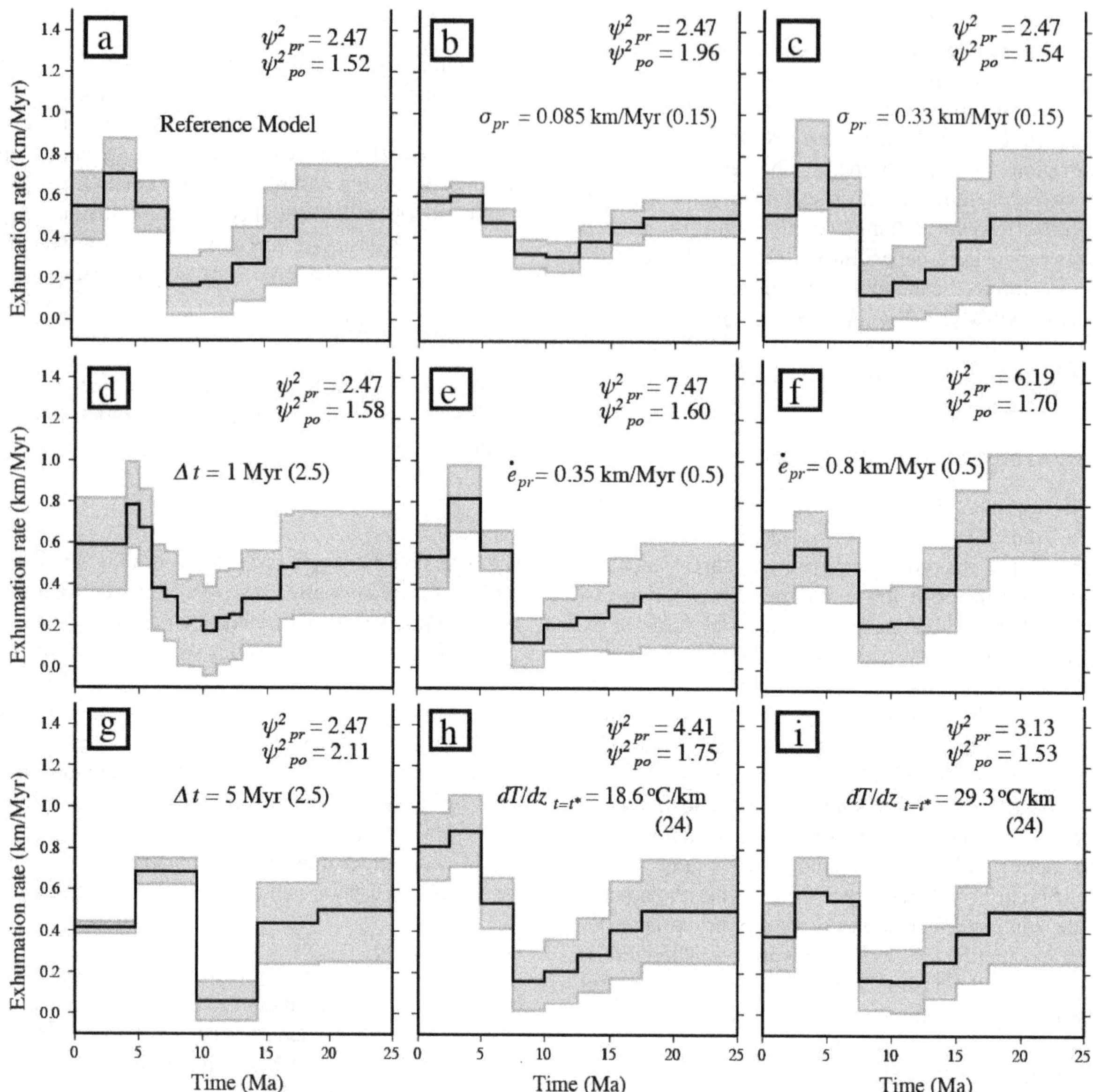

Figure 6. Exhumation rates inferred from age data from Fitzgerald et al. (1995) for different a priori model parameters. **(a)** Reference model; **(b–i)** independent analyses with one parameter changed with respect to the reference model, as indicated. The corresponding parameter value in the reference model is shown in brackets.

weighted average of the exhumation rates evaluated at the data locations. Here the relative weights are defined by the separation distances between the data locations and the point of interest and the correlation function imposed in the inversion.

The resolution matrix still relates the estimated and true values of exhumation rate but now includes temporal and spatial relationships. As a consequence, **R** is difficult to visualize. To simplify, we integrate the resolution values across the spatial dimension for each time interval. An integrated value of one corresponds to an exhumation rate that is resolved independently of other time intervals.

3.3 The effects of topography on a closure isotherm

In Sec. 2.2 we calculated the depth to a closure isotherm from a one-dimensional model that assumed an isothermal boundary condition at the mean topographic elevation. However, if we are to extend the analysis to investigate exhumation rates across a larger region, the topography of Earth's surface needs to be included (Lees, 1910; Bullard, 1938; Jeffreys, 1938; Birch, 1950; Werner, 1985).

Variations in elevation of Earth's surface as well as variations in the surface temperature imply a complex three-dimensional thermal boundary condition. We take an approach that simplifies this problem by reducing surface topography and surface temperature to perturbations of temperature on a horizontal plane located near the Earth's surface. This temperature perturbation at the surface can then be propagated downwards to a specific depth, z_m. In turn the perturbation in temperature at a specific depth can be converted to a perturbation in depth of an isotherm (Turcotte and Schubert, 1982; Stüwe et al., 1994; Mancktelow and Grasemann, 1997).

In Eq. (15) we decomposed the temperature field into two components, a mean temperature that varies with depth, T_m, and a perturbation from this mean, T_d. Mancktelow and Grasemann (1997) derive an expression to calculate temperature perturbations at depth due to a cosine temperature perturbation at the surface in the presence of heat advection. In our notation this expression is

$$T_d(\lambda,z)\big|_{z=z_m} = \exp(\zeta z_m) T_d(\lambda,z)\big|_{z=0}, \tag{23}$$

where $T_d(\lambda,z)\big|_{z=0}$, is a cosine function with known wavelength, λ, and ζ is defined as

$$\zeta = -\left(\frac{\dot{e}}{2\kappa} + \sqrt{\left(\frac{\dot{e}}{2\kappa}\right)^2 + (2\pi k)^2}\right), \tag{24}$$

where k is the wave number $(1/\lambda)$.

However, the depth perturbation of a closure isotherm is required and not simply the temperature perturbation at a specific depth calculated from the thermal model, T_m, as in Sect. 2.2. As closure depth evolves through time due to the advection of heat, the mean closure depth for each thermochronometric system is used to calculate a perturbation for each system.

The elevation of the surface of the model is chosen as the mean elevation of the area covered by the analysis. A temperature perturbation at this elevation is a function of the projection of the surface temperature (surface topography and atmospheric lapse rate) and the local geothermal gradient (Bullard, 1938; Jeffreys, 1938),

$$T_d(\lambda,z)\big|_{z=0} = -h(\lambda)(\gamma_0 - \gamma_a), \tag{25}$$

where $h(\lambda)$ is cosine function representing Earth's topography and γ_a is the atmospheric lapse rate. γ_0 is the geothermal gradient at $z = 0$ taken from the transient mean thermal model, T_m, and the specific time in the past.

Similarly, the temperature perturbation at a mean closure depth, z_m, can be written in terms of the isotherm topography and the geothermal gradient at that depth, γ_{z_m}:

$$T_d(\lambda,z)\big|_{z=z_m} = -p(\lambda)\gamma_{z_m}, \tag{26}$$

where $p(\lambda)$ is the perturbation of the closure isotherm about z_m. At this point we have expressions for T_d at $z = 0$ as a function of the topography that can be easily calculated, and T_d at $z = z_m$ as a function of the topography of a closure isotherm. These expressions, Eq. (26) and Eq. (25), can be combined in

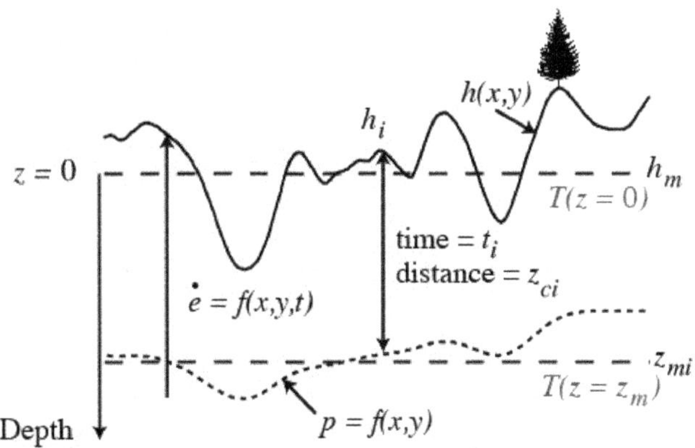

Figure 7. Surface topography, h, and a closure isotherm, p. The mean elevation, h_m, is defined relative to sea level, but subsequent depths and elevations are taken from this datum, thus defined to be $z = 0$. The mean closure depth z_{mi}, calculated for a specific age i, is shown as long dashed line, and p is the closure isotherm. An age, τ, from elevation, h_i, records the time taken to exhume from the closure depth, z_{ci}.

Eq. (23) to give the perturbation of closure depth. The resulting expression for a specific wavelength of the topography, $h(\lambda)$, is

$$p(\lambda) = A_0 \exp(\zeta z_m) h(\lambda), \tag{27}$$

where A_0 contains the atmospheric lapse rate and geothermal gradients evaluated from T_m,

$$A_0 = \left(\frac{\gamma_0 - \gamma_a}{\gamma_{z_m}}\right). \tag{28}$$

Finally, as any complex topography in one or two dimensions can be described as a infinite sum of periodic functions, and the principle of superposition applies, this analysis can be extended to account for any topography. We calculate $p(x,y)$ in the frequency domain (e.g Ducruix et al., 1974; Blackwell et al., 1980; Blakely, 1996). The distance to a closure isotherm for a single age is given by

$$(z_c)_i = (z_m)_i - p_i + h_i \tag{29}$$

where p_i is the value of $p(x,y)$ at the spatial location of the age, as illustrated in Fig. 7.

4 Case study from the Dabie Shan, SE China

We illustrate the ability of our inversion procedure to resolve spatially variable exhumation rates based on data from the Dabie Shan in southeastern China, Fig. 8. The Dabie Shan is a 200 km wide mountain belt with subdued relief. The last major phase of tectonics was during the early to middle Cretaceous. Reiners et al. (2003) reported ages from a vertical transect at the core of the range and from isolated samples collected around the flanks of the range, using three thermochronometric systems, (U-Th)/He ages from

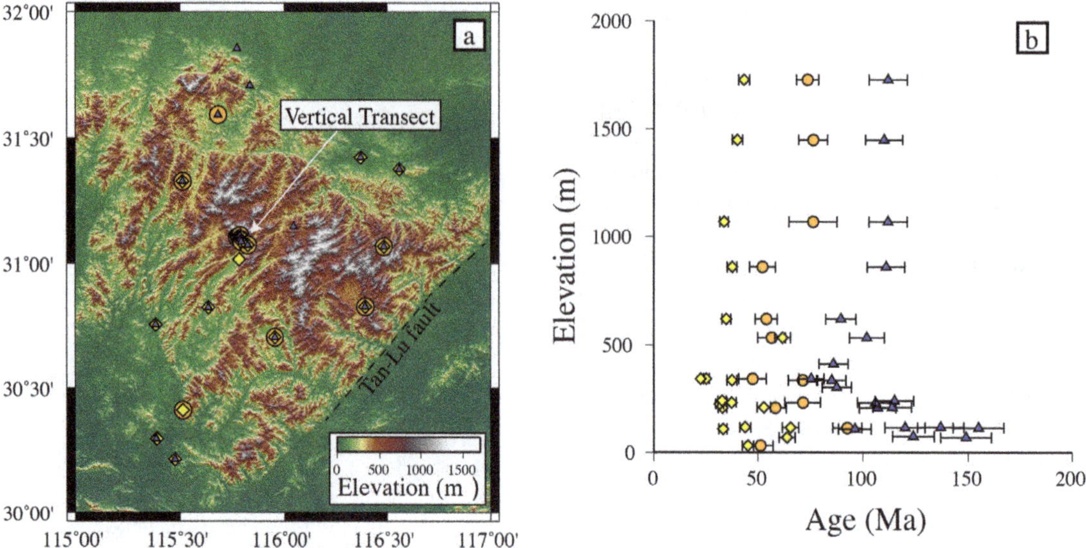

Figure 8. **(a)** Topographic map of the Dabie Shan, SE China, showing zircon (U-Th)/He, apatite fission track and apatite (U-Th)/He ages in blue triangles, orange circles and yellow diamonds, respectively, from Reiners et al. (2003). **(b)** The ages plotted as a function of elevation.

zircon (ZHe) and from apatite (AHe), and AFT dating (Reiners et al., 2003). The ages across the region show a positive AER in the core of the range and older ages around the flanks (Reiners et al., 2003). They concluded that exhumation rate has been slow, $0.06 \pm 0.01\,\mathrm{km\,Myr^{-1}}$, since the Cretaceous, with a possible increase in exhumation rate (up to $0.2\,\mathrm{km\,Myr^{-1}}$) between 80–40 Ma. There appear to be few tectonic structures responsible for this observed pattern of cooling ages, suggesting decay of old topography. Braun and Robert (2005) attributed the exhumation pattern to an isostatic response of relief reduction in the core of the range.

We use the Dabie Shan topography and data distribution to demonstrate and validate two key components of our inversion methodology: first, the approximation for the influence of topography on a closure isotherm; and second, the importance of the correlation structure of the parameter covariance matrix. As a test of the inversion scheme, before applying the algorithm to the measured ages, we apply the analysis to a suite of known synthetic ages produced with a thermo-kinematic model.

4.1 Example of the closure isotherm approximation

We use the topography of the Dabie Shan to demonstrate how the perturbation of a closure isotherm is calculated. The mean closure depths for AHe and ZHe are estimated from a one-dimensional thermal model representative of the Dabie Shan, T_m. The upper boundary of the thermal model is set at the average elevation, inferred from SRTM data (Farr et al., 2007), over the region shown in Fig. 9, 149.1 m.a.s.l.; the temperature at this elevation is set and held at 14.1 °C. Other parameters used in the thermal model are defined in Table 3. This initial model results in geothermal gradient of $22.7\,°\mathrm{C\,km^{-1}}$, which increases to $25.3\,°\mathrm{C\,km^{-1}}$ after 120 Myr of erosion at $0.06\,\mathrm{km\,Myr^{-1}}$.

From T_m we obtain the average values for the parameters required to calculate the closure isotherms, Table 4.

The topography with respect to the mean elevation of the area is shown in Fig. 9a. The two-dimensional discrete Fourier transform of the topography is computed using standard methods (e.g., Press et al., 1992) and we compute $p(x,y)$ in the frequency domain. The perturbation of the closure isotherms of AHe and ZHe about the mean closure depths are shown in Fig. 9b and 9c, respectively.

4.2 Testing the closure isotherm approximation

The analytical method for calculating the topographic effects on isotherm depth involve a number of approximations, foremost of which is the conversion of the Earth's surface into a temperature perturbation on a flat plane. To compare the perturbation of an isotherm about a mean depth, we initially calculate a value for the mean depth and temperature, as in Sect. 2.2. We also calculate T_m by applying a constant temperature at depth 900 °C at $z = 31.5$ km so that the models are compatible. We evaluate a mean closure depth using T_m for AHe closure at the mean age of AHe ages, $\tau^{\mathrm{AHe}} = 40.9$ Ma, at a temperature of 57.44 °C. To test our approximation for the thermal structure of the crust we generate a three-dimensional thermal field, using a finite-element solution to the advection–conduction equation, **Pecube** (Braun, 2003). Using identical thermal parameters and the topographic surface, the depth of the 57.44 °C isotherm is calculated using the three-dimensional thermal model and the

Table 3. Parameters used to calculate the temperature field for the Dabie Shan, as described in Sect. 4.1.

Parameter	Value	Units
h_m	149.1	m.a.s.l
T_0	14.1	°C
κ	35	$km^2\,Myr^{-1}$
k_l	2.6	$Wm^{-1}K^{-1}$
l	51.5	km
t^*	120	Ma
$\bar{\dot{e}}_{pr}$	0.08	$km\,Myr^{-1}$

method presented in Sect. 3.3. The misfit in depth between the isotherms calculated with the two different approaches is small, −1.68 m.

4.3 Testing resolution of the data

In this section, we conduct a resolution test to see how well the data from the Dabie Shan could be expected to resolve a spatially variable exhumation pattern. This test is made by generating synthetic age data from a known exhumation function, then analyzing these data with our inversion scheme in order to see how well the known exhumation rates are recovered. The procedure provides a confirmation of our inversion algorithm, but is primarily a test of the resolving capability of the Dabie Shan data number and location.

Synthetic ages are produced by integrating the temperature along material paths in the 3-D finite-element code **Pecube** (Braun, 2003). We specify a vertical velocity with an exhumation rate of 0.07 $km\,Myr^{-1}$ within a rectangular region in the center of the model domain and 0.05 $km\,Myr^{-1}$ outside of this rectangle. We predict AHe, AFT and ZHe ages with the Dodson approximation, at the same spatial locations as in Reiners et al. (2003). We assign the same measurement errors as in the reported data.

For the inversion analysis, we used an a priori exhumation rate of 0.06 $km\,Myr^{-1}$ with an a priori standard deviation of 0.02 $km\,Myr^{-1}$. We used an exponential spatial correlation function with a length scale parameter of 28 km, although we investigate the effects of this parameter later. The time interval length for the model is 30 Myr. We use the same thermal parameters and boundary conditions as were used to generate the synthetic ages.

Figure 10 shows the resulting inferred exhumation rates. Results are shown for each of the four time intervals of the inverse model. During the time interval from 120 Ma to 90 Ma, only the oldest ages influence the estimate of the exhumation rate and these are all found in the peripheral, slow-exhumation rate region. We normalize the a posteriori variance by the a priori variance as this provides a measure of the information content of the data. A normalized variance value of one and a temporal resolution value of zero highlights where the model is not resolved. The normalized a

Table 4. Parameters used to calculate closure isotherms for the Dabie Shan, as described in Sect. 2.3.1.

Parameter	ZHe value	AHe value	Units
z_m	6.4	1.7	km
T_c	167.1	56.3	°C
$\gamma(0,t)$	24.0	25.2	$°C\,km^{-1}$
$\gamma(z_m,t)$	23.7	25.1	$°C\,km^{-1}$

posteriori variance and the temporal resolution show that the solution is poorly resolved during this time interval.

From 90–60 Ma, there are also age data within the block of fast exhumation rate, and near these data, the exhumation rate is partially resolved, as indicated by the reduction of the a posteriori variance and the increase of temporal resolution. By 60–30 Ma, the exhumation rate is resolved and accurately estimated; however, due to the smoothness imposed in the model covariance matrix, the distinct boundaries of the block are “blurred”. This pattern is also observed in the final time interval, 30–0 Ma; however, the exhumation rates are slightly lower than the true values in the core of the block. Where there is an absence of data, for example in the south-east corner, the exhumation rate never deviates from the a priori value and shows a normalized variance close to one, whereas values of temporal resolution remain close to zero, as shown in Fig. 10.

The blurred nature of the result is due to the smooth correlation function used for the parameter covariance matrix. With additional ages located close to the boundaries of the block, we would be more likely to resolve this discrete step in exhumation rate.

As a further complication, it is evident that the blurring in space during one time interval also influences the entire exhumation rate history for the region. The exhumation rate during the oldest time interval demonstrates this effect. During this time interval there are only ages where the exhumation rate is slow. These slow rates are correctly inferred outside of the block, but are also inferred within the block. The degree to which exhumation rate outside of the block depends on exhumation rate within the block is a function of the correlation scale, and secondarily of the a priori variance; the larger the a priori variance, the easier it is for distant data to influence the result.

In addition, as this spatial averaging results in a low estimate of exhumation rate in the block, the advective heat transport and geothermal gradients are also underestimated and closure depths overestimated. This effect is similar to the example shown in Sect. 2.3.3.

4.4 Effect of the correlation function

As the imposed correlation function influences our results, we test two covariance functions as well as the correlation

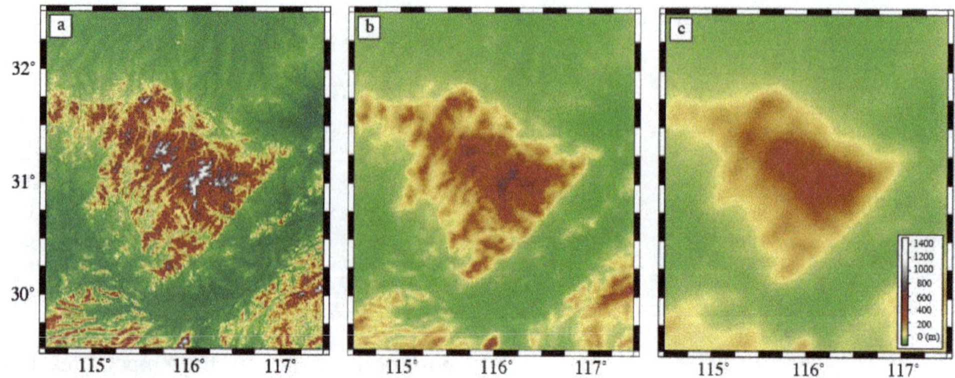

Figure 9. **(a)** The surface topography of Dabie Shan, **(b)** the topography of the closure isotherm for the AHe system and **(c)** the topography of the closure isotherm for the ZHe system. The isotherms are plotted as perturbations about the mean closure depths for the two systems. Refer to Sect. 3.3 and Sect. 4.1 for a full description.

length scale parameter, Eq. (21). Figure 11 shows how the a posteriori misfit, ψ^2_{po}, changes as a function of the correlation length scale, ϕ. Two correlation functions are used, an exponential function and a Gaussian function. If the correlation length, ϕ, is low, in the extreme this becomes equivalent to inverting each age independently with no ability to average out noise in the data or to identify changes in exhumation rate through time. Furthermore, deviations of the a posteriori rate from the a priori exhumation rate are suppressed (see Fig. 11). In contrast, when ϕ is large, samples which record different exhumation rates are forced to correlate and so an overly smooth exhumation pattern is obtained, as can be seen in Fig. 11.

4.5 Exhumation history of the Dabie Shan

We now apply our method to the measured ages of Reiners et al. (2003). As with the synthetic data, the model is initiated at 120 Ma and the exhumation history is discretized into 30 Myr time intervals. The a priori exhumation rate is $0.08 \pm 0.03\,\mathrm{km\,Myr^{-1}}$ based on Al-in-hornblende geobarometer estimates of intermediate calc-alkaline plutons and orthogneisses from within the Dabie Shan (Ratschbacher et al., 2000). Our thermal model predicts an increase of surface geothermal gradients from the initial value at the onset of exhumation, $22\,°\mathrm{C\,km^{-1}}$ to the present day value of $25\,°\mathrm{C\,km^{-1}}$ through time (Hu et al., 2000). The closure depths are calculated as described in Sect. 4.1. Prior to about 115 Ma, exhumation rates were very high, $\sim 2\,\mathrm{km\,Myr^{-1}}$ (Liu et al., 2010), and may have perturbed the thermal regime, but we assume that this effect does not influence the late thermal history (Ratschbacher et al., 2000). We impose a correlation length scale parameter of $\phi = 28$ km.

Estimated exhumation rates in space and time are shown in Fig. 12. During the time interval of 120–90 Ma, there is low spatial resolution due to the limited number of old ages (see Fig. 12). At the eastern end of the range (where the ages are oldest), exhumation rates of $\sim 0.09\,\mathrm{km\,Myr^{-1}}$ are resolved, as indicated by the low a posteriori variance. In the core of the range, the exhumation rates are slightly lower, $\sim 0.07\,\mathrm{km\,Myr^{-1}}$. The north and south flanks of the range are not well resolved, as indicated by the high normalized variance and low resolution values.

From 90–60 Ma we see high exhumation rates ($> 0.1\,\mathrm{km\,Myr^{-1}}$) near the Tan-Lu Fault, and a gradual decrease towards the northwest, possibly supporting activity on this fault during this time interval (Grimmer et al., 2002). The northeastern extent of the range continues to exhume at $\sim 0.09\,\mathrm{km\,Myr^{-1}}$.

The time interval from 60–30 Ma shows slower exhumation rates at the front of the range close to the Tan-Lu fault, in agreement with Reiners et al. (2003). There is also a decrease in exhumation rate in the eastern region. In contrast, we observe a slight increase in the core of the range. This pattern of exhumation rate, with the core of the range exhuming faster than the flanks, is consistent with an isostatic response to a decrease in relief and reduction of topography.

During the final time interval of exhumation, 30–0 Ma, a similar structure as in the previous time interval is inferred, with the core of the range exhuming faster than the flanks. However, the magnitude of the exhumation rates has reduced, most noticeably in the core of the range from ~ 0.08 to $\sim 0.06\,\mathrm{km\,Myr^{-1}}$.

5 Discussion

We have presented a method for deriving time and spatially variable exhumation rates based on a collection of thermochronometric ages from different systems, locations and elevations. The analysis can be regarded as an extension of age–elevation plots in that it capitalizes on the different distances traveled by rocks at the surface of high relief topography with respect to closure depth. In addition, by combining ages obtained from different thermochronomet-

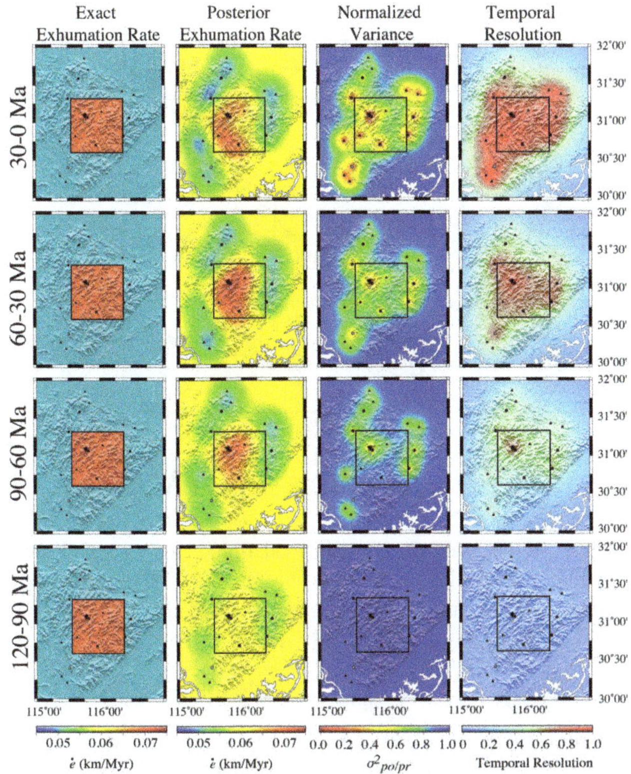

Figure 10. Results of the inversion of synthetic data with locations at the black points; thermochronometric system of each datum is consistent with the measurements of Reiners et al. (2003). Each row corresponds to a different time interval. The red boxes in the left column define a region with an exhumation rate of 0.07 km $\mathrm{Myr^{-1}}$; background rate is 0.05 m $\mathrm{Myr^{-1}}$. The second column shows exhumation rates inferred through inversion of these data. The third column shows the parameter variances normalized by the a priori variance. The right column displays the temporal resolutions.

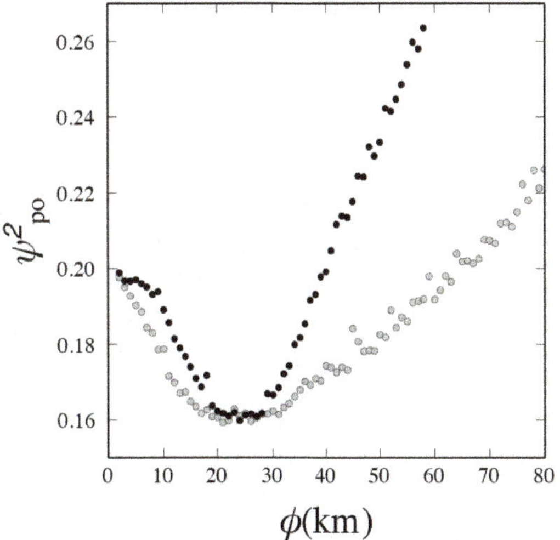

Figure 11. Sum of the squared misfit between predicted and observed ages as a function of the correlation length scale, ϕ. The gray circles show the results assuming an exponential correlation function ($k = 1$, Eq. (21)); the black circles use a Gaussian correlation function ($k = 2$, Eq. (21)).

ric systems, we resolve cooling rates in time by fitting the travel time between closure isotherms as well as final cooling to the surface.

The innovation of our method is that it combines many of the common methods of analyzing thermochronometric ages to derive exhumation rate. Least-squares fitting of ages distributed in elevation, thermal modeling of depth between closure temperatures for different mineral systems, and thermal modeling of individual ages can all be considered subsets of our analysis method. The problem with many of these traditional methods is that one must assume that all the ages distributed across a landscape have a common exhumation history, which is often not the case. Here we address this problem explicitly by imposing a spatial correlation, thereby requiring a common exhumation history only for points within a distance defined by a correlation length scale.

A major assumption of our method is that the kinetics of all thermochronometers are governed by a linear first order Arrhenius process, which enables us to use Dodson's approximation to estimate the depth of closure. However, a large body of work shows that the kinetics of AFT or AHe can become nonlinear due to effects such as radiation damage or multi-compositional annealing (Carlson et al., 1999; Ketcham et al., 1999, 2007; Shuster et al., 2006; Flowers et al., 2009). Fortunately, expanding the use of Dodson's parameters to represent the kinetics of closure can still include some of these complexities. For example, where compositional information is available for AFT ages, specific populations of ages can be modeled with specific sets of kinetic parameters. In this case, retentive apatites and non-retentive apatites would be modeled with separate kinetic parameters and thus closure temperatures. Similarly, grain size or radiation damage in (U-Th)/He ages can be accounted for by redefining the corresponding first order kinetic parameters (Reiners and Brandon, 2006; Shuster et al., 2006). In the extreme case, defining specific kinetic parameters independently for each measured age is possible and would require no modifications to our method as defined here.

In certain scenarios the linear assumption may break down. As a result, the uncertainties associated with the inferred exhumation rates may not be Gaussian, and the resulting parameter correlations may be more complex than reported. This is a common limitation of linear inverse methods, and a nonlinear method may be more appropriate in scenarios where the full range of models consistent with thermochronometric data are required. The linear assumption would be particularly restrictive if the extremes in parameter values were required. However, we expect nonlinear effects to be small over the parameter range we are interested in.

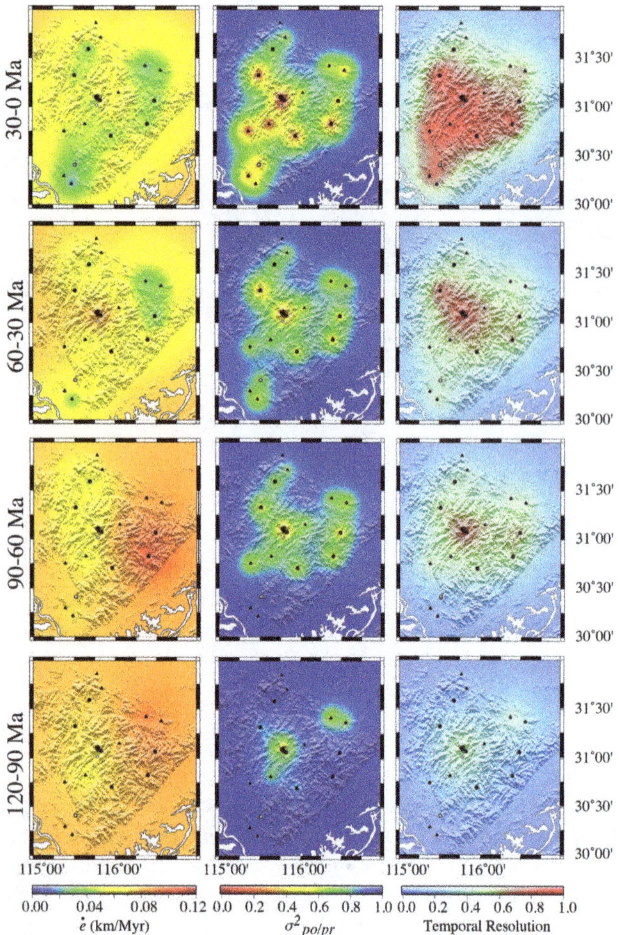

Figure 12. Exhumation rate history of the Dabie Shan. The left column shows the a posterior exhumation rates, the center column shows the a posteriori variances normalized by the a priori variance, and the right-hand column shows the temporal resolutions.

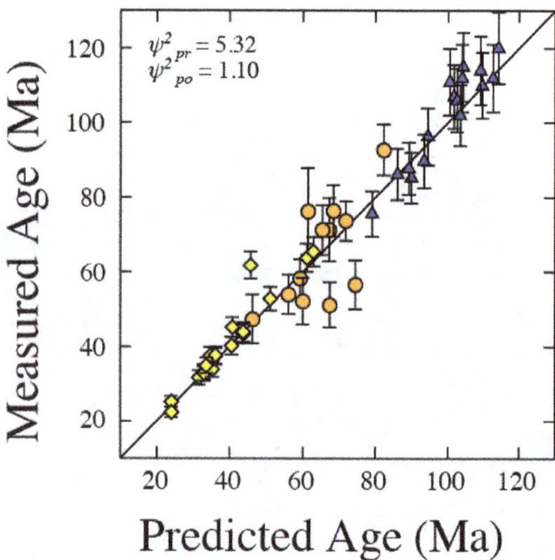

Figure 13. Comparison between model-predicted ages and measured ages. The solid black line is the 1 : 1 line. The error bars are the reported standard deviation of the measured ages.

Our method is best used for regional studies where the exhumation history is relatively simple. This is, first, because we assume that rocks only experienced monotonic cooling, implying that complex reheating has not occurred. Second, exhumation rates are smooth in space and are not strongly affected by surface-breaking faults. This latter complication can be easily accounted for, where these are well-identified, by building them into the correlation structure. In such cases, samples from either side of a fault could follow independent exhumation histories. In scenarios where ages record complex cooling and reheating histories, our approach would not be suitable because Dodson's approximation is invalid. Fortunately, complex cooling histories can be identified through additional information such as the analysis of fission track length distributions (Gallagher, 1995; Willett, 1997; Ketcham et al., 2007; Gallagher et al., 2005).

The thermal model, although simpler than solving a full three-dimensional problem, includes the major components of heat transfer necessary to solve this problem (i.e., conduction and advection). This combined with a spectral method enables us to include the effects of topography on the shape of underlying isotherms. It should be successful in most cases, but it might fail in regions of very high relief or very high exhumation rates where the approximation of topography into temperatures on a plane might not be accurate. Furthermore, where exhumation rate is described by 2- or 3-D kinematics, our methodology may be less successful. For example, orogens with high rates of horizontal displacement, which is associated with large scale thrust faulting, should be treated with caution. Our implementation could lead to an apparent change in exhumation rate in time, which would correspond to a change in the trajectory of rocks traveling through an orogenic wedge or up a thrust ramp. Likewise, the thermal model that we have implemented will not account for exhumation in situations where extensional unroofing dominates, unless it is very shallow.

The choice of initial and boundary conditions on the thermal model can have significant implications on the near-surface geothermal gradient and, therefore, the estimated exhumation rates. Although we have demonstrated the inference is robust when good age–elevation data exist or several thermochronometric systems are available, the thermal model can have a significant impact on the exhumation rates during the latest stages of exhumation. In this case, the thermal model will entirely control the distance the rock has to travel from its closure isotherm to the surface over the time defined by its thermochronometric age. It is therefore important to test several models using different initial and boundary conditions and assess how robust the estimated exhumation rates are.

One of the advantages of the formal inversion we adopt is in the ability to assess noise propagation from data to model parameters and to establish resolution. As the results of the analysis vary due to changes in the imposed parameterization and thermal model, shown in Fig. 6, the inferred exhumation rate uncertainty (obtained from the a posteriori covariance matrix) does not reflect the true model uncertainty. Therefore, a range of models with different imposed parameterization are required to convey the true uncertainty. We focus on interpreting the resolution in time; spatial resolution can be inferred from the spatial distribution of an age data set. It is computed, by definition, using the data error and parameter covariance matrices and, in our case, integrated spatially. It indicates how well exhumation rates at each time interval can be resolved independently of exhumation rate in other time intervals. The examples we report clearly show that resolution degrades back in time, which implies that we are more likely to resolve more recent exhumation rates and their variations.

As the problem as we have defined it is underdetermined, we require additional information to determine exhumation rate. This is in the form of a mean exhumation rate, $\dot{e}_{\mathrm{pr}}$, and an expected variance about this mean, σ^2 (along with a model for spatial covariance for the spatial case). In the majority of cases the a priori exhumation rate is an estimate of the average exhumation rate derived from any available information, although in principle it should be independent of the thermochronometric data. This could be in the form of sediment flux into neighboring sedimentary basins, geodetic derived rock uplift rates, or paleo–barometry estimates. Additional complexities, in the form of spatial and temporal variations, could be built into the a priori exhumation rate. However, where data exhibit good elevation distributions, results are relatively independent of the a priori exhumation rate.

For the spatial case, the parameter covariance matrix we implement largely controls the spatial resolution. It is defined through a correlation length scale and an a priori variance on the a priori exhumation rates. It serves as a trade-off parameter, trading solution resolution in space against the need to average out noise and to combine ages from different elevations or with different closure temperatures in order to resolve exhumation rates in time. Therefore, it is key to choose a correlation length scale that is appropriate. In most cases, it can be defined using statistical methods, for example, through the computation of a semi-variogram (e.g., Matheron, 1963). Similarly, the a priori variance must be chosen carefully. Its primary influence is as a weighting factor for the data uncertainty, and is chosen based on a trial-and-error approach.

Application of our method requires that exhumation histories be discretized into a predefined, carefully chosen, number of time intervals. Time interval lengths should be short enough to suitably represent the temporal variations of interest without attempting to infer too many model parameters. With a short time interval length, parameter resolution is reduced and the a priori information dominates. Conversely, as the time interval length increases, parameter resolution increases. However, the ability of the method to resolve changes in exhumation rate decreases. Therefore, time interval length should be chosen based on the age distribution.

6 Conclusions

The linear Gaussian inversion method presented here provides a practical, yet powerful tool to convert age information into exhumation rates. Our method generalizes the correlated concept of travel–time from a closure depth to the Earth's surface that is implicit in age–elevation relationships. In addition, it permits the simultaneous analysis of ages from different thermochronometric systems, provided their retention characteristics can be expressed in terms of first-order kinetics.

In addition to providing an estimate of exhumation rates, the formalisms of inverse theory provide associated measures of the quality of the exhumation rate estimates. These are in the form of covariance and resolution matrices. These matrices show that data variance, their geographic locations, a priori knowledge of exhumation rate, and the temporal distribution of the ages all play an important role in inferring exhumation rates.

We propose that our method is best suited to regional studies, where a general model for space and time variations in exhumation rate is desired. Given that the approach is linear, and the isotherms are calculated using an analytical solution, we can efficiently estimate exhumation rates across a range of wavelengths and timescales.

Acknowledgements. We would like to thank Jean Braun and Peter van der Beek for stimulating discussions throughout the development of the work. Thanks to Mark Brandon for his help with the topographic perturbation to the closure isotherms. Jeff Moore and Rebecca Reverman are thanked for reviews of the manuscript. Figures were prepared using the Generic Mapping Tools (Wessel and Smith, 1998).

Edited by: J. Braun

References

Backus, G. and Gilbert, F.: The resolving power of gross earth data, Geophys. J. Roy. Astron. Soc., 16, 169–205, 1968.

Backus, G. and Gilbert, F.: Uniqueness in the inversion of inaccurate gross earth data, Philosophical Transactions of the Royal Society of London, Series A, Mathemat. Phys. Sci., 266, 123–192, 1970.

Batt, G. E. and Braun, J.: On the thermomechanical evolution of compressional orogens, Geophys. J. Internat., 128, 364–382, 1997.

Batt, G. E. and Braun, J.: The tectonic evolution of the Southern Alps, New Zealand: insights from fully thermally coupled dynamical modelling, Geophys. J. Internat., 136, 403–420, 1999.

Berger, A. L., Spotila, J. A., Chapman, J. B., Pavlis, T. L., Enkelmann, E., Ruppert, N. A., and Buscher, J. T.: Architecture, kinematics, and exhumation of a convergent orogenic wedge: A thermochronological investigation of tectonic/climatic interactions within the central St. Elias orogen, Alaska, Earth Planet. Sci. Lett., 270, 13–24, 2008.

Birch, F.: Flow of heat in the Front Range, Colorado, Geol. Soc. Am. Bull., 61, 567–630, 1950.

Blackwell, D. D., Steele, J. L., and Brott, C. A.: The terrain effect on terrestrial heat flow, J. Geophys. Res.h, 85, 4757–4772, 1980.

Blakely, R. J.: Potential theory in gravity and magnetic applications, Cambridge University Press, Oxford, 1996.

Brandon, M. T., Roden-Tice, M. K., and Garver, J. I.: Late Cenozoic exhumation of the Cascadia accretionary wedge in the Olympic Mountains, northwest Washington State, Geol. Soc. Am. Bull., 110, 985–1009, 1998.

Braun, J.: Quantifying the effect of recent relief changes on age-elevation relationships, Earth Planet. Sci. Lett., 200, 331–343, 2002.

Braun, J.: Pecube: a new finite-element code to solve the 3D heat transport equation including the effects of a time-varying, finite amplitude surface topography, Comput. Geosci., 29, 787–794, 2003.

Braun, J. and Robert, X.: Constraints on the rate of post-orogenic erosional decay from low-temperature thermochronological data: application to the Dabie Shan, China, Earth Surf. Proc. Landforms, 30, 1203–1225, 2005.

Braun, J., van der Beek, P., Valla, P., Robert, X., Herman, F., Glotzbach, C., Pedersen, V., Perry, C., Simon-Labric, T., and Prigent, C.: Quantifying rates of landscape evolution and tectonic processes by thermochronology and numerical modeling of crustal heat transport using PECUBE, Tectonophysics, 524, 1–28, 2012.

Bullard, E.: The Disturbance of the Temperature Gradient in the Earth's Crust by Inequalities of Height, Geophys. J. Internat., 4, 360–362, 1938.

Campani, M., Herman, F., and Mancktelow, N.: Two-and three-dimensional thermal modeling of a low-angle detachment: Exhumation history of the Simplon Fault Zone, central Alps, J. Geophys. Res., 115, B10420, doi:10.1029/2009JB007036, 2010.

Carlson, W. D., Donelick, R. A., and Ketcham, R. A.: Variability of apatite fission-track annealing kinetics; I, Experimental results, Am. Mineral., 84, 1213–1223, 1999.

Clark, S. P. and Jäger, E.: Denudation rate in the Alps from geochronologic and heat flow data, Am. J. Sci., 267, 1143–1160, 1969.

Dadson, S. J., Hovius, N., Chen, H., Dade, W. B., Hsieh, M. L., Willett, S. D., Hu, J. C., Horng, M. J., Chen, M. C., Stark, C. P., et al.: Links between erosion, runoff variability and seismicity in the Taiwan orogen, Nature, 426, 648–651, 2003.

Dodson, M. H.: Closure temperature in cooling geochronological and petrological systems, Contr. Mineral. Petrol., 40, 259–274, 1973.

Ducruix, J., Mouel, J. L. L., and Courtillot, V.: Continuation of Three-dimensional Potential Fields Measured on an Uneven Surface, Geophys. J. Roy. Astronom. Soc., 38, 299–314, 1974.

Ehlers, T. A., Willett, S. D., Armstrong, P. A., and Chapman, D. S.: Exhumation of the central Wasatch Mountains, Utah: 2. Thermokinematic model of exhumation, erosion, and thermochronometer interpretation, J. Geophys. Res.-Sol. Earth, 108, 2156–2202, 2003.

England, P. and Richardson, S.: The influence of erosion upon the mineral fades of rocks from different metamorphic environments, J. Geol. Soc., 134, 201–213, 1977.

Farley, K. A.: Helium diffusion from apatite: General behavior as illustrated by Durango fluorapatite, J. Geophys. Res., 105, 2903–2914, 2000.

Farr, T. G., Rosen, P. A., Caro, E., Crippen, R., Duren, R., Hensley, S., Kobrick, M., and Paller, M.: The Shuttle Radar Topography Mission, Rev. Geophys., 45, doi:10.1029/2005RG000183, 2007.

Fitzgerald, P. G. and Gleadow, A. J. W.: Fission-track geochronology, tectonics and structure of the Transantarctic Mountains in northern Victoria Land, Antarctica, Chem. Geol., 73, 169–198, 1988.

Fitzgerald, P. G., Sorkhabi, R. B., Redfield, T. F., and Stump, E.: Uplift and denudation of the central Alaska Range: A case study in the use of apatite fission track thermochronology to determine absolute uplift parameters, J. Geophys. Res., 100, 20175–20192, 1995.

Flowers, R. M., Ketcham, R. A., Shuster, D. L., and Farley, K. A.: Apatite (U-Th)/He thermochronometry using a radiation damage accumulation and annealing model, Geochimica et Cosmochimica Acta, 73, 2347–2365, 2009.

Franklin, J. N.: Well-posed stochastic extensions of ill-posed linear problems, J. Mathemat. Analys. Appl., 31, 682–716, 1970.

Gallagher, K.: Evolving temperature histories from apatite fission-track data, Earth Plane. Sci. Lett., 136, 421–435, 1995.

Gallagher, K.: Transdimensional inverse thermal history modeling for quantitative thermochronology, J. Geophys. Res., 117, 2156–2202, 2012.

Gallagher, K., Stephenson, J., Brown, R., Holmes, C., and Fitzgerald, P.: Low temperature thermochronology and modeling strategies for multiple samples 1: Vertical profiles, Earth Planet. Sci. Lett., 237, 193–208, 2005.

Gleadow, A. and Duddy, I.: A natural long-term track annealing experiment for apatite, Nuclear Tracks, 5, 169–174, 1981.

Glotzbach, C., van der Beek, P. A., and Spiegel, C.: Episodic exhumation and relief growth in the Mont Blanc massif, Western Alps from numerical modelling of thermochronology data, Earth Planet. Sci. Lett., 304, 417–430, 2011.

Grasemann, B. and Mancktelow, N. S.: Two-dimensional thermal modelling of normal faulting: the Simplon Fault Zone, Central Alps, Switzerland, Tectonophysics, 225, 155–165, 1993.

Green, P., Duddy, I. R., Laslett, G. M., Hegarty, K. A., Gleadow, A. J. W., and Lovering, J. F.: Thermal annealing of fission tracks in apatite 4. Quantitative modelling techniques and extension to geological timescales, Chem. Geol., 79, 155–182, 1989.

Green, P. F., Duddy, I. R., Gleadow, A. J. W., Tingate, P. R., and Laslett, G. M.: Fission-track annealing in apatite: track length measurements and the form of the Arrhenius plot, Nucl. Tracks Radiat. Measurem., 10, 323–328, 1985.

Grimmer, J. C., Jonckheere, R., Enkelmann, E., Ratschbacher, L., Hacker, B. R., Blythe, A. E., Wagner, G. A., Wu, Q., Liu, S., and Dong, S.: Cretaceous-Cenozoic history of the southern Tan-Lu fault zone: apatite fission-track and structural constraints from the Dabie Shan (eastern China), Tectonophysics, 359, 225–253, 2002.

Harrison, T. M., Copeland, P., Kidd, W., Yin, A., et al.: Raising tibet, Science, 255, 1663–1670, 1992.

Harrison, T. M., Ryerson, F. J., Le Fort, P., Yin, A., Lovera, O. M., and Catlos, E. J.: A Late Miocene-Pliocene origin for the Central Himalayan inverted metamorphism, Earth Planet. Sci. Lett., 146, 1–7, 1997.

Herman, F., Rhodes, E., Braun, J., and Heiniger, L.: Uniform erosion rates and relief amplitude during glacial cycles in the Southern Alps of New Zealand, as revealed from OSL-thermochronology, Earth Planet. Sci. Lett., 297, 183–189, 2010.

House, M. A., Wernicke, B. P., and Farley, K. A.: Dating topography of the Sierra Nevada, California, using apatite (U-Th)/He ages, Nature, 396, 66–69, 1998.

Hu, S., He, L., and Wang, J.: Heat flow in the continental area of China: a new data set, Earth Planet. Sci. Lett., 179, 407–419, 2000.

Jackson, D. D.: Interpretation of Inaccurate, Insufficient and Inconsistent Data, Geophys. J. Roy. Astronom. Soc., 28, 97–109, 1972.

Jeffreys, H.: The Disturbance of the Temperature Gradient in the Earth's Crust by Inequalities of Height, Geophys. J. Internat., 4, 309–312, 1938.

Ketcham, R. A., Donelick, R. A., and Carlson, W. D.: Variability of apatite fission-track annealing kinetics: III. Extrapolation to geological time scales, Am. Mineral., 84, 1235–1255, 1999.

Ketcham, R. A., Carter, A., Donelick, R. A., Barbarand, J., and Hurford, A. J.: Improved modeling of fission-track annealing in apatite, Am. Mineral., 92, 799–810, 2007.

Lees, C. H.: On the shapes of the isogeotherms under mountain ranges in radio-active districts, Proceedings of the Royal Society of London. Series A, Containing Papers of a Mathematical and Physical Character, 83, 339–346, http://www.jstor.org/stable/92976, 1910.

Liu, S., Zhang, G., Ritts, B. D., Zhang, H., Gao, M., and Qian, C.: Tracing exhumation of the Dabie Shan ultrahigh-pressure metamorphic complex using the sedimentary record in the Hefei Basin, China, Geol. Soc. Am. Bull., 122, 198–218, 2010.

Mancktelow, N. and Grasemann, B.: Time-dependent effects of heat advection and topography on cooling histories during erosion, Tectonophysics, 270, 167–195, 1997.

Matheron, G.: Principles of geostatistics, Econom. Geol., 58, 1246–1266, 1963.

McQuarrie, N., Horton, B. K., Zandt, G., Beck, S., and DeCelles, P. G.: Lithospheric evolution of the Andean fold–thrust belt, Bolivia, and the origin of the central Andean plateau, Tectonophysics, 399, 15–37, 2005.

Molnar, P. and England, P.: Late Cenozoic uplift of mountain ranges and global climate change: chicken or egg?, Nature, 346, 29–34, 1990.

Moore, M. A. and England, P. C.: On the inference of denudation rates from cooling ages of minerals, Earth Planet. Sci. Lett., 185, 265–284, 2001.

Omar, G. I., Steckler, M. S., Buck, W. R., and Kohn, B. P.: Fission-track analysis of basement apatites at the western margin of the Gulf of Suez rift, Egypt: evidence for synchroneity of uplift and subsidence, Earth Planet. Sci. Lett., 94, 316–328, 1989.

Press, W. H., Teukolsky, S. A., Vetterling, W. T., and Flannery, B. P.: Numerical recipes in FORTRAN. The art of scientific computing, Cambridge: University Press, c1992, 2nd Edn., 1992.

Ratschbacher, L., Hacker, B. R., Webb, L. E., McWilliams, M., Ireland, T., Dong, S., Calvert, A., Chateigner, D., and Wenk, H. R.: Exhumation of the ultrahigh-pressure continental crust in east central China: Cretaceous and Cenozoic unroofing and the Tan-Lu fault, J. Geophys. Res., 105, 13303–13338, 2000.

Reiners, P. W. and Brandon, M. T.: Using thermochronology to understand orogenic erosion, Annu. Rev. Earth Planet. Sci., 34, 419–466, 2006.

Reiners, P. W., Zhou, Z., Ehlers, T. A., Xu, C., Brandon, M. T., Donelick, R. A., and Nicolescu, S.: Post-orogenic evolution of the Dabie Shan, eastern China, from (U-Th)/He and fission-track thermochronology, Am. J. Sci., 303, 489–518, 2003.

Reiners, P. W., Spell, T. L., Nicolescu, S., and Zanetti, K. A.: Zircon (U-Th)/He thermochronometry: He diffusion and comparisons with $^{40}Ar/^{39}Ar$ dating, Geochimica et Cosmochimica Acta, 68, 1857–1887, 2004.

Schildgen, T. F., Balco, G., and Shuster, D. L.: Canyon incision and knickpoint propagation recorded by apatite $^{4}He/^{3}He$ thermochronometry, Earth Planet. Sci. Lett., 293, 377–387, 2010.

Shuster, D. L., Flowers, R. M., and Farley, K. A.: The influence of natural radiation damage on helium diffusion kinetics in apatite, Earth Planet. Sci. Lett., 249, 148–161, 2006.

Shuster, D. L., Cuffey, K. M., Sanders, J. W., and Balco, G.: Thermochronometry Reveals Headward Propagation of Erosion in an Alpine Landscape, Science, 332, 84, 2011.

Stephenson, J., Gallagher, K., and Holmes, C. C.: Low temperature thermochronology and strategies for multiple samples: 2: Partition modelling for 2D/3D distributions with discontinuities, Earth Planet. Sci. Lett., 241, 557–570, http://www.sciencedirect.com/science/article/B6V61-4J022T3-1/2/723715442bc0c979bec22f09aa135035, 2006.

Stüwe, K. and Hintermuller, M.: Topography and isotherms revisited: the influence of laterally migrating drainage divides, Earth Planet. Sci. Lett., 184, 287–303, 2000.

Stüwe, K., White, L., and Brown, R.: The influence of eroding topography on steady-state isotherms. Application to fission track analysis, Earth Planet. Sci. Lett., 124, 63–74, 1994.

Sutherland, R., Gurnis, M., Kamp, P. J. J., and House, M. A.: Regional exhumation history of brittle crust during subduction initiation, Fiordland, southwest New Zealand, and implications for thermochronologic sampling and analysis strategies, Geosphere, 5, 409, 2009.

Tarantola, A.: Inverse problem theory and methods for model parameter estimation, Society for Industrial Mathematics, 2005.

Tarantola, A. and Nercessian, A.: Three-dimensional inversion without blocks, Geophys. J. Roy. Astronom. Soc., 76, 299–306, 1984.

Tarantola, A. and Valette, B.: Generalized nonlinear inverse problems solved using the least squares criterion, Rev. Geophys. Space Phys., 20, 219–232, 1982.

Thomson, S. N., Brandon, M. T., Reiners, P. W., Zattin, M., Isaacson, P. J., and Balestrieri, M. L.: Thermochronologic evidence for orogen-parallel variability in wedge kinematics during extending convergent orogenesis of the northern Apennines, Italy, Bull. Geol. Soc. Am., 122, 1160–1176, 2010.

Turcotte, D. L. and Schubert, G.: Geodynamics applications of continuum physics to geological problems, New York, NY (US); John Wiley and Sons, Inc., 1982.

Valla, P. G., van der Beek, P. A., Shuster, D. L., Braun, J., Herman, F., Tassan-Got, L., and Gautheron, C.: Late Neogene exhumation and relief development of the Aar and Aiguilles Rouges massifs (Swiss Alps) from low-temperature thermochronology modeling and 4He/3He thermochronometry, J. Geophys. Res., 117, F01004, doi:10.1029/2011JF002043, 2012.

Van den Haute, P.: Fission-track ages of apatites from the Precambrian of Rwanda and Burundi: relationship to East African rift tectonics, Earth Planet. Sci. Lett., 71, 129–140, 1984.

Vernon, A. J., van der Beek, P. A., Sinclair, H. D., and Rahn, M. K.: Increase in late Neogene denudation of the European Alps confirmed by analysis of a fission-track thermochronology database, Earth Planet. Sci. Lett., 270, 316–329, 2008.

Wagner, G. A. and Reimer, G. M.: Fission track tectonics: the tectonic interpretation of fission track apatite ages, Earth Planet. Sci. Lett., 14, 263–268, 1972.

Wagner, G. A., Miller, D. S., and Jäger, E.: Fission track ages on apatite of Bergell rocks from Central Alps and Bergell boulders in Oligocene sediments, Earth Plan. Sci. Lett., 45, 355–360, 1979.

Werner, D.: Geothermal problems in mountain ranges (Alps), Tectonophysics, 121, 97–108, 1985.

Wessel, P. and Smith, W. H. F.: New, improved version of Generic Mapping Tools released, Eos Transactions, 79, 579–579, 1998.

Willett, S. D.: Stochastic inversion of thermal data in a sedimentary basin: resolving spatial variability, Geophys. J. Internat., 103, 321–339, 1990.

Willett, S. D.: Inverse modeling of annealing of fission tracks in apatite 1: A controlled random search method, Am. J. Sci., 297, 939–969, 1997.

Willett, S. D. and Brandon, M. T.: On steady states in mountain belts, Geology, 30, 175–178, 2002.

Willett, S. D. and Brandon, M. T.: Some analytical methods for converting thermochronometric age to erosion rate, Geochem., Geophys., Geosystems, 2013.

Zeitler, P. K.: Cooling history of the NW Himalaya, Pakistan, Tectonics, 4, 127–151, 1985.

5

The impact of particle shape on the angle of internal friction and the implications for sediment dynamics at a steep, mixed sand–gravel beach

N. Stark[1,*], A. E. Hay[1], R. Cheel[1], and C. B. Lake[2]

[1]Dalhousie University, Department of Oceanography, Halifax, Canada
[2]Dalhousie University, Department of Civil Engineering, Halifax, Canada
[*]now at: Virginia Tech, Department of Civil and Environmental Engineering, Blacksburg, VA, USA

Correspondence to: N. Stark (ninas@vt.edu)

Abstract. The impact of particle shape on the angle of internal friction, and the resulting impact on beach sediment dynamics, is still poorly understood. In areas characterized by sediments of specific shape, particularly non-rounded particles, this can lead to large departures from the expected sediment dynamics. The steep slope (1 : 10) of the mixed sand–gravel beach at Advocate Harbour is stable in large-scale morphology over decades, despite a high tidal range of 10 m or more, and intense shore-break action during storms. The Advocate sand ($d < 2$ mm) was found to have an elliptic, plate-like shape (Corey Shape Index, CSI $\approx$ 0.2–0.6). High angles of internal friction of this material were determined using direct shear, ranging from $\phi \approx 41$ to $49°$, while the round to angular gravel was characterized as $\phi = 33°$. The addition of 25 % of the elliptic plate-like sand-sized material to the gravel led to an immediate increase in friction angle to $\phi = 38°$. Furthermore, re-organization of the particles occurred during shearing, characterized by a short phase of settling and compaction, followed by a pronounced strong dilatory behavior and an accompanying strong increase of resistance to shear and, thus, shear stress. Long-term shearing (24 h) using a ring shear apparatus led to destruction of the particles without re-compaction. Finally, submerged particle mobilization was simulated using a tilted tray submerged in a water-filled tank. Despite a smooth tray surface, particle motion was not initiated until reaching tray tilt angles of 31° and more, being $\geq 7°$ steeper than for motion initiation of the gravel mixtures. In conclusion, geotechnical laboratory experiments quantified the important impact of the elliptic, plate-like shape of Advocate Beach sand on the angles of internal friction of both pure sand and sand–gravel mixtures. The resulting effect on initiation of particle motion was confirmed in tilting tray experiments. This makes it a vivid example of how particle shape can contribute to the stabilization of the beach face.

1 Introduction

Subaqueous sediment dynamics play a major role in coastline, river and lake development, as well as scour around submerged structures, and coastal hazards such as submarine landslides (Kuehl et al., 1996; Simons and Şentürk, 1992; Bradley and Stolt, 2006; Masson et al., 2006). Despite the widespread interest and ongoing research in this field, the complex system consisting of the many diverse factors governing subaqueous sediment dynamics and beach dynamics (hydrodynamics, morphology and sediment properties) is still far from being fully understood. Focusing on the sediment properties, the friction angle is known to be a major factor controlling the critical shear stress required to initiate particle motion (Middleton and Southard, 1984; Bagnold, 1988; Kirchner et al., 1990; Soulsby, 1997). The friction angle depends on grain size, sorting, density, particle arrangement, and particle shape (Schanz and Vermeer, 1996; Das, 1990). In particular, the importance of particle shape with regard to the friction angle and initiation of subaqueous

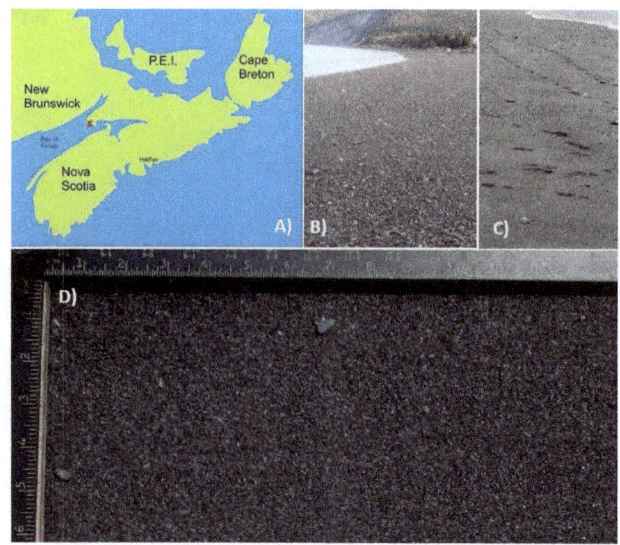

Figure 1. **(a)** The location of Advocate Harbour at the Bay of Fundy between Nova Scotia and New Brunswick, Canada. **(b)** Photograph of the beach face at low water during calm weather conditions. **(c)** Photograph of the beach face at low water after a storm event. **(d)** Photograph of the surficial sediments after a storm event (scale in inches).

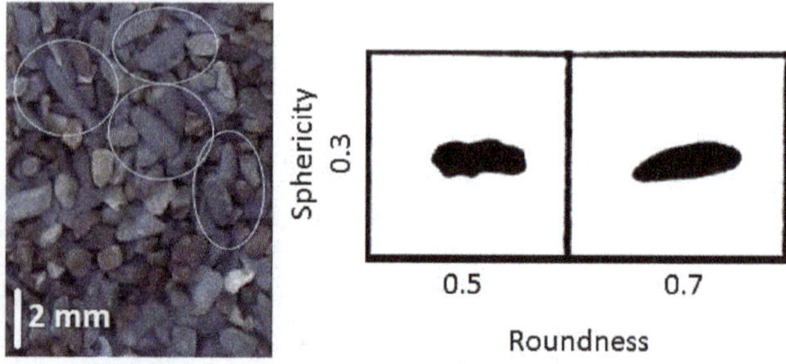

Figure 2. Left panel: example of flat, elliptic sand particles. Right panel: particle shapes which reached the highest friction angles in the study of Cho et al. (2006) (modified after Santamarina and Cho, 2004).

sediment motion was pointed out by Kirchner et al. (1990) for river sediments. Generally, there is still a certain lack of data pertaining to the impact of particle shape – particularly non-rounded shapes – on subaqueous sediment dynamics and beach morphodynamics.

The mixed sand–gravel beach near Advocate Harbour (Fig. 1) was investigated in the framework of a sediment dynamics experiment in May 2012. Hydrodynamics, large- and small-scale morphology, and sediment distribution were monitored over 3 weeks. With regard to the latter, a distinct cross-shore zonation varying with the lunar tidal cycle and weather/hydrodynamic conditions was observed (Stark et al., 2014b). However, despite the energetic hydrodynamics at the beach during wind-wave forcing events in particular, the beach slope has remained constant at about 6.4° for decades. Also – and rather surprisingly – following a significant storm wave event, the beach face was dominated by sandy sediments (Fig. 1). As documented in a companion paper (Hay et al., 2014), the fine-grained surficial sediments observed during active wave-forced conditions were associated with a very particular size-sorting process involving the development of orbital-scale (or vortex) ripples on the beach face during the rising tide. The sandy sediments ($d < 2$ mm) were characterized by a strong variations in particle shape, and a high abundance of flat, elliptic particles (Fig. 2). As a result, it was hypothesized that the flat, elliptic shape of the sand-sized particles impacts the friction angle and, by doing so, contributes to the stability of the beach.

This hypothesis is in agreement with findings by, e.g., Miller and Byrne (1966). Instead of the geometric approach to estimate the average friction angle by Eagleson and Dean (1961), they proposed the following equation:

$$\overline{\phi} = \alpha_{\mathrm{mb}} \left(d/\overline{K}\right)^{-\beta}, \tag{1}$$

where $\overline{\phi}$ is the average friction angle, d is the test grain diameter (tests were conducted with one single loose particle on a fixed rough bed), $\overline{K}$ is the average diameter of the bed grains, and α_{mb} and β are parameters fitted by regression to the data. Regarding the latter, Miller and Byrne (1966) found that α_{mb} was sensitive to the particle shape, and increased with decreasing sphericity. Li and Komar (1986) specified that α increased with particle flatness (i.e., with the decrease of the ratio of the smallest and intermediate axial diameters S/I). Additionally, these authors found that β was influenced by particle shape when comparing spheres to smooth ellipsoidal pebbles. However, Kirchner et al. (1990) argued that there is no consistent relationship between α_{mb} and grain shape in the literature, and that more controlled experiments are required. Nevertheless, these authors also found a decrease of median friction angle by 10–15° by eliminating grain roughness, meaning approaching spherical particles. Buffington et al. (1992) stressed the importance of testing friction angles of the native sediments to estimate critical shear stresses instead of relying on the grain size and neglecting particle shape. They presented results from testing naturally formed gravel from a streambed with Corey Shape Indices CSI = 0.57–068, indicating compact but not yet rounded particles (Illenberger, 1991). The results supported the discussion by Kirchner et al. (1990) that the empirical relationship obtained by Miller and Byrne (1966) cannot be generally applied using coefficients α and β determined from other samples.

Riley and Mann (1972) focused particularly on flat particles (i.e., "flakes"). They found that the angle of repose increased in the order of spheres, cylinders, angular particles and flakes for glass particles in the same size. Flakes with a particle size of 16 mm reached angles of repose of ~ 38.27°, but the authors argued that they expected even higher angles

Table 1. Summary of sediment properties and shear test result terms of median grain size d_{50}, modality (unimodal vs. bimodal), maximum shear stress measured for the highest load level of the respective sample in the direct shear boxes τ_{max}, internal angle of friction ϕ_{max}, maximum observed vertical displacement during shearing in the respective direct shear box (dilation is positive) and tray tilt angle when sediment started moving α_{tray}.

Sample	d_{50} (mm)	Modality	τ_{max} (kPa)	ϕ (°)	Dilation (mm)	α_{tray} (°)
Sand: reef	0.28	unimodal	28	41	0.2	
Sand: fine sand patch	0.27	unimodal	30	42	0.4	
Sand: Pos 182 (yd 131)	0.78	unimodal	32	42	0.2	
Sand: Pos 184 (yd 131)	0.92	unimodal	38	49	0.6	
Sand: Pos 186 (yd 125)	1.0	unimodal	35	44	0.1	31
Sand: Pos 186 (yd 131)	1.1	unimodal	36	49	0.5	
Sand: Pos 187 (yd 125)	1.2	unimodal	35	46	0.6	
Gravel: Pos 181 (yd 125)	12.5	unimodal	89	32	−1.4	24
75 % gravel and 25 % sand:	12.5 and 1.0	bimodal	113	38	−4.7	
50 % gravel and 50 % sand:	12.5 and 1.0	bimodal	113	38	−4.6	

of repose, and that such values were probably not reached because the flakes always packed with their flat faces horizontally, making slipping easier. This conclusion highlights the need for investigations of shearing behavior dependence on particle shape to fully understand sediment transport.

The detailed investigation of shearing behavior is traditionally a research focus in geotechnical engineering and soil mechanics, and the impact of particle shape on shearing behavior and angle of internal friction has been recognized for some time (Terzaghi, 1943; Taylor, 1948; Craig, 1974; Das, 1990). More recently, Santamarina and Cho (2004) and Cho et al. (2006) discussed the role of particle shape in soil behavior. Among other results, they showed that a decrease in particle regularity that included sphericity and/or roundness leads to an increase in the constant volume critical state friction angle, with the constant volume critical state referring to a state of static equilibrium between volume reduction as a consequence of chain collapse and volume dilation during shearing (Santamarina and Cho, 2004). In the Cho et al. (2006) experiments, constant volume critical state friction angles of up to 41° were documented for a crushed sand with a median grain size of 0.48 mm, a high angularity and a medium sphericity (Fig. 2).

In this study, standard geotechnical laboratory shear tests (direct uniaxial shear and ring shear test apparatuses) were applied (i) to test the applicability of standard geotechnical shearing experiments for the investigation of shearing processes of beach sediments, and (ii) to test the hypothesis of whether the flat, elongated particle shape of the sand-size fractions of the Advocate Beach sediments potentially contributes to the stabilization of the beach face. The study was initially motivated by the high abundance of surprisingly fine particles on the beach face after energetic storm wave events. The geotechnical results were complemented by a simple physical simulation of sediment mobilization along a tilted tray that contributed to the understanding of particle transport of the Advocate Beach sediments. More technical details and results from the latter laboratory study are presented in Stark et al. (2014a).

2 Methods

Beach samples (Table 1) were collected along a cross-shore transect in the vicinity of an instrumented frame that was installed for the full 3 weeks of the experiment. The frame was equipped with a number of different acoustic and other devices estimating flow velocity, bed-load transport velocity, wave orbital velocity, wave height and small-scale morphological variations (Hay et al., 2014). Sampling locations along this cross-shore transect reached from the berm down to the low water level, 10 m apart from one another (position 181 (berm) through position 188 (low water level)). Additionally, samples from two fully submerged sites in about 1.5 m water depth at low water were collected at sand patches, and close to a reef-like assembly of boulders and rocks. These samples represent surficial sediment samples, and were taken using a small shovel.

2.1 Geotechnical laboratory experiments

Three different shear devices were used: a small direct shear box (100 mm × 100 mm) for sandy sediments only, a large direct shear box (300 mm × 300 mm) for gravel and sand–gravel mixtures, and a ring shear device for 24 h tests of the sand samples. The direct shear test is the oldest and simplest shear test arrangement in geotechnical engineering (Das, 1990). The specimen is placed in a metal box that is split horizontally in halves. In the case of fully or partially saturated samples, porous plates are placed on top and below of the specimen to allow free drainage and avoid pore pressure buildup. A normal force is applied via a loading plate on the specimen. The minimum normal load equals the weight of the loading plate and the porous plate, while loads can be as great as ~ 1000 kPa (Das, 1990). Vertical displacement of the specimen is measured by recording the vertical motion of the loading plate using a mechanical displacement sensor. The shear tests could theoretically be performed without load, loading plate and top porous plate, but then measurements of vertical displacement would be difficult. After establishing the desired normal force, the lower half of the box is displaced horizontally at a chosen shear rate. The specimen experiences shear stress along the shear plane between the two halves of the box measured using a load cell. The results are commonly expressed as horizontal displacement versus vertical displacement, expressing dilation or contraction of the specimen during the shearing process, and as horizontal displacement versus shear stress. Different profiles are expected depending on the packing of the specimen. For example in the case of loose sand, the resisting shear stress increases with horizontal displacement until an approximately constant value τ_f (failure shear stress) is reached, while compaction dominates the vertical displacement (Terzaghi, 1943; Poulos, 1971; Das, 1990). In the case of densely packed sand, the shear stress increases to a maximum, the peak shear stress τ_f, while the specimen undergoes a compaction followed by dilation until reaching τ_f. Subsequently, the shear stress gradually decreases until reaching the critical state shear stress τ_{cv} (Terzaghi, 1943; Poulos, 1971; Das, 1990). No significant vertical displacement is observed in this phase.

Specimens of the same samples are then tested at various normal stresses, and plotting the normal stresses versus the corresponding shear stress allows an estimate of the angle of internal friction by application of the Mohr–Coulomb failure criterion (Terzaghi, 1943; Poulos, 1971; Das, 1990):

$$\tau_f = c + \sigma \tan \phi, \qquad (2)$$

with c the cohesion, σ the normal stress and ϕ the angle of internal friction. With this approximation, ϕ determined from tests with a range of normal stresses, including relatively high normal stresses, would still be applicable for the low range of normal stresses. Thus, the shear test results can be applied to beach surface sediments which are exposed to a minimal normal stress.

Different denotations can be found regarding ϕ. Terzaghi (1996) referred to the effective-stress friction angle, the angle of internal friction or the angle of shearing resistance ϕ' when plotting the normal stress versus the peak shear stress τ_f. Taylor (1948) and Rowe (1962) denoted this as ϕ_{max}. Plotting the normal stress versus the critical state shear stress allows one to extract the critical state or constant-volume friction angle ϕ_{cv} (Taylor, 1948; Rowe, 1962; Terzaghi, 1996). In the case of loose sand, it can be assumed that $\phi_{max} \approx \phi_{cv}$. For dense sands $\phi_{max} > \phi_{cv}$ applies as a result of the impact of dilation on shear strength (Taylor, 1948; Rowe, 1962; Terzaghi, 1996). Regarding application to the beach environment, dilation or compaction of the beach surface sediment under initiation of motion and shearing is expected due to particle rearrangement. Thus, the determination of the peak shear stress and the angle of internal friction is desired here. In the following, we will denote the angle of internal friction by ϕ.

2.1.1 Direct shear box

For the direct shear tests of the sand-sized fractions, a standard-sized small shear box (surface area = 36 cm^2) was used, and the sediment was filled in loosely and water-saturated. Each sample was tested at three different normal stresses: 3, 32, and 64 kPa. This is significantly lower than normal stresses usually applied for subsoil testing to account for low normal stresses at the beach-face sediment surface. Between the tests, the samples were stirred to ensure a loose particle arrangement at the start of shearing.

The larger direct shear apparatus was used for shearing the gravel and gravel–sand mixtures. Mixtures of 100 % gravel, 75 % gravel mixed with 25 % sand, and 50 % gravel mixed with 50 % sand were tested at normal stresses of 51, 99, and 149 kPa, which were the lowest normal stresses feasible with the large direct shear apparatus. Similar to the small direct shear box, samples were loosely installed, and stirred after each test.

The direct shear tests were used to monitor the development of shear stress with horizontal shearing of the sample to determine the angle of internal friction. Furthermore, any compression or expansion of the samples during shearing was recorded.

2.1.2 Ring shear test

Shearing behavior over 24 h was tested using a ring shear apparatus (also called annular direct shear apparatus). Here, an annular specimen (sand or finer sediments only) is sheared, under a given normal stress (92 kPa, the lowest normal stress the apparatus allowed; 461; 922 kPa), on a horizontal plane by the rotation of the annular sample relative to a stationary lid (Craig, 1974). This test was mainly conducted to observe long-term dilatory behavior.

2.2 Physical simulation of sediment remobilization

A smooth tray with a hopper feeding sediment onto the tray was arranged in a tank filled with water, and was tilted to angles ranging between 20 and 40° (Stark et al., 2014a). The angle at which sediment started moving α_{tray}, and average particle velocities u were determined via video observations, and via a prototype wide-band coherent Doppler profiler (MFDop) (Hay et al., 2008, 2012a, b). In this study, the main interest was to determine how easily the gravel and sand can be mobilized in comparison to each other, and other gravel types.

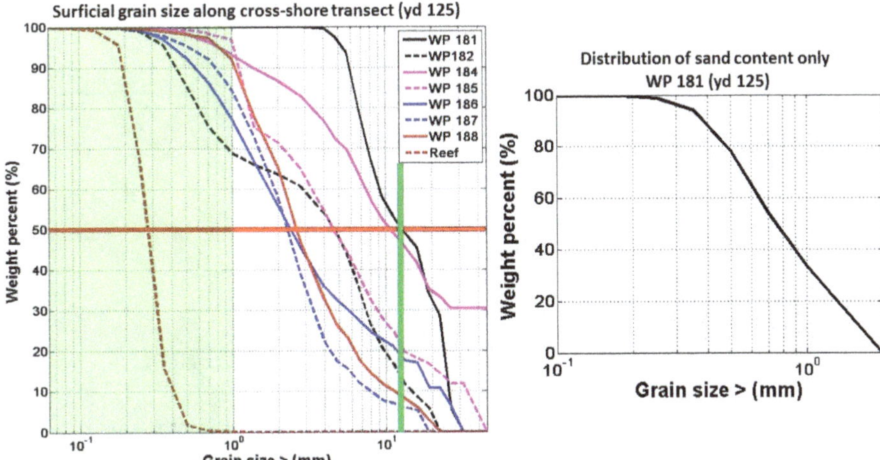

Figure 3. Left panel: representative grain size distributions for beach surface samples collected along a cross-shore profile with position WP 188 being the low tide level and WP 181 at the berm on year day 125. The green bars indicate the grain size range which has been tested in the laboratory shear tests. The red bar indicates the 50 % mark at which median grain size d_{50} is determined. Right panel: representative distribution of grain size of sand fraction only (< 2 mm) extracted from the sample taken at position WP 181 on year day 125. This sample has been tested in the direct shear box.

3 Results

3.1 Sediment description

The sampled sediment showed a strong variation in grain size distributions ($d_{50} = 0.3$–18.5 mm) along the cross-shore transect, as well as depending on the hydrodynamic conditions (Table 1 and Fig. 3). The finest sediments were found at the fully submerged sites ($d_{50} = 0.3$ mm), while the most fine-grained samples from the beach face were characterized by $d_{50} \approx 1$ mm. Figure 3 shows representative grain size distributions along a cross-shore transect, and a detailed grain size distribution of a representative sand sample used for the laboratory testing.

Within the sand-sized fractions ($d < 2$ mm), strong variations in particle shape with a high abundance of flat, elliptic particles were observed (Fig. 2). Grains of a sediment sample from a representative location at the beach surface (Pos. 184) were measured with regard to longest axis length (L), intermediate axis length (I) and shortest axis length (S) to determine disc-rod index ($\mathrm{DRI} = (L - I)/(L - S)$) and Corey Shape Index ($\mathrm{CSI} = S/(\sqrt{I\,L})$) after Illenberger (1991). Most of the particles were rather elongated and ranged from extremely flat to compact (Fig. 4). Compact particles were predominantly of gravel size. Low DRIs in the case of the sand particles can likely be associated with broken particles.

The samples were dominated by sandstone, in accordance with geological studies of the region by, e.g., Amos and Long (1980), Amos et al. (1991) and Dalrymple et al. (1990). Particle densities of randomly chosen samples were consistently 2.4–2.5 g cm^{-3}.

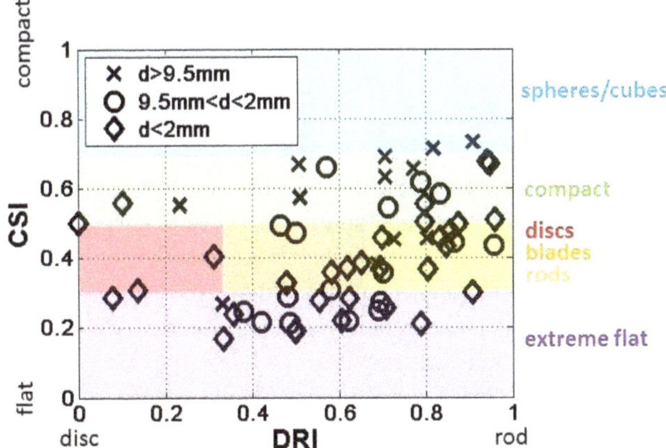

Figure 4. Sample measurements of particle shape expressed as disc-rod index (DRI) and Corey Shape Index (CSI). Description and schematic modified after Illenberger (1991).

3.2 Geotechnical laboratory experiments

3.2.1 Direct shear box

The plots of horizontal displacement versus shear stress (Fig. 5) illustrate the different response to applied shear stress for the different loading stages. At the lowest applied normal stress, sample failure occurred at a low value of the applied shear stress. At the second loading stage, we found some stress-vs.-horizontal displacement curves which matched a shear stress path expected in the case of loose sediments, and some which already showed a tendency towards a peak shear strength, as expected for denser sand. In some cases, a step-like feature was observed in the stress-vs.-horizontal displacement curves, likely corresponding to processes of

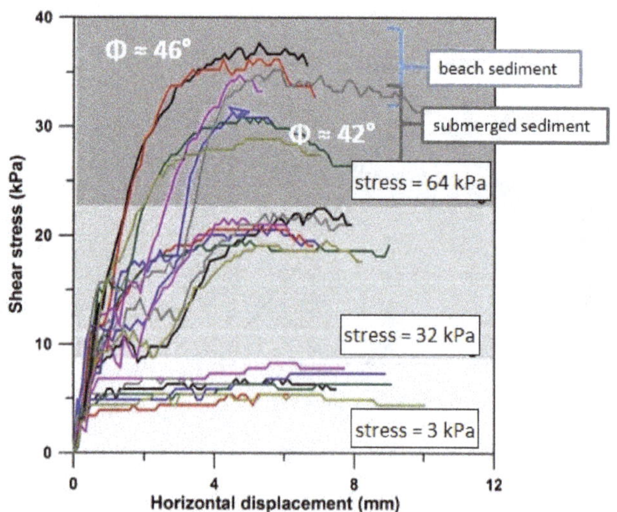

Figure 5. Horizontal displacement versus shear stress of samples with a grain size $d \leq 2$ mm measured in the small direct shear box. The different stages of normal stress are indicated by different gray shades. At the highest loading stage, a difference in the shear stress path between the beach-face sediments and samples from permanently submerged sites can be observed, leading to differences in the internal angle of friction ϕ.

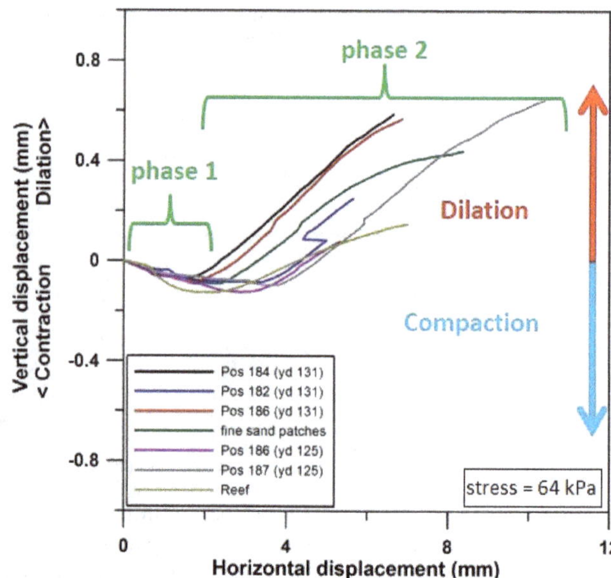

Figure 6. Horizontal displacement versus vertical displacement of samples with a grain size $d \leq 2$ mm measured in the small direct shear box under the highest normal stress. First a short phase of compaction was observed, before a strong dilation was monitored.

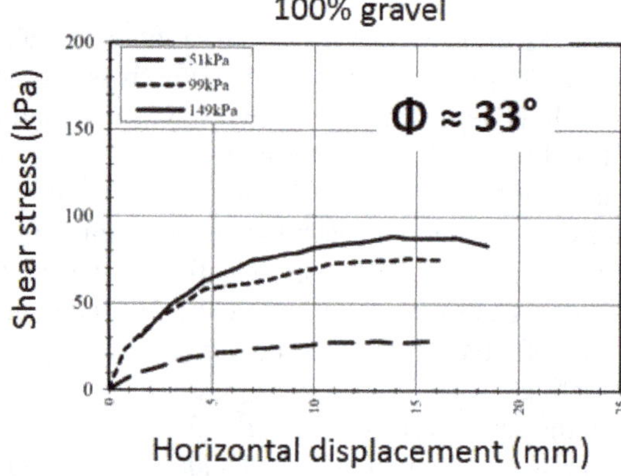

Figure 7. Horizontal displacement versus vertical displacement of gravel measured in the large direct shear box.

re-arrangement of the particles. The same was true for the highest normal load. Here, we additionally observed clear differences between the samples from the beach face and those from the permanently submerged sites. Analysis of the profiles led to $\phi \approx 46°$ for the beach-face sands and $\phi \approx 42°$ for the sand from the permanently submerged sites.

The vertical displacement was characterized by a two-phase behavior that was approximately similar for all sands independent of their origin (Fig. 6). First, a phase of specimen compaction, i.e., a negative vertical displacement, was observed, corresponding to the range of horizontal displacement when also the first plateau in the step-like shear stress profiles was noted. This observation supports the hypothesis of particle re-arrangement and alignment during this phase. After this compaction phase, a strong dilation behavior is characteristic of the Advocate Beach samples, reaching significantly larger positive vertical displacements than the previous negative vertical displacement (Fig. 6).

The gravel tests in the large direct shear box revealed typical shear stress profiles, as expected for loose granular material (Fig. 7), and led to $\phi \approx 33°$ for the pure gravel sample, while the addition of sand resulted in an increase to $\phi \approx 38°$. The vertical displacement was characterized by compaction: no dilation was observed.

3.2.2 Ring shear test

The vertical displacement during shearing was investigated in more detail in the 24 h ring shear tests (Fig. 8). Under all three values of normal stress, the specimen expressed the previously described phases of compaction and stronger dilation. However, under the much higher normal stresses tested in the ring shear, it stood out that the magnitude of compaction as well as dilation is governed by the normal stress. Nevertheless, a noticeable compaction followed by a strong dilation was confirmed by the low normal stress measurements in the ring shear apparatus, likely the test closest to the in situ beach-face conditions where normal stresses are low. Afterwards, a stagnation is reached in which the vertical displacement level remains constant, before it comes to another negative vertical displacement, governed in magnitude by the normal stress stage again. Furthermore, it was observed that over the long-term shearing the sand experienced a process

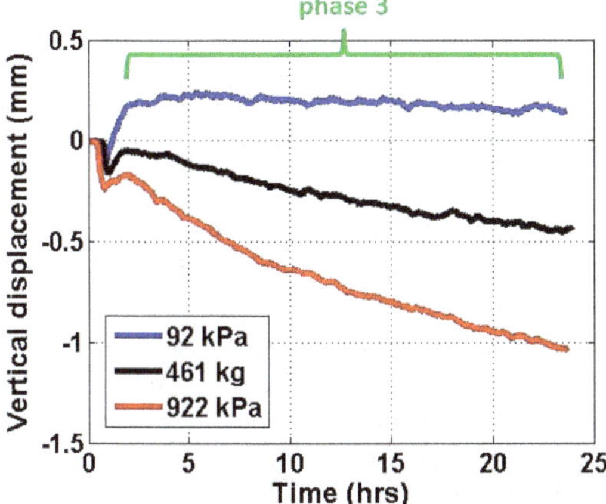

Figure 8. Vertical displacement versus time for a sand sample measured in the ring shear apparatus. After the previously described phases 1 and 2 of vertical displacement, the specimen shows a third phase of behavior: a return to negative vertical displacement. The magnitude is governed by the normal stress. Also, the intensity of dilation in phase 2 seemed to be restricted by the normal stress.

such as grinding. The particle sizes at the shear face were reduced and, in the case of the highest normal stress, down to a silt size level.

3.3 Physical simulation of sediment remobilization

In a simple physical simulation, we tested the initiation of particle motion of sand and gravel from Advocate in comparison to each other, and to other commercially available gravel. The test series was part of a larger effort investigating the suitability of a new wideband coherent acoustic Doppler profiler for the investigation of coarse sediment transport. Detailed results and discussion can be found in Stark et al. (2014a). The tray tilt angle α_{tray} at which first particle motion was correlated to the respective transport velocities reached approximately the center of the tray and was measured using the acoustic Doppler profiler MFDop and video observations. The Advocate gravel showed a behavior similar to more angular, as well as more rounded commercially available gravel. Deviations depending on shape and size were observed, but will not be discussed in detail in this article. A more distinctively different behavior was observed in the case of the sand. Initiation of particle motion of the Advocate sand did not start until tray tilt angles of 31° were reached, while all gravel size particles started moving at tray tilt angles of 21–24°. This trend matched the observation of the angle of internal friction determined by direct shearing; however, the exact values did not match. The difference can be explained with the experimental setup. The majority of the sediment was fed from a hopper which was mounted to the tray. Nevertheless, it has to be considered that the predominant "failure" or slipping plane was the tray–sediment interface, likely leading to the initiation of particle motion at lower inclinations.

4 Discussion

The application of geotechnical laboratory experiments to investigate the behavior of elliptically shaped beach sediment under shear stress proved to be a suitable and useful approach to study the response to applied shear under controlled conditions. Nevertheless, some issues have to be considered. First, the collected samples hardly represented the in situ texture at the beach face, in particular in a submerged state, and under active flow and wave action. This is a well-recognized issue regarding sediment sampling in the field of subaqueous sediment dynamics (Blomqvist, 1991; Larson et al., 1997; Edwards and Glysson, 1988). Specifically loosely arranged surface samples have to be considered significantly disturbed after retrieval, transport and storage. Instead of trying to preserve the original state, we decided to account for this by installing the sample in the shear boxes in a very loose, not consolidated and fully water saturated state, aiming for conditions representative of submerged beach-face surface sediments which are frequently re-arranged by bed-load transport processes. As we observed the re-arrangement of the loose, unaligned particles during shearing, we argue that the laboratory-prepared samples mimicked fairly well the in situ conditions of recently and loosely deposited sediments which are exposed to increasing shear stress in the swash and surf zone.

A second issue is the applied normal stress. Despite the fact that the lowest normal stress possible, dependent on the specific shear apparatus, was applied in the shear tests, there is no doubt that all tested normal stresses exceeded the normal stresses beach surface sediments are exposed to, a function of the particle weight of the mobile sediment layer. In the case of the large direct shear box and the ring shear, the use of lower normal stresses than standard would have impacted on the operation of the shear boxes, and so was not possible. In the case of the small direct shear box, it was feasible to test at lower normal stresses. According to the Mohr–Coulomb failure criterion, the angle of internal friction ϕ derived from direct shear tests in which a range of different normal stresses σ was tested is valid for the minimal normal stress scenario as well (see Sect. 2.1). However, Mohr's criterion allows for a curved shape of the failure envelope, representing a nonlinear relationship $\tau_{\mathrm{f}} = f(\sigma)$ (Mohr, 1900; Labuz and Zang, 2012). Additionally, there are concerns as to whether failure stress and friction angles can be properly determined at low stress levels in a shear box (Bruton et al., 2007). For the case of fine sands, Lehane and Liu (2013) demonstrated that a conventional direct shear can indeed be applied at low stress levels, but that corrections might be required, particularly at normal stresses below 10 kPa. Their corrections led to a decrease in ϕ by $\sim 4°$ for $\sigma = 4$ kPa, by 1–2° for $\sigma = 10$ kPa, and to no significant changes when $\sigma \geq 30$ kPa. Based on these findings, no corrections were applied in this study because most of the tests were performed at normal stresses significantly above 10 kPa. Furthermore, $R^2 \geq 0.95$ regarding the measurements and the fitted Mohr–Coulomb failure criterion line from which the angles of internal friction were determined, and repetition of measurements for specimen from the same sample material delivered consistent results (up to three repetitions per sample). This indicates that the shear box results are reliable and that the angle of internal friction is unlikely significantly biased by the comparably low normal stress measurements.

The question remains of whether the results are applicable to particle remobilization of the beach surface. Bagnold (1966) addresses the principle of solid friction with regard to bed-load transport. He argues that the limiting angle of repose α of a mass of cohesionless granular solids relates the required shear force T to initiate motion under a respective normal force P in the following manner:

$$T/P = \tan\alpha, \tag{3}$$

consistent with the Mohr–Coulomb failure criterion. This supports the applicability of the performed shear tests for the investigation of the shearing processes of the beach sediments, and suggests that the observed shearing behavior in the laboratory can be qualitatively transferred to the beach environment. However, some major differences arise resulting from sampling, sample positioning and sample preparation. The issue of sample retrieval has been discussed in a previous paragraph. Other factors are, e.g., beach slope (Rowe, 1962) and the characteristics of hydrodynamic forces exerting shear stress onto the bed. These aspects exceed the scope of this article, but the latter is addressed in Hay et al. (2014). The impact of sample density and packing on the angle of internal friction and the relation to the angle repose however will be discussed in more detail here.

It has been well known since the mid-20th century that the angle of internal friction highly depends on sample density and packing and that it decreases with looser packing (Taylor, 1948; Poulos, 1971). While Bagnold (1966) in his review on the principle of solid friction neglects the possible difference in particle arrangement of a mass of cohesionless granular solids and, thus, assumes that the angle of internal friction approximates the angle of repose, Metcalf (1966) investigates the angle of internal friction in comparison to the angle of repose in more detail. Contrary to the common assumption that the angle of repose equals the angle of internal friction for the loosest particle arrangement, Metcalf (1966) showed that the angle of repose is the angle of internal friction after a first consolidation following the initiation of shearing. For example, for a washed quartz sand he found $\alpha = 37°$, $\phi_{\text{loose}} = 32°$ and $\phi_{\text{consolidated}} = 37°$. He also observed that the angle of solid friction equaled 37°, consistent with Bagnold (1966). Based on these findings and the negligible deviations between our shear stress measurements for low and larger normal stresses from the Mohr–Coulomb failure criterion approximation, the application of the direct shear test derived angles of internal friction on the bed shear stress scenario at the beach face seems to be justified. Thus, the derived angle of internal friction can be applied as an estimate of angle of repose based on previous literature.

In this study we determined angles of internal friction of $\phi \approx 41$–$49°$ for the flat, elliptic sand from Advocate Beach (Fig. 4) using direct shear (Fig. 7). Considering that the tests were conducted on loose sand samples, these angles of internal friction were significantly larger than what was expected from the literature (i.e., loose sand (rounded): $\phi \approx 27$–$30°$; loose sand (angular): $\phi \approx 30$–$35°$; Das, 1990). Instead, these loose samples fall in the high range of dense sand: $\phi \approx 40$–$45°$ (Craig, 1974; Das, 1990). Similarly high friction angles were reported for, e.g., angular rock, metal cubes and elliptic particles by Carson (1977), Carrigy (1970), Cho et al. (2006) and Frette et al. (1996), respectively, highlighting the importance of particle shape. Particularly, the elliptic rice particles had characteristic angles of repose ranging from 46.6 to 51.1° (Frette et al., 1996).

The threshold stress τ_t of the Advocate sand was estimated using the direct shear box results and the equation suggested by Chepil (1959) and adapted for elliptic particles by Komar and Li (1986):

$$\tau_t = k'(\rho_s - \rho)\, g\, I\, \frac{\tan\phi}{1 + 0.75\tan\phi}, \tag{4}$$

with $k' = 1$ an empirical coefficient, the grain density ρ_s, water density ρ, gravitational acceleration g, the length of the intermediate grain axis I and the angle of internal friction ϕ. The estimated range of τ_t for the sandy fractions has been indicated by the dark gray patch in Fig. 9 based on a diagram presented by Komar and Li (1986) plotting the grain diameter versus the threshold stress for a variety of particle shapes. Our estimates agree well the curves suggested by Komar and Li (1986) particularly for imbricated, angular to ellipsoid particles. In addition to the above-mentioned high angles of repose for elliptic rice particles, this favorable agreement between our results and the study by Komar and Li (1986) strengthens the argument that the direct shear tests are applicable to the beach environment and are likely to be useful for the investigation of subaqueous sediment transport.

The friction angles measured for gravel matched the range that has been previously observed for loose gravel (Craig, 1974; Das, 1990; Simoni and Houlsby, 2006). However, mixing the Advocate sand with the gravel led to an immediate increase in the friction angle, expressing the opposite behavior to that observed in other studies in which rounded to sub-angular sand was mixed with gravel. Both Simoni and Houlsby (2006) and Yagiz (2001) described a steady decrease in friction angles with the addition of sand to the gravel sample, while the Advocate sand led to an immediate increase of friction angle after adding 25 % of sand, but remained constant when adding 50 % of sand. This supported the first observation that the Advocate sand withstood exceptionally high shear stresses in comparison to similar particle sizes of rounded or angular shape before failure. Furthermore, this observation may motivate a more detailed

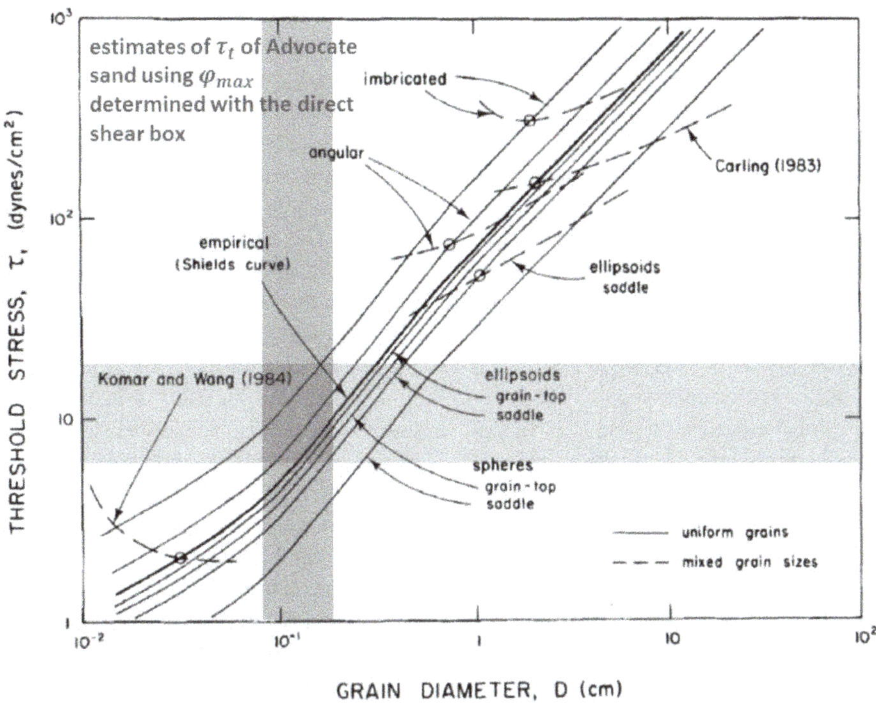

Figure 9. Estimated range of threshold stress based on the shape measurements and direct shear tests of the Advocate sands plotted as the dark gray shade into a scheme by Komar and Li (1986). Our results agree well with their measurements of imbricated, angular and ellipsoid particles in the same size range.

investigation of this issue. Our results suggest that the mixing of elliptic plate-shaped sand with gravel can lead to an increase in the internal friction angle compared to the characteristic internal friction angle of the gravel and that this behavior is related to the particle shape and characteristics of the sand. More detailed geotechnical laboratory investigations are required to test this hypothesis. For beach and other aquatic environments, this finding might imply that specific sands may strengthen a sand–gravel mixture against shear forcing, while often the finer sediments are considered the weak spot in cohesionless mixtures. Investigators have studied the impact of bimodality of sand–gravel mixtures on sediment entrainment and remobilization (Wilcock, 1993; Wilcock and McArdell, 1993; Shvidchenko et al., 2001), but the behavior of the overall mixture and the impact of specific sand characteristics has rarely been the focus of such studies. This research could potentially have a large impact on beach nourishment strategies.

The observed dilatory behavior is another main result of this study. It contributes to the understanding of particle response to shearing in the soil matrix, and offers a conceptual scheme explaining high angles of internal friction and strong resistance to shear of the sand fraction at Advocate Beach (Fig. 10). Again the difference in normal stresses and location of the shear plane must be discussed here. At the beach face, shear is applied to the uppermost sediment surface and failure occurs between the uppermost particle layers. In the shear box the failure plane is approximately in the vertical center of the box. Thus, compaction of the specimen is supported, while dilation is hampered by normal stress in the shear box. Nevertheless, we observed a strong dilatory behavior of the Advocate sand specimen after an initial settlement associated with particle rearrangement during initiation of shear force. The samples dilated and remained at a certain

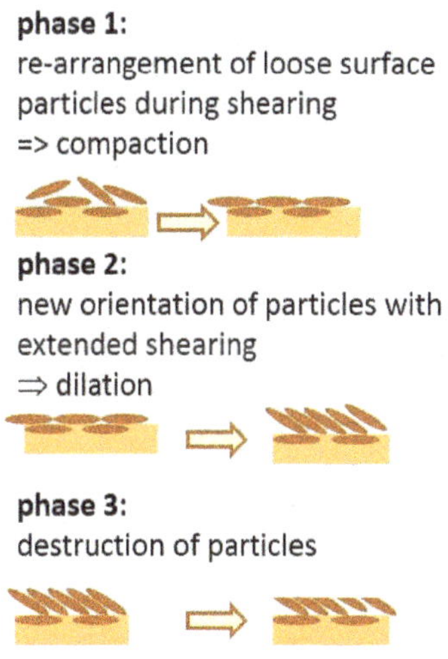

phase 1:
re-arrangement of loose surface particles during shearing
=> compaction

phase 2:
new orientation of particles with extended shearing
⇒ dilation

phase 3:
destruction of particles

Figure 10. Conceptual sketch of particle arrangement explaining the observed time dependence of vertical displacement during shear.

level of dilation until the end of the test. Even during long-term shearing in the ring shear, the specimen remained in a state of maximum dilation until particles were ground into silt-sized material (Fig. 8). This led to the hypothesis that the Advocate sand re-arranges in a dilatory manner under shearing and that this rearrangement supports a large shear resistance (Fig. 10). Under the negligible normal stresses in the beach scenario, dilation is even easier, allowing a rapid and easy rearrangement of the particle matrix. However, the impact of particle protrusion and lift must be considered here. If one of the elongated particles moves into a more protruded position during the process of particle arrangement, shear and lift forces will increase significantly and will possibly lead to the mobilization of the particle (Bagnold, 1988). Resulting gaps in the sediment matrix and the resulting increase in bed roughness could then lead to a chain reaction of particle mobilizations. These processes are entirely ignored in the shear box tests.

The vertical displacement plots, determined using the small direct shear box and the ring shear apparatus, indicated that the samples underwent different phases of arrangement during shearing. In the following, we propose a concept of particle re-arrangement during the shearing process that may explain the observed vertical displacement and failure shear stresses (Fig. 10). Under first shearing, the loose and unaligned particles rearrange and align corresponding to their shape. This would lead to a denser state of particle packing, explaining a first increase in shear stress and negative vertical displacement. With increasing shear stress, the particles may be erected, again guided by their elongated shape, allowing for higher shear stresses and a strong dilatory behavior (Fig. 10). This arrangement was so strong and final (possibly in an arrangement similar to shingles on a roof) that particles were rather destroyed by grinding than re-arranged again (Fig. 10) during long-term shearing.

A simple physical simulation in which the sand was exposed to the tilting of a smooth tray was used as another independent test of sediment mobilization. The experiment was part of a larger investigation of particle velocity and particle motion in bed-load transport of different particles sizes (Stark et al., 2014a). The tray tilt angle at which mobilization of the sand particles occurred did not match the determined angles of internal friction, which can be explained by the smooth surface of the tray. Thus, the results may only serve for a qualitative comparison. Nevertheless, a clearly delayed mobilization at much higher angles than the tested gravel was observed, confirming that the Advocate sand was resistant to the initiation of particle transport. For future experiments, the impact of the tray surface for testing the angle of repose in comparison to the angle of internal friction can be significantly improved by preparing the tray with a fixed layer particles representing the tray surface as has been demonstrated by several investigators (Miller and Byrne, 1966; Buffington et al., 1992).

In summary, the geotechnical laboratory experiments suggested that the sand fraction at the mixed sand–gravel beach in Advocate, Nova Scotia, has a high resistance towards shear forces and even increases the overall shear resistance when mixed with native gravel. This finding can be explained with the flat and elongated shapes of the sand-sized particles. For the beach environment, it can be concluded that higher shear forces – meaning stronger flows and wave action – are required to mobilize particles than would be assumed from the particle size. When particles are moved, sliding along the flat faces of the particles would be the favored transport behavior, and likely occurs before particle arrangement under increased shearing. When particle arrangement finally takes place, the particle matrix is strengthened by the new arrangement, and more transport and particularly entrainment is hampered. This is in good agreement with the observations at the beach (Hay et al., 2014). No significant change was observed at the beach surface under calm and moderate hydrodynamic conditions. Under storm wave action, sands were shifted and formed ripples with the passing surf zone of the flood tide. During the ebb tide, the ripples were washed out by the retreating surf and swash, leading to a thin fine sand cover of the beach face (Hay et al., 2014). Bed-load transport dominated, while major entrainment of particles appeared unlikely. Only the surf and swash flow conditions were sufficient to trigger significant sediment transport, as no major migration of the ripples was observed. Detailed findings and discussion with respect to flow conditions can be found in Hay et al. (2014).

5 Conclusions

Three different types of laboratory tests were performed to assess the behavior of sediments from Advocate Beach during shear: geotechnical direct shear tests and ring shear tests, as well as simple physical simulations of sediment transport initiation. The sediments ranged in size from sandy particles to gravel, and the sand-sized fractures were characterized by a high abundance of flat, elliptic grains. The study was part of a larger effort targeting beach dynamics at Advocate Beach, Bay of Fundy, Nova Scotia.

With regard to the research objectives of this study, the following conclusions can be drawn. (i) It was found that the geotechnical laboratory methods offer important insight into the soil mechanical processes under shear stress and sediment resistance to shear, with potentially important implications for the sediment dynamical behavior at the beach face. (ii) Particularly, it was proposed that the flat, elliptic shape of the Advocate sand undergoes a specific process of particle re-arrangement and alignment that results in a significant increase in angles of internal friction. This strengthens the sand against shearing processes, potentially contributing to the stability of the sand at the beach face against wave action.

Further investigation is nevertheless needed to better determine how standard geotechnical laboratory experiments should be utilized for the investigation of subaqueous and beach sediment dynamics. The results of this study encourage further pursuit of the issue. This study has demonstrated the importance of particle shape – in particular plate-like shapes – in water-saturated sand-sized sediments subjected to shear stress in the laboratory using a standard geotechnical shear apparatus. In particular, the measured friction angles are very high compared to the values for more rounded particles, a result which we suggest may potentially contribute to the long-term stability of the 1 : 10 slope of the beach face at Advocate.

Acknowledgements. We would like to thank the Natural Sciences and Engineering Research Council of Canada, the Atlantic Innovation Fund, and Nortek for funding; the Advocate field team (L. Zedel, D. Barclay, M. Hatcher, T. Guest and J. Hare) for support in the field; and T. Morton and R. Jamshidi for support during the direct shear experiments. We would also like to thank M. Kleinhans, G. Coco and an anonymous reviewer for suggestions and comments which contributed significantly to the improvement of the article.

Edited by: G. Coco

References

Amos, C. L. and Long, B.: The sedimentary character of the Minas Basin, Bay of Fundy, in: Proc. Coastlines Canada Conf., Halifax, NS, 1980.

Amos, C. L., Tee, K. T., and Zaitlin, B. A.: The post-glacial evolution of Chignecto Bay, Bay of Fundy, and its modern environment of deposition, Clastic Tidal Sedimentology-memoirs 16, Canadian Society of Petroleum Geologists, Special Publications, 59–89, 1991.

Bagnold, R.: An approach to the sediment transport problem from general physics, US Geol. Surv. Prof. Paper 422, in: The Physics of Sediment Transport by Wind and Water: A collection of hallmark papers, ASCE Publications, Reston, USA, 231–291, 1966.

Bagnold, R.: Beach and nearshore processes part I. Mechanics of marine sedimentation, in: The Physics of Sediment Transport by Wind and Water: A Collection of Hallmark Papers, ASCE Publications, Reston, USA, 188–231, 1988.

Blomqvist, S.: Quantitative sampling of soft-bottom sediments: Problems and solutions, Mar. Ecol.-Prog. Ser., 72, 295–304, 1991.

Bradley, M. P. and Stolt, M. H.: Landscape-level seagrass–sediment relations in a coastal lagoon, Aquat. Bot., 84, 121–128, 2006.

Bruton, D., Carr, M., and White, D.: The influence of pipe-soil interraction on lateral buckling and walking of pipelines-the safebuck JIP, Offshore Site Investigation and Geotechnics, Confronting New Challenges and Sharing Knowledge, 6th International Conference on Offshore Site Investigation and Geotechnics, Confronting new challenges and sharing knowledge, London, UK, 2007.

Buffington, J. M., Dietrich, W. E., and Kirchner, J. W.: Friction angle measurements on a naturally formed gravel streambed: Implications for critical boundary shear stress, Water Resour. Res., 28, 411–425, 1992.

Carrigy, M. A.: Experiments on the Angles of Repose of Granular Materials, Sedimentology, 14, 147–158, 1970.

Carson, M.: Angles of repose, angles of shearing resistance and angles of talus slopes, Earth Surf. Proc., 2, 363–380, 1977.

Chepil, W.: Equilibrium of soil grains at the threshold of movement by wind, Soil Sci. Soc. Am. J., 23, 422–428, 1959.

Cho, G.-C., Dodds, J., and Santamarina, J. C.: Particle shape effects on packing density, stiffness, and strength: natural and crushed sands, J. Geotech. Geoenviron. Eng., 132, 591–602, 2006.

Craig, R. F.: Soil Mechanics, E and FN Spon Press, London, UK, 1974.

Dalrymple, R. W., Knight, R., Zaitlin, B. A., and Middleton, G. V.: Dynamics and facies model of a macrotidal sand-bar complex, Cobequid Bay-Salmon River Estuary (Bay of Fundy), Sedimentology, 37, 577–612, 1990.

Das, B. M.: Principles of geotechnical engineering, PWS, Boston, USA, 1990.

Eagleson, P. and Dean, R.: Wave-induced motion of bottom sediment particles, T. Am. Soc. Civ. Eng., 126, 1162–1185, 1961.

Edwards, T. K. and Glysson, G. D.: Field methods for measurement of fluvial sediment, Department of the Interior, US Geological Survey, pubs.usgs.gov, 1988.

Frette, V., Christensen, K., Malthe-Sørenssen, A., Feder, J., Jøssang, T., and Meakin, P.: Avalanche dynamics in a pile of rice, Nature, 379, 49–52, 1996.

Hay, A. E., Zedel, L., Craig, R., and Paul, W.: Multi-frequency, pulse-to-pulse coherent Doppler sonar profiler, in: Current Measurement Technology, 2008, CMTC 2008, IEEE/OES 9th Working Conference, Charlston, USA, 25–29, 2008.

Hay, A. E., Zedel, L., Cheel, R., and Dillon, J.: Observations of the vertical structure of turbulent oscillatory boundary layers above fixed roughness beds using a prototype wideband coherent Doppler profiler: 1. The oscillatory component of the flow, J. Geophys. Res., 117, C03005, doi:10.1029/2011JC007113, 2012a.

Hay, A. E., Zedel, L., Cheel, R., and Dillon, J.: Observations of the vertical structure of turbulent oscillatory boundary layers above fixed roughness using a prototype wideband coherent Doppler profiler: 2. Turbulence and stress, J. Geophys. Res., 117, C03006, doi:10.1029/2011JC007114, 2012b.

Hay, A. E., Zedel, L., and Stark, N.: Sediment dynamics on a steep, megatidal, mixed sand-gravel-cobble beach, Earth Surf. Dynam. Discuss., 2, 117–152, doi:10.5194/esurfd-2-117-2014, 2014.

Illenberger, W. K.: Pebble shape (and size!), J. Sediment. Res., 61, 756–767, 1991.

Kirchner, J. W., Dietrich, W. E., Iseya, F., and Ikeda, H.: The variability of critical shear stress, friction angle, and grain protrusion in water-worked sediments, Sedimentology, 37, 647–672, 1990.

Komar, P. D. and Li, Z.: Pivoting analyses of the selective entrainment of sediments by shape and size with application to gravel threshold, Sedimentology, 33, 425–436, 1986.

Kuehl, S. A., Nittrouer, C. A., Allison, M. A., Faria, L. E. C., Dukat, D. A., Jaeger, J. M., Pacioni, T. D., Figueiredo, A. G., and Underkoffler, E. C.: Sediment deposition, accumulation, and seabed dynamics in an energetic fine-grained coastal environment, Cont. Shelf Res., 16, 787–815, 1996.

Labuz, J. F. and Zang, A.: Mohr–Coulomb failure criterion, Rock Mech. Rock Eng., 45, 975–979, 2012.

Larson, R., Morang, A., and Gorman, L.: Monitoring the coastal environment; part II: Sediment sampling and geotechnical methods, J. Coast. Res., 13, 308–330, 1997.

Lehane, B. and Liu, Q.: Measurement of shearing characteristics of granular materials at low stress levels in a shear box, Geotech. Geol. Eng., 31, 329–336, 2013.

Li, Z. and Komar, P. D.: Laboratory measurements of pivoting angles for applications to selective entrainment of gravel in a current, Sedimentology, 33, 413–423, 1986.

Masson, D., Harbitz, C., Wynn, R., Pedersen, G., and Løvholt, F.: Submarine landslides: Processes, triggers and hazard prediction, Philos. T. Roy. Soc. A, 364, 2009–2039, 2006.

Metcalf, J.: Angle of repose and internal friction, in: International Journal of Rock Mechanics and Mining Sciences & Geomechanics Abstracts, vol. 3, Elsevier, Oxford, UK, 155–161, 1966.

Middleton, G. V. and Southard, J. B.: Mechanics of sediment movement, SEPM, Tulsa, USA, 1984.

Miller, R. L. and Byrne, R. J.: The angle of repose for a single grain on a fixed rough bed, Sedimentology, 6, 303–314, 1966.

Mohr, O.: Welche Umstände bedingen die Elastizitätsgrenze und den Bruch eines Materials, Z. Verein. Deut. Ing., 46, 1524–1530, 1900.

Poulos, S. J.: The Stress-Strain Curves of Soils, Geotechnical Engineers Incorporated, Winchester, USA, 1971.

Riley, G. and Mann, G.: Effects of particle shape on angles of repose and bulk densities of a granular solid, Mater. Res. Bull., 7, 163–169, 1972.

Rowe, P. W.: The stress-dilatancy relation for static equilibrium of an assembly of particles in contact, P. Roy. Soc. Lond. A, 269, 500–527, 1962.

Santamarina, J. and Cho, G.: Soil behaviour: The role of particle shape, in: Advances in geotechnical engineering: The skempton conference, vol. 1, Thomas Telford, London, UK, 604–617, 2004.

Schanz, T. and Vermeer, P.: Angles of friction and dilatancy of sand, Geotechnique, 46, 145–152, 1996.

Shvidchenko, A. B., Pender, G., and Hoey, T. B.: Critical shear stress for incipient motion of sand/gravel streambeds, Water Resour. Res., 37, 2273–2283, 2001.

Simoni, A. and Houlsby, G. T.: The direct shear strength and dilatancy of sand–gravel mixtures, Geotech. Geol. Eng., 24, 523–549, 2006.

Simons, D. B. and Şentürk, F.: Sediment transport technology: Water and sediment dynamics, Water Resources Publication, Littleton, USA, 1992.

Soulsby, R.: Dynamics of marine sands: A manual for practical applications, Thomas Telford, London, UK, 1997.

Stark, N., Hay, A., Cheel, R., Zedel, L., and Barclay, D.: Laboratory Measurements of Coarse Sediment Bedload Transport Velocity Using a Prototype Wideband Coherent Doppler Profiler (MFDop), J. Atmos. Ocean. Tech., 31, 999–1011, 2014a.

Stark, N., Hay, A., and Zedel, L.: Sedimentological and morphological variations at a macro-tidal sandy gravel beach, Mar. Geol., in preparation, 2014b.

Taylor, D. W.: Fundamentals of soil mechanics, Soil Sci., 66, 83–163, 1948.

Terzaghi, K.: Theoretical soil mechanics, Wiley & Sons, Hoboken, USA, 1943.

Terzaghi, K.: Soil mechanics in engineering practice, John Wiley & Sons, Hoboken, USA, 1996.

Wilcock, P. R.: Critical shear stress of natural sediments, J. Hydraul. Eng., 119, 491–505, 1993.

Wilcock, P. R. and McArdell, B. W.: Surface-based fractional transport rates: Mobilization thresholds and partial transport of a sand-gravel sediment, Water Resour. Res., 29, 1297–1312, 1993.

Yagiz, S.: Brief note on the influence of shape and percentage of gravel on the shear strength of sand and gravel mixtures, Bull. Eng. Geol. Environ., 60, 321–323, 2001.

6

Dynamics and mechanics of bed-load tracer particles

C. B. Phillips and D. J. Jerolmack

Earth and Environmental Science, University of Pennsylvania, Hayden Hall, 240 South 33rd st., Philadelphia, PA 19104, USA

Correspondence to: C. B. Phillips (colinbphillips@gmail.com)

Abstract. Understanding the mechanics of bed load at the flood scale is necessary to link hydrology to landscape evolution. Here we report on observations of the transport of coarse sediment tracer particles in a cobble-bedded alluvial river and a step-pool bedrock tributary, at the individual flood and multi-annual timescales. Tracer particle data for each survey are composed of measured displacement lengths for individual particles, and the number of tagged particles mobilized. For single floods we find that measured tracer particle displacement lengths are exponentially distributed; the number of mobile particles increases linearly with peak flood Shields stress, indicating partial bed load transport for all observed floods; and modal displacement distances scale linearly with excess shear velocity. These findings provide quantitative field support for a recently proposed modeling framework based on momentum conservation at the grain scale. Tracer displacement is weakly negatively correlated with particle size at the individual flood scale; however cumulative travel distance begins to show a stronger inverse relation to grain size when measured over many transport events. The observed spatial sorting of tracers approaches that of the river bed, and is consistent with size-selective deposition models and laboratory experiments. Tracer displacement data for the bedrock and alluvial channels collapse onto a single curve – despite more than an order of magnitude difference in channel slope – when variations of critical Shields stress and flow resistance between the two are accounted for. Results show how bed load dynamics may be predicted from a record of river stage, providing a direct link between climate and sediment transport.

1 Introduction

Understanding landscape denudation and its relation to climate requires an understanding of how a flood hydrograph drives the sediment mass flux leaving the system through rivers. While suspended sediment represents the largest fraction of mass exiting the landscape (Milliman and Syvitski, 1992; Willenbring et al., 2013), it is coarse bed load transport that sets the limiting rate of landscape incision through its control on bedrock erosion and channel geometry in gravel rivers (Sklar and Dietrich, 2004; Snyder et al., 2003; Parker et al., 2007). The rate of bed load transport is known to vary both spatially and temporally due to turbulence and granular phenomena such as clustering, bed forms, bed compaction, grain protrusion/hiding, and collective motion (Gomez, 1991; Kirchner et al., 1990; Schmeeckle et al., 2001; Strom et al., 2004; Ancey et al., 2008; Zimmermann et al., 2010; Marquis and Roy, 2012; Heyman et al., 2013), which makes predictions difficult (Recking et al., 2012) and point measurements highly variable (e.g., Gray et al., 2010). Bed load is especially difficult to predict near the threshold of motion (Recking et al., 2012), where transport is highly intermittent, often resulting in partial bed load transport, in which only a fraction of the bed is mobilized during a transporting event (Wilcock and McArdell, 1997). Further confounding predictions is that many gravel streams adjust their geometry to an effective discharge (Wolman and Miller, 1960), which occurs at a flow slightly above (1.2–1.4 times) the threshold of motion for the median grain size (Parker, 1978; Parker et al., 1998, 2007), indicating that partial transport may be the dominant transport regime within gravel rivers. The spatially variable and highly intermittent flux during partial transport (Wilcock and McArdell, 1997; Haschenburger and Wilcock, 2003), compounded with the added difficulty of a varying sediment supply, necessitates long-term observations to decipher bed load dynamics in the field.

Passive tracer particles, in particular passive integrated transponder radio-frequency identification (PIT RFID) tagged particles, are becoming an attractive low-cost and low-maintenance method of measuring bed load particle dynamics. The application of passive tracer particles has taken various forms, such as exotic lithologies (Houbrechts et al., 2011), painted bed material (Wilcock, 1997b), magnetic (Hassan et al., 1991), radioactive (Sayre and Hubbell, 1965; Bradley et al., 2010), and RFID (Lamarre et al., 2005; Bradley and Tucker, 2012; Phillips et al., 2013; Schneider et al., 2014). A benefit of RFID-equipped tracer particles is that each particle is uniquely identified, which allows its position to be measured at longer timescales with high recovery rates (Bradley and Tucker, 2012; Phillips et al., 2013). An advantage of long-term (multi-flood to multi-annual) observations of tracer particles is that they sample over temporal variations in fluid stress and spatial heterogeneity in the river bed, and thus present an integrated picture of bed load transport dynamics.

In this paper we present the results of a 2-year deployment of several populations of RFID tracer cobbles within alluvial and bedrock sections of a river, for single-flood and yearly timescales. Throughout this manuscript, we define bedrock channels following the definition of Turowski et al. (2008) as channels that cannot substantially alter their geometry without eroding bedrock; conversely, alluvial channels are channels that are able to freely adjust their geometry without eroding bedrock. At the individual flood scale we examine the tracer displacement distributions, and the fraction of tracers mobilized. We show that tracer displacements and the fraction mobile are consistent with results from a recent momentum conservation framework (Charru et al., 2004; Lajeunesse et al., 2010). We employ a recently developed dimensionless impulse framework (Phillips et al., 2013) to account for unsteadiness in the hydrograph, which allows us to apply a fluid momentum conservation approach to long-term tracer displacement data. For flows within the partial transport regime, we demonstrate that tracer displacements are short and close to the limit of one step per flood. Furthermore, we demonstrate the generality of the long-term tracer displacement results within the main channel with a smaller deployment of tracers in a step-pool bedrock tributary. We show that, by accounting for flow resistance and differences in the threshold of motion, displacement dynamics in the step-pool and main channels are similar. Lastly, we analyze and compare the sorting of tracer particles with that of the river to show that the emerging sorting patterns are consistent with a size-selective transport sorting model.

2 Theory

In the following sections we present the relevant theoretical background that guides the analysis and interpretations of our tracer particle results. The theoretical background is intended as a brief introduction to the topics of sediment transport mechanics and dynamics, quantifying hydrologic forcing, and the downstream sorting of sediment by particle size.

2.1 Sediment transport at the particle scale

Under a wide range of bed load transport conditions, coarse sediment particles undergo short steps separated by longer periods of rest, which leads to probabilistic descriptions of particle motion (Einstein, 1937). A particle step is defined as the distance the particle is transported from entrainment to deposition, and the rest duration is the time between deposition and subsequent entrainment. The motion of particles in bed load transport is comprised of sliding, rolling, or short hops called saltations (Drake et al., 1988), where the travel time is generally much shorter than the rest duration (Lajeunesse et al., 2010; Martin et al., 2012; Furbish et al., 2012b, a; Roseberry et al., 2012). For near-threshold bed load transport, in which only bed surface particles are mobile, bed load flux may be described as the product of the particle velocity and surface density (particles/area) of moving grains (Bridge and Dominic, 1984; Wiberg and Smith, 1989; Parker et al., 2003; Lajeunesse et al., 2010; Furbish et al., 2012b), or similarly the product of the particle entrainment rate and the average particle step length (Einstein, 1950; Wilcock, 1997a; Wong et al., 2007; Ganti et al., 2010; Furbish et al., 2012b). The combination of particle velocity, number of mobile surface particles, depth of the mobile layer, and a threshold stress typically result in a nonlinear relationship between bed load flux and the fluid shear stress (Meyer-Petter and Muller, 1948; Fernandez Luque and Van Beek, 1976; Wong and Parker, 2006; Furbish et al., 2012b). Here it should be noted that the above formulations for bed load particle flux both require averaging the measured quantities over yet-undetermined timescales (Ancey, 2010; Furbish et al., 2012b). For steady turbulent flows in the laboratory, the particle velocity and step length have been shown to scale linearly with the excess shear velocity ($U_* - U_{*c}$) (Fernandez Luque and Van Beek, 1976; Lajeunesse et al., 2010; Roseberry et al., 2012; Martin et al., 2012), where $U_* = \sqrt{\rho \tau_b}$ is the shear velocity ($\mathrm{m\,s^{-1}}$), τ_b is the basal shear stress, ρ is the fluid density ($1000\,\mathrm{kg\,m^{-3}}$), and U_{*c} ($\mathrm{m\,s^{-1}}$) is the threshold shear velocity for initiation of sediment motion. Specifically, Lajeunesse et al. (2010) found that the modal particle step length scales as

$$X/D = C\,(U_* - U_{*c})\,/V_s, \tag{1}$$

where X is the transport distance (m), D is the particle median axis (m), $C = 70$ is an empirically determined constant, and $V_s = \sqrt{R\,g\,D}$ is the settling velocity in the limit of large particle Reynolds numbers, where R is the submerged specific gravity of the particles and g is the acceleration due to gravity ($\mathrm{m\,s^{-2}}$). The surface density of moving particles was found to increase linearly with the Shields stress (τ_*) (Lajeunesse et al., 2010), where $\tau_* = \tau_b/(\rho_s - \rho) g D$, and ρ_s is the sediment density ($2650\,\mathrm{kg\,m^{-3}}$ for quartz). That the dependencies of the step length, particle velocity, and mobile surface density on shear velocity have been recently validated for both unimodal and bimodal grain size distributions under turbulent flow (Lajeunesse et al., 2010; Houssais and Lajeunesse, 2012) encourages us to extend these results to interpret tracer particle data at the field scale.

Treating the particle behavior probabilistically, we focus on the distributions of particle steps and rests. In the laboratory, the distribution of particle step lengths for a given stress and grain size have been observed to follow exponential or gamma-like distributions (Lajeunesse et al., 2010; Hill et al., 2010; Martin et al., 2012; Roseberry et al., 2012). However, for mixed grain size distributions, heavy-tailed statistics can emerge due to a summation of exponential step lengths for each size group (Hill et al., 2010). Examining passive tracers in the field introduces an ambiguity; one measures particle displacement – i.e., the distance a particle travels between successive surveys of its position – but this displacement is composed of an unknown number of steps and rests. Displacement length distributions measured for individual floods, and at longer timescales over many floods, typically follow exponential or gamma-like distributions (Hassan et al., 1991; Schmidt and Ergenzinger, 1992; Habersack, 2001; Lamarre and Roy, 2008; Bradley and Tucker, 2012; Hassan et al., 2013; Phillips et al., 2013). We propose two simple limits for particle displacement during a flood: (1) the lower limit is that a particle executes a single step, with a characteristic length scale predicted by Eq. (1), and (2) the upper limit is continuous particle transport, with no rests, for the duration of the flow that exceeds the threshold entrainment stress. We explore tracer displacements within the context of these two limits.

Upon deposition, the rest duration before subsequent entrainment is constrained by two criteria: first the stress must exceed the threshold of motion locally, and second the particle must be exposed to the flow (Martin et al., 2012). The stochastic erosion and deposition of the river-bed surface acts to bury and excavate particles, and recent laboratory results suggest that this produces heavy-tailed particle rest durations (Martin et al., 2012, 2014). Although these rest durations cannot be measured from passive tracers in the field, our previous work used the dispersion of the tracer plume to infer similar behavior to laboratory experiments (Phillips et al., 2013). Accordingly, we will not consider the particle rest duration or tracer dispersion in this article.

2.2 Dimensionless impulse

Flows in natural coarse-grained rivers are inherently unsteady: from the microscopic scale of variations in turbulence, to macroscopic fluctuations in discharge within a flood, to the rise and fall of the hydrograph throughout a series of floods. At the smallest relevant scales of turbulence, the threshold of motion is determined by the impulse, the product of shear stress magnitude, and duration (Diplas et al., 2008). Due in part to the difficulties in measuring tracer particle motion and near-bed stresses during floods, the fluid shear stress is commonly quantified through use of a bulk-flow parameter such as the depth–slope product, $\tau_b = \rho g h S$ (Church and Hassan, 1992; Hassan et al., 1991; Ferguson and Wathen, 1998; Haschenburger and Church, 1998; Lenzi, 2004; Haschenburger, 2011), where h is the flow depth (m) and S is channel slope. For coarse-grained streams this simplification is perhaps more reasonable, as particle inertial timescales are large and thus coarse particles are insensitive to a range of turbulent stress fluctuations (Diplas et al., 2008; Celik et al., 2010; Valyrakis et al., 2010, 2013). Although some readers may object to the assumption of steady and uniform flow for a flood, the flow may be considered quasi-steady so long as the hydrograph varies slowly compared to the grain inertial timescale (on the order of several seconds), and quasi-uniform so long as water surface slope remains approximately constant. Thus for large ensembles of particles over many floods, the idea of employing a normal flow approximation becomes tenable. Accordingly, Phillips et al. (2013) introduced the dimensionless impulse

$$I_* = \int_{t_s}^{t_f} (U_* - U_{*c})\,\mathrm{d}t / D_{50}, \quad U_* > U_{*c} \tag{2}$$

to quantify the time-integrated fluid momentum in excess of threshold, assuming normal flow. Here D_{50} represents the median grain size of the tracers, t_s represents the start of a flood, and t_f represents the end of a flood of interest. The integral is only calculated over the record of $U_* > U_{*c}$, as sub-threshold flows do not transport sediment. We note that I_* in this study represents a cumulative metric of reach-averaged fluid momentum.

2.3 Downstream sediment sorting

The spatial pattern of diminishing grain size with increasing distance from the headwaters is near universal among gravel rivers, and results from a combination of size-selective sorting and particle abrasion (e.g., Paola et al., 1992; Kodama, 1994; Paola and Seal, 1995; Ferguson et al., 1996; Gasparini et al., 1999, 2004; Fedele and Paola, 2007; Jerolmack and Brzinski, 2010). For tracer particles, the relatively short distances and timescales involved preclude abrasion as a mechanism for observed downstream fining (Ferguson et al., 1996). Thus we further explore the mechanisms and implications of size-selective transport as it pertains to tracer particles and the river bed. We look to laboratory experiments to inform the following analysis, as it is difficult to generalize the rate at which tracer particles sort from previous field studies. Flume experiments with a heterogeneous sediment input show that particles rapidly segregate by size to achieve an equilibrium profile, and that subsequent transport results in an elongation (stretching) of this sorting profile (Paola et al., 1992; Paola and Seal, 1995; Seal et al., 1997; Toro-Escobar et al., 2000). The self-similar sorting profile means that longitudinal profiles collapse onto a single curve when downstream distance (X) is cast as a dimensionless extraction length $X_* = X/L$, where L (m) is the distance at which 100 % of the coarse material in transport has been extracted (deposited). In laboratory experiments and in natural rivers, L is taken as the distance from the input/source to the gravel-sand transition (Paola and Seal, 1995); $X_* = 1$ indicates that all gravel particles are deposited upstream of this location. In the absence of a well-defined gravel front the extraction length can be difficult to determine, though Toro-Escobar et al. (2000) suggest that the location where 90 or 95 % of the source material has been extracted (deposited) is a suitable proxy for L. In the case of tracer particles, this represents the distance downstream from the source to the location of 95 % recovery. Laboratory and modeling results (Fedele and Paola, 2007) demonstrate that the downstream sorting of a gravel mixture can be described by its initial variance and mean in the following formulations:

$$\sigma(X_*) = \sigma_o e^{-C_1 X_*}, \qquad (3)$$

$$\overline{D}(X_*) = \overline{D}_o + \sigma_o (C_2/C_1)\left(e^{-C_1 X_*} - 1\right), \qquad (4)$$

where σ_o and $\overline{D}_o$ are the standard deviation and mean of the input material, $\overline{D}$ is the mean grain size, σ is the standard deviation of the grain size distribution, and C_1 and C_2 are constants. For selective deposition in gravel rivers, $\overline{D}$ and σ should decrease exponentially at approximately the same proportion, resulting in a constant coefficient of variation ($\sigma/\overline{D}$) (Fedele and Paola, 2007). These results also suggest that, due to the self-similar sorting profile, separate populations of tracer particles with similar initial grain size distributions (e.g., $\overline{D}$ and σ) should exhibit similar dynamics when properly rescaled by X_*.

3 Field site and methods

3.1 Field site

Field deployment of tracer particles took place in the Mameyes River basin, located within the Luquillo Critical Zone Observatory in northeastern Puerto Rico. Coarse-grained tracers equipped with PIT RFID tags were deployed in the main stem of the Mameyes River, and in a steep tributary (Fig. 1a). The Mameyes River drains the center of the Luquillo Mountain Range, which commonly receives more than 4000 mm yr^{-1} of precipitation in the headwaters due to a large orographic effect. Precipitation occurs frequently throughout the year in the form of high-intensity, short-duration events, resulting in frequent flash flooding (Schellekens et al., 2004). The 1.2 km study reach in the Mameyes begins just downstream of where the river exits the mountains (drainage area of 24.21 km^2). At the main channel field site, stage was recorded at 5 min intervals for 40 consecutive days by an In-Situ Level Troll 500 and measured from surveys of high flow debris following the largest floods, which were correlated to discharge (Q) measured 3.5 km upstream by a US Geological Survey (USGS) gage (gage no. 50065500, 15 min resolution) to obtain a reach-averaged depth record for the study period (Fig. 2a). The combination of the automated stage measurements and high flow surveys captures the entire range of discharge for the 2-year study period. For calculations of the frictional resistance we use the hydraulic radius (h_r). This section of the Mameyes River (Fig. 3a) exhibits minimal meandering with nearly constant width (20 m), and has a slope of $S = 7.8 \times 10^{-3}$. The slope was extracted from a lidar digital elevation model (DEM) along the channel center (1 m horizontal and vertical resolution) (Fig. 3b). The smaller headwaters tributary is located in catchment three (drainage area of 0.58 km^2) of the Bisley Experimental Watershed (Bisley 3). The Bisley 3 study reach is characterized as a step-pool stream with a slope of $S = 1.2 \times 10^{-1}$, width ranging from 2 to 4 m, and boulder steps that range from 0.5 to 2 m in height (Fig. 3c). For this reach (Fig. 3c) the slope was determined from a longitudinal profile surveyed in the field (Fig. 3d); due to extremely dense forest canopy, a longitudinal profile extracted from the lidar DEM does not accurately represent the heterogeneity in channel topography (Fig. 3c). It should be noted that a simple linear regression is unlikely to capture the heterogeneity in transport slopes for this step-pool stream; however we use it to remain consistent with our main channel field site. Stage was recorded at 1 min intervals at the field site for 59 consecutive days by an In-Situ Level Troll 500, which was correlated to discharge measured by a US Forest Service (USFS) gage (15 min resolution) located ~ 100 m upstream to obtain a reach-averaged depth record for the duration of the study (Fig. 2b).

To characterize the spatial sorting of the stream bed, we performed pebble counts at 200 m intervals from the start of the Mameyes RFID tracers to the perceived gravel–sand transition. When analyzing the stream sorting we restrict the analysis to the depositional part of the river, and thus we only use measurements downstream of the start of the alluvial plain (approximately 200 m downstream of tracer installation location). The $\overline{D}$ and σ of the tracers were measured at the center of eight linearly spaced bins moving downstream from the initial placement location. The number of bins was determined to ensure enough tracers within each bin for accurate statistics. To determine the extraction length X_* for the stream, we set the basin length (L) as the distance from

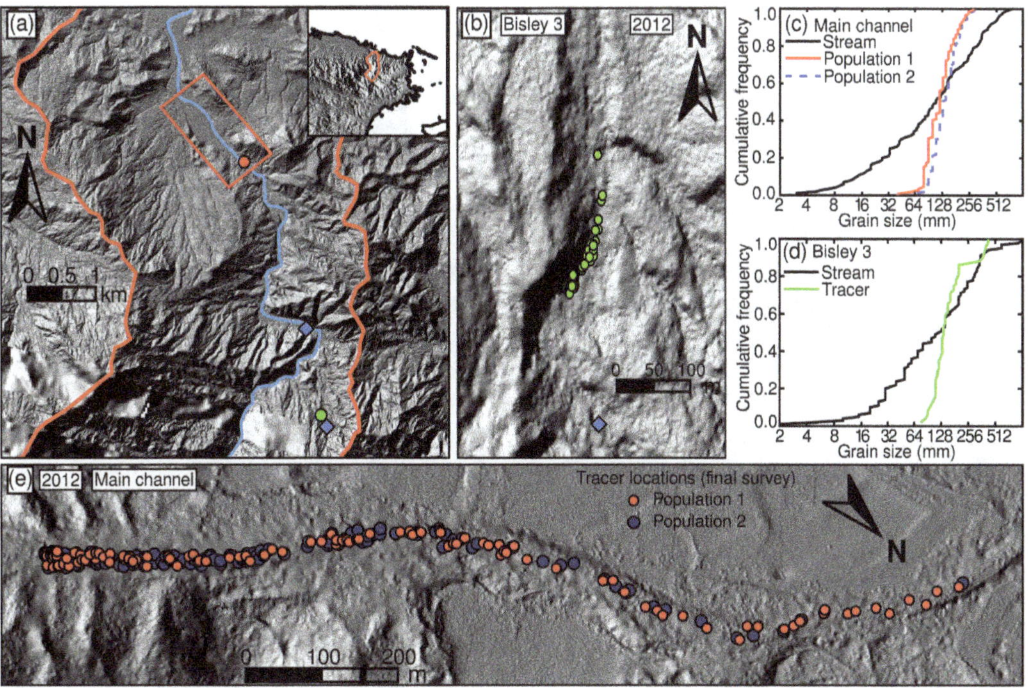

Figure 1. **(a)** DEM of northeastern Puerto Rico (inset) with Mameyes watershed outlined in red. The red and green circles represent the approximate starting locations of tracer particles in the main channel and headwaters stream (Bisley 3), respectively. Blue diamonds represent USGS and USFS stream gages. The blue line is the main channel of the Mameyes River; flow is from south to north. The red bounding rectangle represents the area in panel **(e)**. **(b)** Close-up map of the headwaters stream showing the location of the tracer particles (green circles) at the time of the final survey. **(c)** Grain size distributions for the main channel site determined by Wolman pebble count for the channel (black line), initial population of tracers (red line), and second population of tracers (blue dashed line). **(d)** Grain size distributions for the headwaters site determined by Wolman pebble count for the channel (black line), and population of tracers (green line). **(e)** Close-up map of the main channel field site showing the locations of the first (red circles) and second (blue circles) populations of tracer particles at the time of the final survey. Flow direction is from left to right.

the start of the alluvial plain to the perceived gravel–sand transition. The exact location of the gravel–sand transition in the Mameyes River is obscured in the field due to substantial anthropogenic modification of the river. However, with the use of airborne lidar data we have identified a substantial break in channel slope within the confines of the golf course approximately 3.5 km downstream of the start of the alluvial plain, which we take as the gravel–sand transition. As there is no well-defined front for both populations of tracer particles, we set L as the distance at which 95 % of the tracers recovered remain upstream for each population of tracers at the time of the final survey. Ideally L is the distance at which 100 % of the total (recovered and unrecoverable) tracers remain upstream; however we cannot accurately determine this point. In order to reduce the variability in the surface grain size distributions we determined the D_{84} from an exponential fit to pebble count data collected up- and downstream of each study reach.

3.2 Tracer particles

Two populations of 150 tracers were installed in the summers of 2010 and 2011 in the Mameyes as it exits the mountains near the start of the alluvial plain. A smaller population of 51 tracers was installed in the Bisley 3 stream in the summer of 2010. All three tracer particle populations are composed of cobbles from the stream bed and have narrow grain size distributions centered on the bed D_{50} (Fig. 1c and d). Narrow grain size distributions were selected to facilitate equal mobility (e.g., Wiberg and Smith, 1987) within the tracer populations. The median grain size values for both tracer populations and the river bed at the Mameyes site were 12, 13, and 11 cm, respectively. Median particle diameters for the Bisley 3 stream and tracer particles were 12 and 13.5 cm, respectively. The median grain size for the main channel reach represents the average of three separate pebble counts of 100 particles each (Wolman, 1954). In the Mameyes and Bisley 3 field sites the tracers are fully submerged (average at both sites $h/D_{50} = 7$) during transport for flows above the threshold of motion. Both populations of tracers in the Mameyes River were deployed in the same reach (Fig. 1e) in a 20 m × 20 m grid with 1 m spacing across the width of

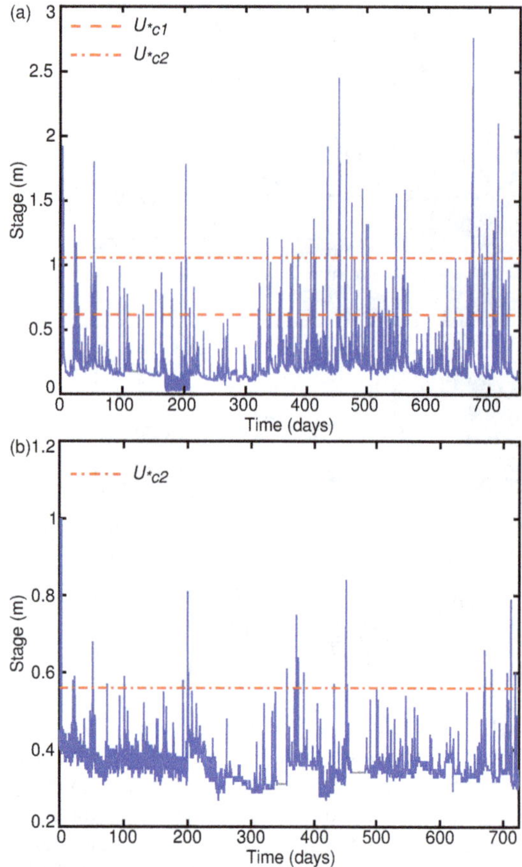

Figure 2. **(a)** Hydrograph for the duration of the study in depth (m) for the main channel of the Mameyes River. The dashed red lines represent two determinations of the critical shear velocity ($m\,s^{-1}$). **(b)** Hydrograph for the duration of the study in depth (m) for the headwaters field site for the duration of the study. The dashed red line represents the critical shear velocity ($m\,s^{-1}$). Gray lines represent missing data.

Figure 3. **(a)** Photograph of the main channel of the Mameyes River looking upstream to the location where the tracer particles were installed. The width of the wetted portion of the channel is approximately 20 m. **(b)** Longitudinal profile extracted from a lidar DEM of the main channel of the Mameyes River. *S* is the slope. **(c)** Photograph of the Bisley 3 stream looking upstream, showing the location of the farthest tracer found downstream. The wetted region in the foreground is approximately 2 m wide. **(d)** Longitudinal profile from field survey; gray crosses represent the location of survey points. *S* is the slope.

the channel. They were surveyed two, three, and one time(s) during the summers of 2010, 2011, and 2012, respectively. Tracer recovery percentages for the first population for the six field surveys were 62, 92.5, 86.6, 88, 86.6, and 93 %. Recovery percentages for the second tracer population for field surveys in 2011 (2) and 2012 (1) were 100, 99, and 94.6 %. The initial recovery rate for population 1 is low as a result of an incomplete survey, which was cut short for safety concerns due to the occurrence of a second large flood. The second population of tracers was placed in the river as two installments of 80 and 70 tracers on two consecutive days due to a small flood, which resulted in a minor amount of burial from fine sediment for the initial 80 tracers installed. Impacts of the initial increased embeddedness on this subset of tracers are not observable at the multi-flood scale. Unrecovered tracers have the potential to bias the mean value, as it is possible that tracers could be buried beyond the detection limit, destroyed, missed, or are farther downstream. However, it was common to find previously missing tracers on subsequent surveys, suggesting that the unrecovered tracers were buried or missed. Only 7 % of installed tracers for the first population were permanently lost or unaccounted for in subsequent surveys. Tracers were surveyed in the Bisley 3 stream two, four, and one time(s) during the summers of 2010, 2011, and 2012, respectively. Tracer recovery percentages for the seven surveys were 91, 91, 93, 98, 98, 100, and 93 %. Final surveyed positions of the tracers can be seen in Fig. 1b and f. Surveyed positions for both field sites were transformed from Cartesian coordinates to a stream-wise normal system using a methodology similar to that developed by Legleiter and Kyriakidis (2007). Tracer particles were located using two wands (manufactured by Oregon RFID) with empirically determined horizontal detection limits of 50 and 20 cm, respectively. PIT RFID tags (32 mm HDX tags supplied by Oregon RFID) were detectable when buried at depths up to 10–20 cm below the river bed for the small wand (depending on tag orientation), and 50 cm for the large wand. The maximum combined survey and detection error is estimated to be 1 m and 45 cm for the large and small wands, respectively (for a discussion of wand detection distances and limitations see Chapuis et al., 2014). All measured tracer motion recorded below the detection threshold was considered to be error and set to zero. Assuming that tracers detectable with the larger wand only were buried beyond the detection limit of the small wand, then only 6 % of the tracers were buried more than 20 cm below the surface. This suggests that the majority of the tracers were near the surface. All tracer data are available online as part of the Luquillo Critical Zone Observatory database (see acknowledgements).

3.3 Hydrologic forcing

For the study reaches, $U_* = \sqrt{g\,h\,S}$ was estimated assuming steady and uniform flow; Shields stress was also estimated for comparison to other studies. Long-term flow records for the Bisley 3 stream are measured near a series of large culverts that artificially truncate the largest floods; this does not affect the calculations of I_* greatly, but does add ambiguity to the distributions examined in following sections. Therefore, we limit our analysis to the hydrograph on the main channel of the Mameyes River, as it represents the highest quality data.

The value of I_* (Eq. 2) was computed for each flood above the threshold of motion (Fig. 2), and also for the cumulative time periods between successive tracer surveys. The calculation of I_* is particularly sensitive to the value of U_{*c}, a parameter that is known to vary both temporally and spatially within a flood and from flood to flood (Kirchner et al., 1990; Charru et al., 2004; Turowski et al., 2011; Marquis and Roy, 2012). We treat U_{*c} as a constant by necessity as we lack a theoretical or empirical methodology with which to account for these effects. We determined the value of U_{*c} in two manners: (1) U_{*c1} was calculated from the fraction of mobile tracers for individual floods, and (2) U_{*c2} was determined as the value that provided the best collapse of the mean tracer displacement data (both methods are used in Sects. 4.2.1 and 4.2.2). Using two definitions of critical shear velocity, the resulting values of I_* for a representative flood are 3815 for U_{*c1} and 806 for U_{*c2} (Fig. 4). The discrepancy between the values of I_* is due to the broadening asymmetric shape of the hydrograph, where a small change in U_{*c} can result in an order of magnitude increase in I_*. When calculated over numerous floods, uncertainty in U_{*c} can result in substantial differences in I_*. However, it should be noted that both approaches produce the same scaling relationships, but differ in the magnitude of the coefficients. A potential drawback of calculating U_* and U_{*c} from a reach-averaged depth–slope product is that large instantaneous values of U_* due to turbulent fluctuations cannot be accounted for. This simplification could result in tracer movement from turbulent fluctuations being attributed to a reach-averaged stress, and hence a biased estimate for the actual threshold value. The upshot is that estimates for I_* are least accurate for low-magnitude, short-duration floods. Data points likely to be affected by this consequence of determining a reach-averaged U_{*c} are included for completeness and indicated in the following figures where appropriate. Due to this drawback we do not recommend using Eq. (2) for individual floods without an independent measure of U_{*c}. For the remainder of the manuscript, except where noted, we use the value of U_{*c} as determined by the fraction of mobile tracers (U_{*c1}).

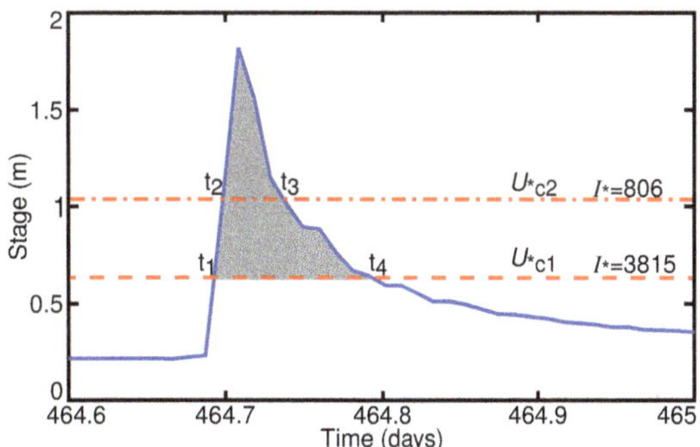

Figure 4. Calculation of the dimensionless impulse (I_*) for two estimates of U_{*c} for a single flood, where the time represents the floods location on the hydrograph in Fig. 2a. The limits of integration for U_{*c1} and U_{*c2} are t_1 to t_4 and t_2 to t_3, respectively. The shaded region represents the region integrated for the calculation of I_* using U_{*c1}.

4 Results

4.1 Hydrology filtered through sediment mechanics and dimensionless impulse

River discharge is the most commonly reported variable in relating long-term sediment dynamics to hydrologic forcing. However, the momentum framework presented in Sect. 2.1 reminds us that fluid stress – rather than water discharge – is the relevant parameter to consider for driving sediment motion. For considerations of bed load transport, the threshold of motion applies a filter to these data; only flows that exceed the critical stress for entrainment are relevant for assessing particle transport. Accordingly, we empirically estimate the threshold of motion by determining the fraction of mobi-

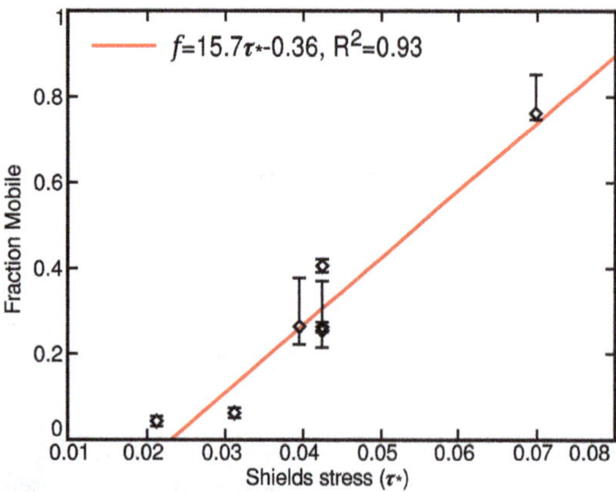

Figure 5. Fraction of mobile tracers (f) for single floods against peak Shields stress (τ_*). The red line represents the best fitting linear relationship, for which the intercept represents the critical Shields stress. See text for discussion of error bars.

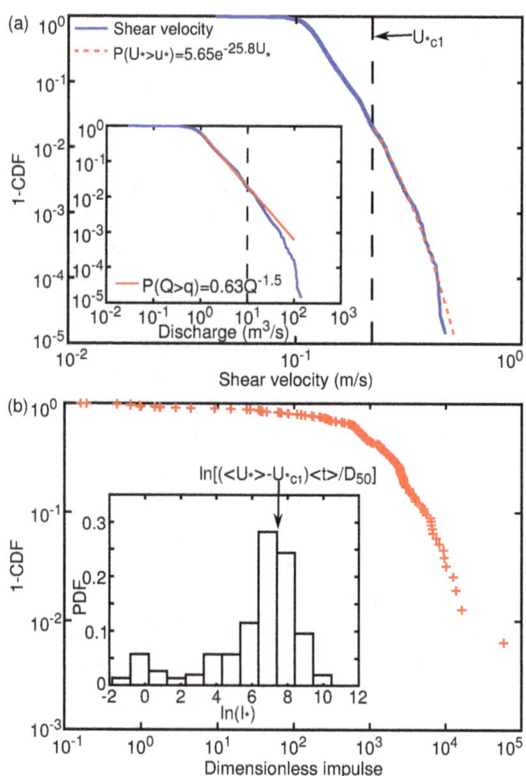

Figure 6. **(a)** Frequency magnitude distribution of shear velocity for the main channel of the Mameyes river. The dashed red line represents an exponential function fit to the distribution for $U_* > U_{*c}$. **(a)** Inset: magnitude frequency distribution of discharge. The red line represents a power-law relationship, and the vertical dashed black line is the location of the threshold of motion (U_{*c1}). **(b)** Magnitude frequency distribution of the dimensionless impulse (I_*). **(b)** Inset: PDF of $\ln(I_*)$, where $\langle U_* \rangle$ is the average value of the distribution of $U_* > U_{*c1}$, $\langle t \rangle$ is the average duration of a flood above the threshold of motion for the duration of the study, and D_{50} is the median grain size of the tracer particles.

lized tracers (f) for several individual floods, where tracers' positions were surveyed before and after the event. Based on a momentum balance approach (cf. Lajeunesse et al., 2010), we anticipate that f should scale linearly with Shields stress (Fig. 5). A linear relation provides a reasonable fit to the data when plotted against the peak Shields stress of the flood for the D_{50} of the tracer population. We treat the intercept as the threshold of motion, determining that $\tau_{*c} = 0.023$ ($U_{*c1} = 0.22\,\mathrm{m\,s^{-1}}$) (Fig. 5). We use the Shields stress at the flood peak – rather than average Shields stress – as it does not require additional information; computing the average stress associated with a flood requires choosing a threshold value for fluid stress. Error bars for f account for the number of missing tracers, where the upper and lower lines indicate the absolute maximum and minimum values for the fraction mobile by assuming that all tracers not recovered moved or did not move, respectively. For all flood events monitored, the fraction mobile remained well below 1 ($f < 1$).

The magnitude–frequency distribution for U_* for the entire period of record (Fig. 6a) presents a fuller picture of the statistical scaling of flow within the hydrograph. For small to intermediate values of U_* the curve appears to be a straight line on a log–log plot (Fig. 6a), indicating potential power-law scaling for this region, a common feature of flood hydrology (Turcotte, 1994; Lague et al., 2005; Molnar et al., 2006). The power-law scaling is even more evident for discharge (Fig. 6a, inset). The magnitude–frequency distribution for both U_* and Q exhibits a truncation to the power-law scaling that occurs at approximately the threshold of motion (Fig. 6a). The upper truncation of the distribution of U_* is well fit by an exponential function, indicating that the shear velocity for flows exceeding threshold is well described by a single average value, $\langle U_* \rangle = 0.27\,\mathrm{m\,s^{-1}}$ ($\langle \tau_* \rangle = 0.033$, $\langle U_* \rangle / U_{*c1} = 1.23$). The exponential decay in probability for shear velocities above critical further indicates that most bed load transport events in the Mameyes do not exceed the threshold of motion by much. Despite a large range in discharge ($Q_{\mathrm{peak}}/Q_{\mathrm{c}} = 14.6$), the peak shear velocity $U_{*\mathrm{peak}} = 0.46\,\mathrm{m\,s^{-1}}$ ($\tau_{*\mathrm{peak}} = 0.1$) observed only exceeded U_{*c1} by a factor of 2. This filtering of the discharge data is particularly strong as the highest discharge values in the 20-year instantaneous record exceed our peak observed flood by a factor of 3.1 ($Q = 444.6\,\mathrm{m^3\,s^{-1}}$), while the highest shear velocity values associated with these extreme events are only 1.28 times ($U_* = 0.59\,\mathrm{m\,s^{-1}}$) as high as the peak observed value in this study.

We analyze the magnitude–frequency distribution of I_* (Fig. 6b) and find that the distribution of I_* is composed of several scaling regions, though it does not appear to be heavy tailed. The smallest values of I_* are artificially truncated by the resolution of the river stage measurements (15 min). The probability density function (PDF) of I_* (Fig. 6b, inset) has a pronounced peak that coincides with the product of the average excess shear velocity and average flood duration, $(\langle U_* \rangle - U_{*c})\langle t \rangle / D_{50} = 2070$. Here $\langle U_* \rangle$ is the average value of the magnitude frequency distribution of $U_* > U_{*c}$ (Fig. 6a), and $\langle t \rangle$ is the average duration of a flood above the threshold of motion for the period of record. The distributions of U_* and I_* calculated over the relatively short study duration (2 years) are the same as those calculated using a longer flow record (20 years available from the USGS) for the Mameyes River. Thus, data indicate that there is a well-defined "characteristic flood" for the Mameyes.

4.2 Mechanics of sediment tracer particles

4.2.1 Individual flood scale

The distribution of particle displacements resulting from several floods was determined from surveys of both populations of tracers at the Mameyes site, in the summers of 2010 and 2011. We normalize each tracer's transport distance (X_i) by its median diameter (D_i). For individual flooding events above the threshold of motion, the majority of the cumulative distribution functions (CDF) of tracer particle displacement are well described by exponential functions (Fig. 7a), except for two tracer displacement CDFs which decay faster than exponential functions (green circles and cyan squares in Fig. 7a). Each tracer displacement CDF is normalized by its mean displacement ($\langle X/D \rangle$) to facilitate plotting all single events in one graph (Fig. 7a). Typical travel distances for individual tracers were on the order of a few meters for each flood. We compare the dimensionless tracer distance for all single events against the peak shear velocity for that event (Fig. 7b), normalized by V_s, and find that the modal tracer displacement is well described by a linear relation, as anticipated by the momentum framework presented earlier. A fit to the modal displacement distances provides another estimate of the threshold stress, $U_{*c} = 0.13$ ($\tau_{*c} = 0.016$), from the intercept. This is likely a lower estimate of U_{*c}, and is not the same value as determined previously for the fraction mobile data. Finally, we plot Eq. (1) over a contour density map of our field data (Fig. 7b), using $k = 70$ from Lajeunesse et al. (2010). Remarkably, the modal step lengths predicted from the laboratory-derived relation of Lajeunesse et al. (2010) (Eq. 1) run through the modal displacement distances measured from our field data in the Mameyes.

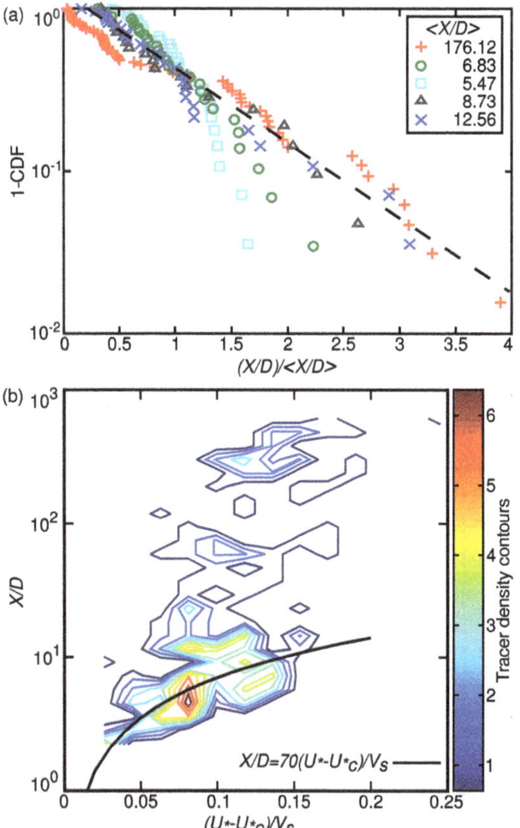

Figure 7. **(a)** Dimensionless displacement distributions for individual floods normalized by the mean ($\langle X/D \rangle$) displacement for that flood. The black dashed line is an exponential distribution. Dimensionless mean displacement lengths for each flood are labeled in the legend. **(b)** Contour density plot of X/D against the excess shear velocity normalized by the settling velocity for each tracer. The contour colors represent the density of tracers within that location. The value of U_* is for the flood peak, while the value of U_{*c} is treated as a fitting parameter. The black line represents the expected linear relationship between the dimensionless shear velocity and the modal tracer step length (Eq. 1).

4.2.2 Multi-flood scale

At the multi-flood scale we analyze the long-term behavior of the tracer particles' displacement. We normalize the CDFs of cumulative travel distance by each survey's mean value, which results in a collapse of the data. This collapse suggests that the mean value is a reasonable descriptor of the dynamics of each tracer population. We note here that the CDFs in Fig. 8 are truncated at the lower end due to measurement accuracy, and at the upper end of the distribution due to unrecovered tracers (see Hassan et al., 2013, for a discussion of the effects of unrecovered tracers on the scaling of the CDF, while for a discussion of the functional distribution for the CDF see Bradley and Tucker, 2012). We therefore analyze the mean cumulative tracer travel distances ($\langle X/D \rangle$) using the dimensionless impulse (I_*) Eq. (2). We find that the $\langle X/D \rangle$ scales linearly with I_* for both populations of tracer particles (Fig. 9). Due to a limited number of repeat surveys we utilize all permutations of tracer surveys; in other words, we determine the $\langle X/D \rangle$ and I_* for all possible sampling intervals. Using all permutations of tracer surveys does require the assumption that the sequence of floods does not exert substantial control on the mechanics of particle displacement. However, flood sequence and particle embeddedness may explain some of the scatter in the $\langle X/D \rangle$ displacement data (Fig. 9). Due to the close agreement between our field data and laboratory results (Fig. 7b) we use Eq. (1) to calcu-

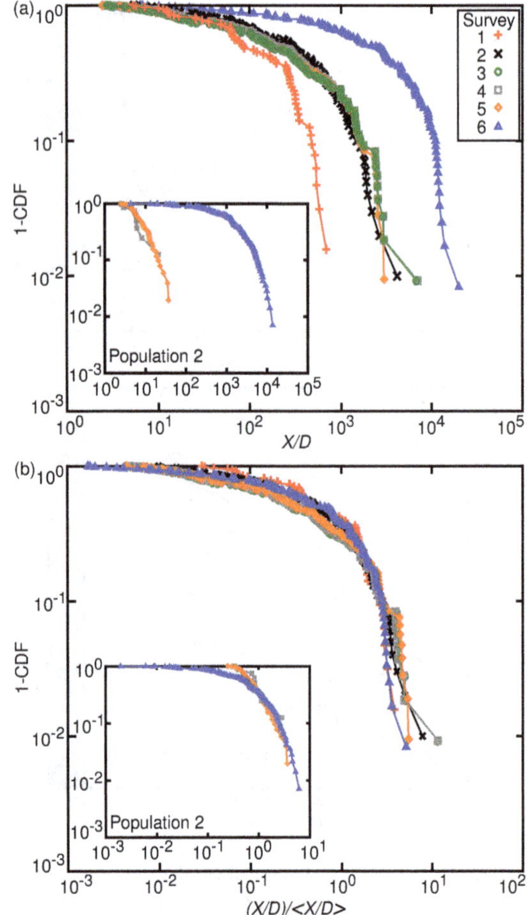

Figure 8. **(a)** Cumulative dimensionless displacement for each tracer survey for the first population of tracers. Survey number is denoted in the legend. **(a)** Inset: cumulative dimensionless displacement for the second population of tracer particles, installed immediately prior to survey 4. **(b)** Cumulative dimensionless displacement data normalized by the mean displacement for each survey for tracer population 1. **(b)** Inset: cumulative dimensionless displacement data normalized by the mean displacement for each survey for tracer population 2.

late the two limits of particle transport discussed in Sect. 2.1. Limit 1, in which entrained particles execute one step per flood, was calculated using Eq. (1) with the peak U_* for each flood for the study duration above U_{*c1}. Limit 2, continuous motion, was calculated as the product of Eq. (1) and t/T_s, where t is the duration of the flood above the threshold of motion, and $T_s = 10.6\sqrt{D_{50}/Rg}$ is the expected particle step duration from Lajeunesse et al. (2010). For limit 2 we use the average shear velocity over the duration of a flood in Eq. (1). When these limits are compared with the tracer particle data, we find that the tracer particles' mean displacement is significantly closer to limit 1 than limit 2 (Fig. 9), consistent with highly intermittent and partial bed load transport.

Due to time limitations in the field we were unable to survey more than three individual floods for the Bisley 3

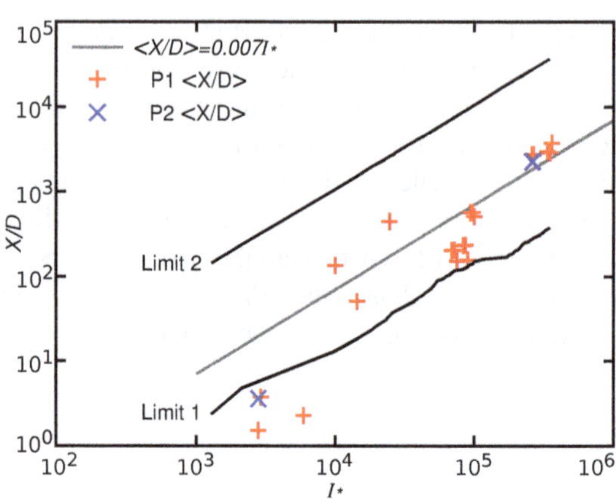

Figure 9. Mean displacement data for the first (red +) and second (blue x) populations of tracer particles vs. dimensionless impulse (I_*). The gray line is the linear relationship determined by Phillips et al. (2013). The black lines represent the upper and lower limits of sediment particle transport (see Sect. 2.1 for discussion of limits). The data plotting below limit 1 have unrealistic values for I_* (see Sect. 3.3 for explanation).

stream, and thus are unable to determine the threshold of motion from the fraction of mobile tracers. In order to compare the Mameyes data with the Bisley 3 tracer data, values for U_{*c} from both sites must be determined using the same methodology (Wilcock, 1988). Therefore we utilize method two, described in Sect. 3, to determine the values of U_{*c2} for this comparison. The values of U_{*c2} that provide the best collapse of the tracer data are 0.28 m s^{-1} ($\tau_{*c2} = 0.038$) and 0.81 m s^{-1} ($\tau_{*c2} = 0.303$) for the Mameyes and Bisley 3 sites, respectively. We find that the Bisley 3 data are also well characterized by a linear relation, with a slope that is 1 order of magnitude lower than that of the Mameyes site (Fig. 10). It is intriguing that the Bisley 3 displacement data form a well-defined linear trend, considering the limited number of tracers and the rough channel geometry (Fig. 3c and d). We note here that the slope of the linear relation computed with U_{*c2} for the Mameyes site is an order of magnitude larger than that determined using U_{*c1}; however, the linear form of the relationship is robust for a wide range of threshold values.

4.3 Tracer particle sorting in the Mameyes

At longer timescales, tracer particle sorting by size is readily apparent in the Mameyes, as observed in other studies (Hassan et al., 1991; Ferguson et al., 1996; Hodge et al., 2011). Sorting may be the result of (1) an inverse relation between particle step length and grain size, and/or (2) differences in entrainment frequency throughout a flood as a function of grain size. To examine differences in entrainment, we separate the tracer particles into two populations: particles that moved at least once for all single floods, and particles that remained immobile. All particle size distributions are well fit by lognormal functions. We aggregate all of

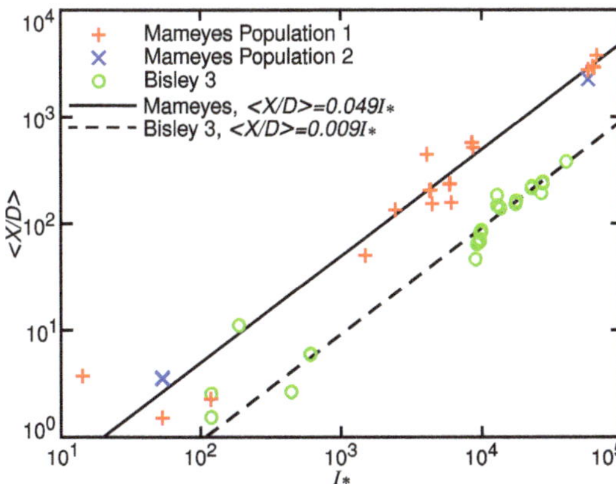

Figure 10. Mean displacement data for the first (red +) and second (blue x) Mameyes tracers, and Bisley 3 tracers (green o). The solid and dashed black lines represent a linear relation between $\langle X/D \rangle$ and I_* for the Mameyes tracers, and the Bisley 3 tracers, respectively. I_* is calculated using U_{*c2}.

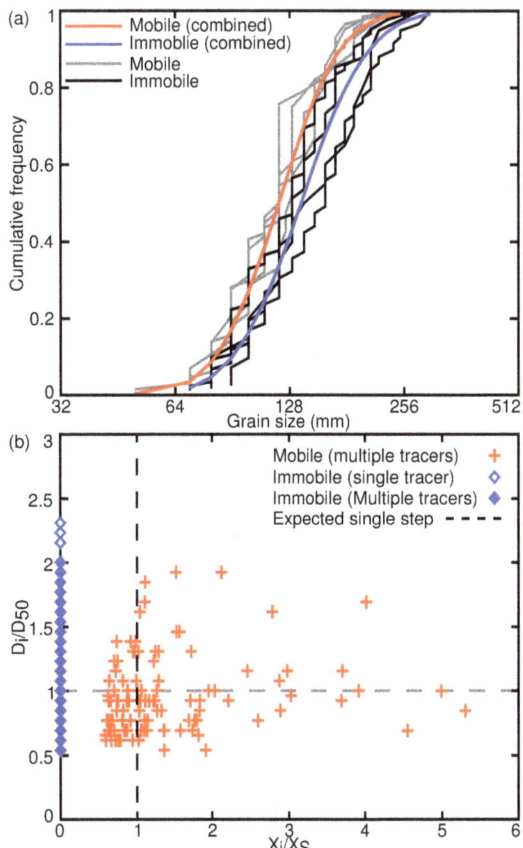

Figure 11. **(a)** CDFs for tracer grain size (mm) for single floods separated by whether the tracer moved (light-gray lines) or remained immobile (black lines). The red and blue lines represent lognormal functions fit to the combined mobile and immobile tracers, respectively. **(b)** Tracer grain size normalized by D_{50} against tracer travel distance normalized by the expected step length (X_S) from Eq. (1). Red crosses represent mobile tracers for all single floods ($n = 108$) near the threshold of motion ($\tau_* < 0.045$), and blue diamonds represent tracers that did not experience movement ($n = 269$) (solid diamonds represent multiple tracers plotted on top of each other). The dashed black line denotes the expected single step length.

the mobile and immobile particles into two combined distributions for statistical purposes (Fig. 11a). The size distribution of mobile particles is finer at the coarse end of the distribution compared to immobile particles (Fig. 11a), though there is a fair amount of overlap between the CDFs. We used a two-sample t test (equal variance, unequal sample size) on the natural-log-transformed distributions to determine that the difference in the mean values ($\langle$mobile$\rangle = 119$ mm, $\langle$immobile$\rangle = 137$ mm) between the mobile and immobile populations is statistically different (t statistic $= 4.02$, degrees freedom $= 345$, p value < 0.001). At the single-flood scale there does not appear to be a significant dependence of displacement length on particle size. When we compare the displacements during a flood to the expected step length calculated from Eq. (1), it becomes apparent that the majority of tracers have displacements that are close to the expected step length, while a significant number of smaller ($D_i < D_{50}$) particles have longer displacements (Fig. 11b).

The cumulative effects of minor grain size sorting at the flood scale (Fig. 11) result in the rapid development of tracer sorting at the annual scale for the Mameyes River. In order to connect the dynamics of tracers to the downstream fining pattern of river bed, we first analyze the downstream grain size trend of the bed surface of the Mameyes River as determined from pebble counts. The starting location of the tracers coincides with where the Mameyes River exits the mountains (at approximately 0 km downstream in Fig. 12a). The start of the alluvial plain begins at approximately 0.5–0.75 km downstream in Fig. 12a. From this point downstream the bed D_{84} fines by roughly a factor of 2, while there is only minimal decrease in the bed D_{50} and D_{16} (Fig. 12a). This suggests that downstream fining of the river occurs through deposition of the coarsest particles. Starting at $X = 300$ m (Fig. 12a), the standard deviation and mean grain size decrease exponentially with initial statistics of $\sigma_o = 185$ mm and $\overline{D} = 185$ mm (Fig. 12b and e). The exponent of the fitted exponential function (C_1 in Eq. 3) is 1.2 for the stream. This results in an approximately constant coefficient of variation $\sigma/\overline{D} = 0.93$ (Fig. 12c). Using the parameters determined in Fig. 12b and c, we apply the full model of Fedele and Paola (2007) (Eq. 4) to the spatial decrease in $\overline{D}$ and find that the model seems to underpredict the mean value (Fig. 12e). Potential reasons for the underfit are given in Sect. 5.2.

Turning to the tracers emplaced in the Mameyes, sorting of both populations by size is readily apparent (Fig. 12a). The spatial decrease in σ for both tracer populations (σ_{P1} and σ_{P2}) are well described by exponential functions (Fig. 12b). The exponents of the fitted exponential functions (C_1 in Eq. 3) for population 1 and 2 are 0.83 and 0.73, respectively. The

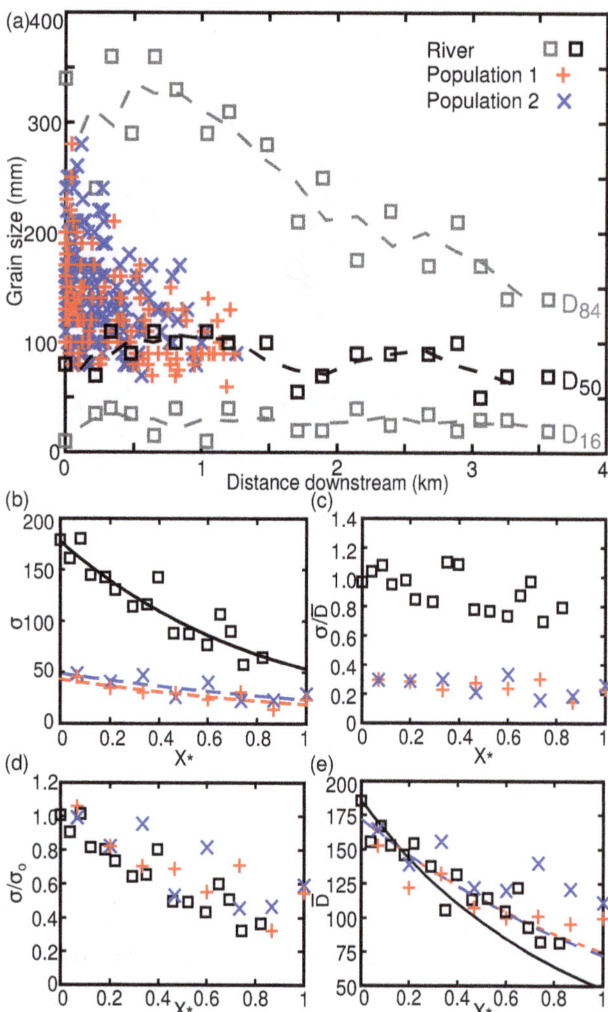

Figure 12. **(a)** Tracer grain size (mm) against distance (km) the particle has traveled for the first (red +) and second (blue x) tracer populations at the final survey. Black and gray squares represent the grain size percentiles at the corresponding distance downstream for the river. The dashed lines are moving averages to guide the eye. **(b)** Spatial standard deviation for the river, and both populations of tracer particles against dimensionless distance (X_*). The lines represent fitted exponential functions with the form of Eq. (3) with coefficients 185, 43, and 49 and exponents 1.2, 0.83, and 0.73 for the stream, population 1, and population 2, respectively. **(c)** Coefficient of variation moving downstream for the river and both populations of tracer particles. **(d)** Collapse of the stream and tracer populations standard deviation data. σ_o represents the coefficients from the fitted equations in **(b)**. **(e)** Downstream decrease in the mean ($\overline{D}$) grain size. Lines represent Eq. (4).

initial standard deviations of tracer populations 1 and 2 at placement are 48 and 41 mm, respectively. These values are very close to the coefficients for the fitted exponential function in Fig. 12a. The initial mean values for tracer populations 1 and 2 are 129 and 140 mm, respectively. In accordance with the model of Fedele and Paola (2007) we find that both populations of tracers have nearly constant coefficients of variation ($\sigma/\overline{D}$) of 0.25 and 0.26, respectively (Fig. 12c). The coefficient of variation for the tracers is expected to be low due to the narrow grain size distribution. Furthermore, we find that the σ data can be reasonably collapsed by normalizing by σ_o from the upstream end of the depositional system (Fig. 12d). Here we apply Eq. (4) using the parameters determined in Fig. 12b and c and find that it underpredicts the rate at which $\overline{D}$ decays. In applying Eq. (4) for the tracers we set $\overline{D}_o = 173$ mm to the value for the stream at the location where they were placed. This is because the model is for a bed with a continuous source where the mean at $X_* = 0$ does not change, while for a finite population of tracers the coarser particles are deposited, and thus over time $\overline{D}$ at $X_* = 0$ will coarsen. Potential reasons for the underfit are given in Sect. 5.2.

5 Discussion

5.1 Sediment mechanics

At the single-flood scale, tracer particle displacements have been shown to be well described by exponential, gamma, and power-law distributions (Phillips et al., 2013; Habersack, 2001; Hassan et al., 2013). Here we find that the majority of observations are well described by exponential distributions, with two exceptions that decay faster than exponentially (Fig. 7a). The distributions that decay faster than exponentially are likely due to undersampling as a result of the small number of mobile tracer particles in these floods. The particle velocity distribution has been shown in laboratory experiments and theoretically to scale exponentially, except for small sample sizes for which the distribution decays faster than exponential (Furbish and Schmeeckle, 2013); this pattern is also likely true for the distribution of particle displacements. Therefore, we might expect that a larger population of tracers would converge to the exponential distribution. The remarkable agreement between the modal tracer displacement data at the single-flood scale for the Mameyes and the laboratory results of Lajeunesse et al. (2010) (Fig. 7b) strongly suggests that the most likely tracer displacement for these floods is a single step length. This further suggests that the observed exponential distributions for tracer displacement (Fig. 7a) represent the distributions of tracer step lengths. The observation of mostly single step displacements holds for all observed individual floods, despite a 3-fold increase in Shields stress ($\tau_*/\tau_{*c1} = 3.04$), demonstrating that partial and intermittent bed load transport occurred under all observed conditions. This is consistent with previous field observations by Mao and Surian (2010), who reported that partial transport occurred over a 3-fold increase in Shields stress. However, tracer displacement length distributions become increasingly skewed toward larger values with increasing flood strength (Fig. 7a and b), indicating that increasing numbers of particles experience more frequent re-entrainment as shear stress increases. Nonetheless, the fraction of mobile tracers remained significantly below 1 for all observed floods (Fig. 5), reinforcing the inference that partial bed load transport is the dominant mode of transport here. Plots of mean tracer displacement length against cumulative impulse show that tracer motion is far from the continuous limit, and much closer to the lower limit of one step per flood (Fig. 9). The linear scaling observed (Fig. 9) for both populations of Mameyes tracers indicates that, to first order, the mean displacement is determined by the total momentum imparted to the stream bed. The agreement in the two populations suggests that effects of embeddedness and initial placement are not evident at the annual scale. Linear scaling with I_* and the use of all possible permutations of tracer surveys indicates that flood sequence is not exerting a first-order control on tracer displacement at annual timescales.

The agreement of observed tracer displacement distributions from the Mameyes with models and laboratory data (Lajeunesse et al., 2010) gives us hope that the results here are general. Our earlier observations of tracer dispersion provided an indirect method for inferring particle rest times (Phillips et al., 2013); thus, results from this tracer study may provide the basis for future probabilistic modeling of long-term bed load transport, for which the distributions of particle steps and rests are required input parameters (Zhang et al., 2012). As pointed out above, although larger floods have occurred in the historical record of instantaneous discharge, the associated stresses were no more than 1.28 times the largest observed flood. From the USGS records of annual maximum flood peaks (1967–2012, discontinuous), the largest flood is only 1.35 times our observed largest value of U_*. The rapid decay of frequency of occurrence for flows above threshold (Fig. 6a) indicates that transport conditions observed during our study are representative of the river system. Partial bed load transport during near-threshold conditions is also consistent with expectations from equilibrium channel theory for gravel rivers (Parker, 1978; Parker et al., 2007). Indeed, the channel depth inferred from the hydrograph for the "characteristic flood" on the Mameyes – i.e., the peak value in the impulse distribution – agrees with an independent estimate of bankfull flow deduced from vegetation markers, channel morphology, and flow frequency analysis (Pike, 2008; Pike and Scatena, 2010).

We now turn to data from the Bisley 3 tracer deployment, which can serve as a critical test of the generality of the impulse framework and tracer displacement results. Because it is currently unknown how many tracer particles are required to produce accurate statistics, we note here that we only analyze the limited number of Bisley 3 tracers alongside the larger set of Mameyes tracers. Until such a number is known we caution that researchers should not solely rely on a limited set of tracers to inform their results. Given the small number of tracers deployed and the particularly variable stream profile (Fig. 3d), it is intriguing that the Bisley 3 data fall on a well-defined linear relation when plotted

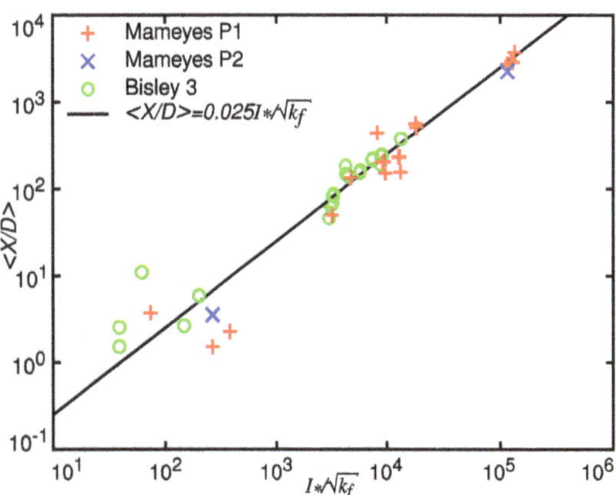

Figure 13. Collapse of the tracer data for the first (red +) and second (blue x) Mameyes tracer populations, and Bisley 3 tracers (green o). The black line represents a linear relationship between the mean particle displacement and the dimensionless impulse over the dimensionless friction factor (Eq. 5). The grouping of the date is a result of surveys in three separate years.

against I_* (Fig. 10). The offset between the two field sites indicates that, for equivalent values of I_*, tracer particles at the Mameyes field site have traveled farther than those in the Bisley 3 stream. This could result from enhanced particle trapping and hiding effects, or greater flow resistance due to the rougher bed. We attempt to collapse the Mameyes and Bisley 3 data onto a single curve by accounting for each of these two effects separately. We assess particle hiding effects using a simple hiding function (Einstein, 1950; Wilcock and Crowe, 2003), which does not produce a collapse of the data. To test the effect of flow resistance, we calculate the dimensionless friction factor k_f using a modified Keulegan equation that was found to provide a reasonable fit to a large compilation of field data (Ferguson, 2007):

$$U/U_* = \sqrt{8/k_\mathrm{f}} = (1/K)\ln(11h_\mathrm{r}/4D_{84}), \tag{5}$$

where U is the flow velocity (m s^{-1}), and $K = 0.41$ the von Karman constant. The h_r (average from three cross sections) for the Mameyes main channel and Bisley 3 sites is 0.9 and 0.29 m, and the D_{84} is 0.31 and 0.55 m, respectively. To reduce variability in the surface grain size counts, the D_{84} is determined from an exponential fit to pebble count data collected up and downstream of each study reach. For the Mameyes and Bisley 3 reaches, k_f is 0.30 and 9.27, respectively. When values for I_* are normalized by $\sqrt{k_\mathrm{f}}$ computed using Eq. (5), the mean tracer displacement data for the Mameyes and Bisley 3 streams collapse onto a single curve (Fig. 13). This collapse indicates that, when one accounts for the difference in relative submergence and its effect on flow resistance, the resulting mean particle transport distance is the same. The morphologic characteristics of these two field sites represent end members for the Mameyes watershed, indicating that the linear function $\langle X/D\rangle = 0.025\, I_*/\sqrt{k_\mathrm{f}}$ may be a general relationship. This conjecture would be well supported should this relationship be found to hold for tracer studies in other regions. We note here that one can achieve a similar collapse of the data using other recently proposed flow resistance equations as well (see Ferguson, 2007; Rickenmann and Recking, 2011; Ferguson, 2012).

A pitfall of the dimensionless impulse is its sensitivity to the determination of U_{*c}. As seen in this manuscript, and in general (Wilcock, 1988), the value of the threshold stress is dependent on the method used to determine it. We have determined two separate values for U_{*c} by using the intercept of the fraction mobile data ($U_{*c1} = 0.22$, $\tau_{*c1} = 0.023$, Fig. 5), and the value that best collapses the long-term mean displacement data ($U_{*c2} = 0.28$, $\tau_{*c2} = 0.038$, Fig. 10). Both of the determined values are within the range reported from field and laboratory data (Buffington and Montgomery, 1997; Mueller et al., 2005; Lamb et al., 2008); however the range of values we have recorded underscores the need for an independent empirical measure of the threshold of motion.

5.2 Sediment sorting

The pattern of smaller particles having larger displacements (Fig. 11b) is near universally observed in tracer studies (e.g., Church and Hassan, 1992; Ferguson and Wathen, 1998; Hodge et al., 2011; Scheingross et al., 2013; Schneider et al., 2014). Here we offer a potential explanation for these observations. Results indicate that the modal displacement length for particles of all sizes is approximately one step (Figs. 7a, b, 9), and that the dimensionless particle step length depends only weakly on particle size (Fig. 11b). The latter result appears to support laboratory experiments that show that the largest particles travel the farthest for rivers with steep to moderate slopes (Solari and Parker, 2000; Hill et al., 2010). This effect has been attributed to particle inertia for narrow unimodal sediment size distributions (Solari and Parker, 2000), and to bed roughness effects for wider grain size distributions (Hill et al., 2010). The observation that the majority of tracer particles' displacements match the expected step length (Fig. 11b) combined with grain-size-dependent entrainment (Fig. 11a) suggests that the relatively larger displacements for smaller particles result from a greater frequency of entrainment events during a flood. In other words, the largest particles appear to take one step during a flood due to a single entrainment event, while small particles may take multiple steps. Our field results suggest that sorting happens through smaller particles possessing a higher probability of re-entrainment, rather than possessing longer step lengths as compared to larger particles. This may be a consequence of near-threshold transport conditions in the Mameyes, but more work is needed to support this hypothesis.

The cumulative sorting results over annual timescales appear to substantiate aspects of the self-similar sorting theory of Fedele and Paola (2007), and are in general agreement with earlier laboratory experiments (Paola et al., 1992; Paola and Seal, 1995; Seal et al., 1997; Toro-Escobar et al., 2000). The two tracer plumes in the Mameyes had significantly different front positions (Fig. 12a) at the end of the study, as they were emplaced in different years; the first population terminated at $X = 1$ km, while the second ended at $X = 1.2$ km. However, the two populations behave dynamically similar when their travel distances are rescaled by their extraction lengths (Fig. 12b–e). The decrease in the standard deviation for tracers is offset from that of the river bed, and results from the narrow grain size distribution of the tracers when compared to the substrate of the river (Fig. 1c). Accordingly, we normalize the tracer and river-bed data by their initial standard deviation at the start of the depositional section of the river ($X = 0$ km); the result is that tracer sorting appears to track that of the river bed when the initial particle size population and distance traveled are taken into account (Fig. 12b–e). Data show that the initial establishment of the sorting profile can be quite rapid, in that the tracer particles behave dynamically similar to the stream despite different residence times within the river. To our knowledge, this tracer study represents the first active field confirmation of the selective deposition theory (Paola et al., 1992; Paola and Seal, 1995; Seal et al., 1997; Toro-Escobar et al., 2000; Fedele and Paola, 2007).

The full model predicting the mean concentration of the sediment plume downstream consistently underpredicts the data further downstream (Fig. 12e). There are several reasons to expect that a finite population of sediment tracers might not follow the model in Eq. (4). The low values of σ at the leading edge ($X_* = 1$) of the tracer plume may have reached a limit where the size differences between tracers and the stream is negligible and sorting ceases for this mixture. For the stream, the underprediction is potentially from undersampling at the downstream end due to an artificial truncation caused by anthropogenic modification of the stream, which results in a higher value of C_1 in Eq. (3). The scaling exponent is also fairly sensitive to the determination of L in calculating X_* in this short system. In the case of the stream bed a small shift in L can steepen or elongate the sorting profile resulting in larger or smaller scaling exponents, respectively. Another factor complicating the scaling of the sorting in the Mameyes River is that the distance from the mountains to the ocean is particularly short (5.95 km), resulting in an abrupt truncation of the sorting profile somewhere within the final kilometer of the river. Given the short time that it took for the tracer particles to sort, and the minimal decline in D_{50} downstream, one might expect that there should be a rapidly prograding gravel front (Parker and Cui, 1998); however this front may be arrested due to Holocene sea level rise (Toscano and Macintyre, 2003).

6 Conclusions

In this paper we have presented field results on bed load tracer displacement data at the event to annual timescales, and used simple theory to rationalize the displacement scaling of tracer particles, show how a heterogeneous population of particles sorts downstream, and explore the implications of these findings on the statistical scaling of the hydrograph. At the scale of single floods, the distribution of particle displacements is well described by an exponential distribution. Close agreement with laboratory data and theory (Lajeunesse et al., 2010) suggests that these displacements represent the scaling of the fundamental particle step length. We infer that, for near-threshold floods, the most probable transport distance is one step length. Cumulative displacement over many floods reinforces this finding, with data showing that tracers remain in the partial transport regime for a range of flow conditions. We test the applicability of the impulse framework using data from two streams of very different morphologies, and find that tracer displacement data collapse onto a single linear relationship when differences in critical Shields stress and flow resistance are accounted for. For particle sorting, we find that downstream fining emerges after a series of floods. Sorting seems to result from a slight difference in size-dependent particle entrainment at the flood scale. We find that tracers sort to the limit of sorting present in the stream bed. Both the tracers and the stream bed have the same scaling when accounting for the distance each has traveled, as well as the initial statistics of the tracer and stream grain size distributions, respectively. These observations serve as an active field validation of the selective deposition sorting model (Paola et al., 1992; Fedele and Paola, 2007). Finally, we show that the magnitude–frequency distribution of flood stress in the Mameyes is exponential for flows exceeding the threshold of motion. The average stress for flows exceeding critical is approximately 1.4 times the critical Shields stress, and represents the stress of maximum geomorphic work. In addition, the distribution of dimensionless impulse has a well-defined peak coincident with the flood of maximum geomorphic work, indicating that the channel is adjusted to a characteristic flood impulse. We believe that tracer dynamics observed in the Mameyes River are characteristic of many gravel rivers, because many gravel streams are adjusted such that bankfull floods exert a stress that is only slightly in excess of the threshold for entrainment (Andrews, 1984; Pitlick and Cress, 2002; Torizzo and Pitlick, 2004; Mueller et al., 2005; Parker et al., 2007). A caveat for all of our results, however, is that caution should be exercised when considering reported numerical values duc to the difficulty in independently determining the threshold of particle motion. We emphasize that this remains one of the most critical problems in determining coarse-grained sediment mechanics in natural rivers.

Acknowledgements. We thank J. Singh, K. Litwin Miller, D. Miller, R. Glade, J. Evaristo, and G. Salter for outstanding field assistance. We thank D. N. Bradley for assistance concerning tracer tracking equipment. We thank A. E. J. Turowski, M. Schmeeckle, M. Chapuis, and the two anonymous reviewers for thoughtful comments that enhanced the clarity of this manuscript. Data for the Mameyes tracers are available at https://www.sas.upenn.edu/lczodata/content/mameyes-rfid-tracer-data. Data for the Bisley 3 tracers are available at https://www.sas.upenn.edu/lczodata/content/bisley-3-rfid-tracer-data. We gratefully acknowledge the Luquillo Critical Zone Observatory (NSF EAR 1331841), the Geological Society of America for a graduate student research grant (2011), and the University of Pennsylvania Benjamin Franklin Fellowship program for logistical and financial support.

Edited by: J. Turowski

References

Ancey, C.: Stochastic modeling in sediment dynamics: exner equation for planar bed incipient bed load transport conditions, J. Geophys. Res., 115, F00A11, doi:10.1029/2009JF001260, 2010.

Ancey, C., Davison, A. C., Bohm, T., Jodeau, M., and Frey, P.: Entrainment and motion of coarse particles in a shallow water stream down a steep slope, J. Fluid Mech., 595, 83–114, doi:10.1017/S0022112007008774, 2008.

Andrews, E. D.: Bed-material entrainment and hydraulic geometry of gravel-bed rivers in Colorado, Geol. Soc. Am. Bull., 95, 371–378, doi:10.1130/0016-7606(1984)95<371:BEAHGO>2.0.CO;2, 1984.

Bradley, D. N. and Tucker, G. E.: Measuring gravel transport and dispersion in a mountain river using passive radio tracers, Earth Surf. Proc. Land., 37, 1034–1045, doi:10.1002/esp.3223, 2012.

Bradley, D. N., Tucker, G. E., and Benson, D. A.: Fractional dispersion in a sand bed river, J. Geophys. Res., 115, F00A09, doi:10.1029/2009JF001268, 2010.

Bridge, J. and Dominic, D.: Bed-load grain velocities and sediment transport rates, Water Resour. Res., 20, 476–490, doi:10.1029/WR020i004p00476, 1984.

Buffington, J. M. and Montgomery, D. R.: A systematic analysis of eight decades of incipient motion studies, with special reference to gravel-bedded rivers, Water Resour. Res., 33, 1993–2029, doi:10.1029/96WR03190, 1997.

Celik, A. O., Diplas, P., Dancey, C. L., and Valyrakis, M.: Impulse and particle dislodgement under turbulent flow conditions, Phys. Fluids, 22, 046601, doi:10.1063/1.3385433, 2010.

Chapuis, M., Bright, C.J., Hufnagel, J., and MacVicar, B.: Detection ranges and uncertainty of passive Radio Frequency Identification (RFID) transponders for sediment tracking in gravel rivers and coastal environments, Earth Surf. Proc. Land., 39, 2109–2120, doi:10.1002/esp.3620, 2014.

Charru, F., Mouilleron, H., and Eiff, O.: Erosion and deposition of particles on a bed sheared by a viscous flow, J. Fluid Mech., 519, 55–80, doi:10.1017/S0022112004001028, 2004.

Church, M. and Hassan, M. A.: Size and distance of travel of unconstrained clasts on a streambed, Water Resour. Res., 28, 299–303, doi:10.1029/91WR02523, 1992.

Diplas, P., Dancey, C. L., Celik, A. O., Valyrakis, M., Greer, K., and Akar, T.: The role of impulse on the initiation of particle movement under turbulent flow conditions, Science, 322, 717–720, doi:10.1126/science.1158954, 2008.

Drake, T. G., Shreve, R. L., Dietrich, W. E., Whiting, P. J., and Leopold, L. B.: Bedload transport of fine gravel observed by motion-picture photography, J. Fluid Mech., 192, 193, doi:10.1017/S0022112088001831, 1988.

Einstein, H. A.: Bed load transport as a probability problem, PhD thesis, ETH Zurich, Zurich, 1937.

Einstein, H. A.: The Bed-load Function for Sediment Transportation in Open Channel Flows, Tech. Bull. No. 1026, US Department of Agriculture, Soil Conservation Service, p. 71, 1950.

Fedele, J. J. and Paola, C.: Similarity solutions for fluvial sediment fining by selective deposition, J. Geophys. Res.-Earth, 112, F02038, doi:10.1029/2005JF000409, 2007.

Ferguson, R. I.: Flow resistance equations for gravel- and boulder-bed streams, Water Resour. Res., 43, W05427, doi:10.1029/2006WR005422, 2007.

Ferguson, R. I.: River channel slope, flow resistance, and gravel entrainment thresholds, Water Resour. Res., 48, W05517, doi:10.1029/2011WR010850, 2012.

Ferguson, R. I., Hoey, T., Wathen, S., and Werritty, A.: Field evidence for rapid downstream fining of river gravels through selective transport, Geology, 24, 179–182, doi:10.1130/0091-7613(1996)024<0179:FEFRDF>2.3.CO;2, 1996.

Ferguson, R. I. and Wathen, S. J.: Tracer-pebble movement along a concave river profile: virtual velocity in relation to grain size and shear stress, Water Resour. Res., 34, 2031–2038, doi:10.1029/98WR01283, 1998.

Fernandez Luque, R. and Van Beek, R.: Erosion and transport of bed-load sediment, J. Hydraul. Res., 14, 127–144, 1976.

Furbish, D. J. and Schmeeckle, M. W.: A probabilistic derivation of the exponential-like distribution of bed load particle velocities, Water Resour. Res., 49, 1537–1551, doi:10.1002/wrcr.20074, 2013.

Furbish, D. J., Haff, P. K., Roseberry, J. C., and Schmeeckle, M. W.: A probabilistic description of the bed load sediment flux: 1. Theory, J. Geophys. Res.-Earth, 117, F03031, doi:10.1029/2012JF002352, 2012a.

Furbish, D. J., Roseberry, J. C., and Schmeeckle, M. W.: A probabilistic description of the bed load sediment flux: 3. The particle velocity distribution and the diffusive flux, J. Geophys. Res.-Earth, 117, F03033, doi:10.1029/2012JF002355, 2012b.

Ganti, V., Meerschaert, M. M., Foufoula-Georgiou, E., Viparelli, E., and Parker, G.: Normal and anomalous diffusion of gravel tracer particles in rivers, J. Geophys. Res., 115, F00A12, doi:10.1029/2008JF001222, 2010.

Gasparini, N. M., Tucker, G. E., and Bras, R. L.: Downstream fining through selective particle sorting in an equilibrium drainage network, Geology, 27, 1079–1082, 1999.

Gasparini, N. M., Tucker, G. E., and Bras, R. L.: Network-scale dynamics of grain-size sorting: implications for downstream fining, stream-profile concavity, and drainage basin morphology, Earth Surf. Proc. Land., 29, 401–421, doi:10.1002/esp.1031, 2004.

Gomez, B.: Bedload transport, Earth-Sci. Rev., 31, 89–132, doi:10.1016/0012-8252(91)90017-A, 1991.

Gray, J., Laronne, J., and Marr, J. D. G.: Bedload-surrogate monitoring technologies, US Geological Survey Scientific Investigations Report 2010-5091, United States Geological Survey, 2010.

Habersack, H. M.: Radio-tracking gravel particles in a large braided river in New Zealand: a field test of the stochastic theory of bed load transport proposed by Einstein, Hydrol. Process., 15, 377–391, doi:10.1002/hyp.147, 2001.

Haschenburger, J. K.: The rate of fluvial gravel dispersion, Geophys. Res. Lett., 38, L24403, doi:10.1029/2011GL049928, 2011.

Haschenburger, J. K. and Church, M.: Bed material transport estimated from the virtual velocity of sediment, Earth Surf. Proc. Land., 23, 791–808, doi:10.1002/(SICI)1096-9837(199809)23:9<791::AID-ESP888>3.0.CO;2-X, 1998.

Haschenburger, J. K. and Wilcock, P. R.: Partial transport in a natural gravel bed channel, Water Resour. Res., 39, 1020, doi:10.1029/2002WR001532, 2003.

Hassan, M. A., Church, M., and Schick, A. P.: Distance of movement of coarse particles in gravel bed streams, Water Resour. Res., 27, 503–511, doi:10.1029/90WR02762, 1991.

Hassan, M. A., Voepel, H., Schumer, R., Parker, G., and Fraccarollo L.: Displacement characteristics of coarse fluvial bed sediment, J. Geophys. Res.-Earth, 118, 155–165, doi:10.1029/2012JF002374, 2013.

Heyman, J., Mettra, F., Ma, H. B., and Ancey, C.: Statistics of bedload transport over steep slopes: separation of time scales and collective motion, Geophys. Res. Lett., 40, 128–133, doi:10.1029/2012GL054280, 2013.

Hill, K. M., DellAngelo, L., and Meerschaert, M. M.: Heavy-tailed travel distance in gravel bed transport: an exploratory enquiry, J. Geophys. Res., 115, F00A14, doi:10.1029/2009JF001276, 2010.

Hodge, R. A., Hoey, T. B., and Sklar, L. S.: Bedload transport in bedrock rivers: the role of sediment cover in grain entrainment, translation and deposition, J. Geophys. Res., 116, F04028, doi:10.1029/2011JF002032, 2011.

Houbrechts, G., Levecq, Y., Vanderheyden, V., and Petit, F.: Long-term bedload mobility in gravel-bed rivers using iron slag as a tracer, Geomorphology, 126, 233–244, doi:10.1016/j.geomorph.2010.11.006, 2011.

Houssais, M. and Lajeunesse, E.: Bedload transport of a bimodal sediment bed, J. Geophys. Res.-Earth, 117, doi:10.1029/2012JF002490, 2012.

Jerolmack, D. J., and Brzinski, T. A.: Equivalence of abrubt grain-size transitions in alluvial rivers and eolian sand seas: a hypothesis, Geology, 38, 719, doi:10.1130/G30922.1, 2010.

Kirchner, J. W., Dietrich, W. E., Iseya, F., and Ikeda, H.: The variability of critical shear stress, friction angle, and grain protrusion in water worked sediments, Sedimentology, 37, 647–672, 1990.

Kodama, Y.: Experimental study of abrasion and its role in producing downstream fining in gravel-bed rivers, J. Sediment. Res., 64, 76–85, 1994.

Lague, D., Hovius, N., and Davy, P.: Discharge, discharge variability, and the bedrock channel profile, J. Geophys. Res.-Earth, 110, F04006, doi:10.1029/2004JF000259, 2005.

Lajeunesse, E., Malverti, L., and Charru, F.: Bed load transport in turbulent flow at the grain scale: experiments and modeling, J. Geophys. Res., 115, F04001, doi:10.1029/2009JF001628, 2010.

Lamarre, H. and Roy, A. G.: A field experiment on the development of sedimentary structures in a gravel-bed river, Earth Surf. Proc. Land., 33, 1064–1081, doi:10.1002/esp.1602, 2008.

Lamarre, H., MacVicar, B., and Roy, A. G.: Using Passive Integrated Transponder (PIT) tags to investigate sediment transport in gravel-bed rivers, J. Sediment. Res., 75, 736–741, doi:10.2110/jsr.2005.059, 2005.

Lamb, M. P., Dietrich, W. E., and Venditti, J. G.: Is the critical Shields stress for incipient sediment motion dependent on channel-bed slope?, J. Geophys. Res.-Earth, 113, F02008, doi:10.1029/2007JF000831, 2008.

Legleiter, C. and Kyriakidis, P.: Forward and inverse transformations between cartesian and channel-fitted coordinate systems for meandering rivers, Math. Geol., 38, 927–958, doi:10.1007/s11004-006-9056-6, 2007.

Lenzi, M. A.: Displacement and transport of marked pebbles, cobbles and boulders during floods in a steep mountain stream, Hydrol. Process., 18, 1899–1914, doi:10.1002/hyp.1456, 2004.

Mao, L. and Surian, N.: Observations on sediment mobility in a large gravel-bed river, Geomorphology, 114, 326–337, doi:10.1016/j.geomorph.2009.07.015, 2010.

Marquis, G. A. and Roy, A. G.: Using multiple bed load measurements: toward the identification of bed dilation and contraction in gravel-bed rivers, J. Geophys. Res., 117, F01014, doi:10.1029/2011JF002120, 2012.

Martin, R. L., Jerolmack, D. J., and Schumer, R.: The physical basis for anomalous diffusion in bed load transport, J. Geophys. Res., 117, F01018, doi:10.1029/2011JF002075, 2012.

Martin, R. L., Purohit, P.K., and Jerolmack, D. J.: Sedimentary bed evolution as a mean-reverting random walk: Implications for tracer studies, Geophys. Res. Lett., 41, 6152–6159, doi:10.1002/2014GL060525, 2014.

Meyer-Petter, E. and Muller, R.: Formulas for bed-load transport, in: Proceedings, Stockholm, Sweden, 39–64, 1948.

Milliman, J. and Syvitski, J.: Geomorphic tectonic control of sediment discharge to the ocean – the importance of small mountainous rivers, J. Geol., 100, 525–544, 1992.

Molnar, P., Anderson, R. S., Kier, G., and Rose, J.: Relationships among probability distributions of stream discharges in floods, climate, bed load transport, and river incision, J. Geophys. Res.-Earth, 111, F02001, doi:10.1029/2005JF000310, 2006.

Mueller, E. R., Pitlick, J., and Nelson, J. M.: Variation in the reference Shields stress for bed load transport in gravel-bed streams and rivers, Water Resour. Res., 41, W04006, doi:10.1029/2004WR003692, 2005.

Paola, C. and Seal, R.: Grain-size patchiness as a cause of selective deposition and downstream fining, Water Resour. Res., 31, 1395–1407, doi:10.1029/94WR02975, 1995.

Paola, C., Parker, G., Seal, R., Sinha, S., Southard, J., and Wilcock, P.: Downstream fining by selective deposition in a laboratory flume, Science, 258, 1757–1760, doi:10.1126/science.258.5089.1757, 1992.

Parker, G.: Self-formed straight rivers with equilibrium banks and mobile bed, Part 2: The gravel river, J. Fluid Mech., 89, 127–146, 1978.

Parker, G. and Cui, Y.: The arrested gravel front: stable gravel-sand transitions in rivers Part 1: Simplified analytical solution, J. Hydraul. Res., 36, 75–100, doi:10.1080/00221689809498379, 1998.

Parker, G., Paola, C., Whipple, K. X., Mohrig, D., Toro-Escobar, C. M., Halverson, M., and Skoglund, T. W.: Alluvial fans formed by channelized fluvial and sheet flow, II: Application, J. Hydraul. Eng., 124, 996–1004, doi:10.1061/(ASCE)0733-9429(1998)124:10(996), 1998.

Parker, G., Seminara, G., and Solari, L.: Bed load at low shields stress on arbitrarily sloping beds: alternative entrainment formulation, Water Resour. Res., 39, 1183, doi:10.1029/2001WR001253, 2003.

Parker, G., Wilcock, P. R., Paola, C., Dietrich, W. E., and Pitlick, J.: Physical basis for quasi-universal relations describing bankfull hydraulic geometry of single-thread gravel bed rivers, J. Geophys. Res., 112, F04005, doi:10.1029/2006JF000549, 2007.

Phillips, C. B., Martin, R. L., and Jerolmack, D. J.: Impulse framework for unsteady flows reveals superdiffusive bed load transport, Geophys. Res. Lett., 40, 1328–1333, doi:10.1002/grl.50323, 2013.

Pike, A. S.: Longitudinal patterns in stream channel geomorphology and aquatic habitat in the Luquillo Mountains of Puerto Rico, PhD thesis, University of Pennsylvania, Philadelphia, 2008.

Pike, A. S. and Scatena, F. N.: Riparian indicators of flow frequency in a tropical montane stream network, J. Hydrol., 382, 72–87, doi:10.1016/j.jhydrol.2009.12.019, 2010.

Pitlick, J. and Cress, R.: Downstream changes in the channel geometry of a large gravel bed river, Water Resour. Res., 38, 34-1–34-11, doi:10.1029/2001WR000898, 2002.

Recking, A., Liébault, F., Peteuil, C., and Jolimet, T.: Testing bed-load transport equations with consideration of time scales, Earth Surf. Proc. Land., 37, 774–789, doi:10.1002/esp.3213, 2012.

Rickenmann, D. and Recking, A.: Evaluation of flow resistance in gravel-bed rivers through a large field data set, Water Resour. Res., 47, W07538, doi:10.1029/2010WR009793, 2011.

Roseberry, J. C., Schmeeckle, M. W., and Furbish, D. J.: A probabilistic description of the bed load sediment flux: 2. Particle activity and motions, J. Geophys. Res., 117, doi:10.1029/2012JF002353, 2012.

Sayre, W. and Hubbell, D.: Transport and dispersion of labeled bed material, North Loup River, Nebraska, US Geological Survey Professional Paper 433-C, US Geological Survey, 63–118, 1965.

Scheingross, J. S., Winchell, E. W., Lamb, M. P., and Dietrich, W. E.: Influence of bed patchiness, slope, grain hiding, and form drag on gravel mobilization in very steep streams, J. Geophys. Res.-Earth, 118, 982–1001, doi:10.1002/jgrf.20067, 2013.

Schellekens, J., Scatena, F. N., Bruijnzeel, L. A., van Dijk, A. I. J. M., Groen, M. M. A., and van Hogezand, R. J. P.: Stormflow generation in a small rainforest catchment in the Luquillo Experimental Forest, Puerto Rico, Hydrol. Process., 18, 505–530, doi:10.1002/hyp.1335, 2004.

Schmeeckle, M. W., Nelson, J. M., Pitlick, J., and Bennett, J. P.: Interparticle collision of natural sediment grains in water, Water Resour. Res., 37, 2377–2391, doi:10.1029/2001WR000531, 2001.

Schmidt, K.-H. and Ergenzinger, P.: Bedload entrainment, travel lengths, step lengths, rest periods – studied with passive (iron, magnetic) and active (radio) tracer techniques, Earth Surf. Proc. Land., 17, 147–165, doi:10.1002/esp.3290170204, 1992.

Schneider, J. M., Turowski, J. M., Rickenmann, D., Hegglin, R., Arrigo, S., Mao, L., and Kirchner, J. W.: Scaling relationships between bed load volumes, transport distances, and stream power in steep mountain channels, J. Geophys. Res.-Earth, 119, 533–549, doi:10.1002/2013JF002874, 2014.

Seal, R., Paola, C., Parker, G., Southard, J., and Wilcock, P.: Experiments on downstream fining of gravel: I. Narrow-channel runs, J. Hydraul. Eng., 123, 874–884, doi:10.1061/(ASCE)0733-9429(1997)123:10(874), 1997.

Sklar, L. S. and Dietrich, W. E.: A mechanistic model for river incision into bedrock by saltating bed load, Water Resour. Res., 40, W06301, doi:10.1029/2003WR002496, 2004.

Snyder, N. P., Whipple, K. X., Tucker, G. E., and Merritts, D. J.: Importance of a stochastic distribution of floods and erosion thresholds in the bedrock river incision problem, J. Geophys. Res.-Sol. Ea., 108, 2117, doi:10.1029/2001JB001655, 2003.

Solari, L. and Parker, G.: The curious case of mobility reversal in sediment mixtures, J. Hydraul. Eng., 126, 185–197, doi:10.1061/(ASCE)0733-9429(2000)126:3(185), 2000.

Strom, K., Papanicolaou, A. N., Evangelopoulos, N., and Odeh, M.: Microforms in gravel bed rivers: formation, disintegration, and effects on bedload transport, J. Hydraul. Eng.-ASCE, 130, 554–567, doi:10.1061/(ASCE)0733-9429(2004)130:6(554), 2004.

Torizzo, M. and Pitlick, J.: Magnitude-frequency of bed load transport in mountain streams in Colorado, J. Hydrol., 290, 137–151, doi:10.1016/j.jhydrol.2003.12.001, 2004.

Toro-Escobar, C., Paola, C., Parker, G., Wilcock, P., and Southard, J.: Experiments on downstream fining of gravel, II: Wide and sandy runs, J. Hydraul. Eng., 126, 198–208, doi:10.1061/(ASCE)0733-9429(2000)126:3(198), 2000.

Toscano, M. A. and Macintyre, I. G.: Corrected western Atlantic sea-level curve for the last 11,000 years based on calibrated ^{14}C dates from Acropora palmata framework and intertidal mangrove peat, Coral Reefs, 22, 257–270, doi:10.1007/s00338-003-0315-4, 2003.

Turcotte, D. L.: Fractal theory and the estimation of extreme floods, J. Res. Natl. Inst. Stan., 99, 377–389, 1994.

Turowski, J. M., Hovius, N., Wilson, A., and Horng, M. J.: Hydraulic geometry, river sediment and the definition of bedrock channels, Geomorphology, 99, 26–38, doi:10.1016/j.geomorph.2007.10.001 2008.

Turowski, J. M., Badoux, A., and Rickenmann, D.: Start and end of bedload transport in gravel-bed streams, Geophys. Res. Lett., 38, L04401, doi:10.1029/2010GL046558, 2011.

Valyrakis, M., Diplas, P., Dancey, C. L., Greer, K., and Celik, A. O.: Role of instantaneous force magnitude and duration on particle entrainment, J. Geophys. Res.-Earth, 115, F02006, doi:10.1029/2008JF001247, 2010.

Valyrakis, M., Diplas, P., and Dancey, C. L.: Entrainment of coarse particles in turbulent flows: an energy approach, J. Geophys. Res.-Earth, 118, 42–53, doi:10.1029/2012JF002354, 2013.

Wiberg, P. L. and Smith, J. D.: Calculations of the critical shear stress for motion of uniform and heterogeneous sediments, Water Resour. Res., 23, 1471–1480, doi:10.1029/WR023i008p01471, 1987.

Wiberg, P. L. and Smith, J. D.: Model for calculating bed load transport of sediment, J. Hydraul. Eng., 115, 101–123, 1989.

Wilcock, P. R.: Methods for estimating the critical shear stress of individual fractions in mixed-size sediment, Water Resour. Res., 24, 1127–1135, doi:10.1029/WR024i007p01127, 1988.

Wilcock, P. R.: The components of fractional transport rate, Water Resour. Res., 33, 247–258, doi:10.1029/96WR02666, 1997a.

Wilcock, P. R.: Entrainment, displacement and transport of tracer gravels, Earth Surf. Proc. Land., 22, 1125–1138, doi:10.1002/(SICI)1096-9837(199712)22:12<1125::AID-ESP811>3.0.CO;2-V, 1997b.

Wilcock, P. R. and Crowe, J. C.: Surface-based transport model for mixed-size sediment, J. Hydraul. Eng., 129, 120–128, 2003.

Wilcock, P. R. and McArdell, B. W.: Partial transport of a sand/gravel sediment, Water Resour. Res., 33, 235–245, doi:10.1029/96WR02672, 1997.

Willenbring, J. K., Codilean, A. T., and McElroy, B.: Earth is (mostly) flat: apportionment of the flux of continental sediment over millennial time scales, Geology, 41, 343–346, doi:10.1130/G33918.1, 2013.

Wolman, M. G.: A method of sampling coarse river-bed material, Trans. Am. Geophys. Union, 35, 951–956, 1954.

Wolman, M. G. and Miller, J. P.: Magnitude and frequency of forces in geomorphic processes, J. Geol., 68, 54–74, 1960.

Wong, M. and Parker, G.: Reanalysis and correction of bed-load relation of Meyer-Peter and Muller using their own database, J. Hydraul. Eng.-ASCE, 132, 1159–1168, doi:10.1061/(ASCE)0733-9429(2006)132:11(1159), 2006.

Wong, M., Parker, G., DeVries, P., Brown, T. M., and Burges, S. J.: Experiments on dispersion of tracer stones under lower-regime plane-bed equilibrium bed load transport, Water Resour. Res., 43, W03440, doi:10.1029/2006WR005172, 2007.

Zhang, Y., Meerschaert, M. M., and Packman, A. I.: Linking fluvial bed sediment transport across scales, Geophys. Res. Lett., 39, L20404, doi:10.1029/2012GL053476, 2012.

Zimmermann, A., Church, M., and Hassan, M. A.: Step-pool stability: testing the jammed state hypothesis, J. Geophys. Res.-Earth, 115, F02008, doi:10.1029/2009JF001365, 2010.

A two-sided approach to estimate heat transfer processes within the active layer of the Murtèl–Corvatsch rock glacier

M. Scherler[1,*], S. Schneider[1], M. Hoelzle[1], and C. Hauck[1]

[1]Departement of Geosciences, University of Fribourg, Chemin du Musée 4, 1700 Fribourg, Switzerland
[*]now at: Swiss Federal Research Institute WSL, Zürcherstrasse 111, 8903 Birmensdorf, Switzerland

Correspondence to: M. Scherler (martin.scherler@wsl.ch)

Abstract. The thermal regime of permafrost on scree slopes and rock glaciers is characterized by the importance of air flow driven convective and advective heat transfer processes. These processes are supposed to be part of the energy balance in the active layer of rock glaciers leading to lower subsurface temperatures than would be expected at the lower limit of discontinuous high mountain permafrost. In this study, new parametrizations were introduced in a numerical soil model (the Coup Model) to simulate permafrost temperatures observed in a borehole at the Murtèl rock glacier in the Swiss Alps in the period from 1997 to 2008. A soil heat sink and source layer was implemented within the active layer, which was parametrized experimentally to account for and quantify the contribution of air flow driven heat transfer on the measured permafrost temperatures. The experimental model calibration process yielded a value of about 28.9 $\mathrm{Wm^{-2}}$ for the heat sink during the period from mid September to mid January and one of 26 $\mathrm{Wm^{-2}}$ for the heat source in the period from June to mid September. Energy balance measurements, integrated over a 3.5 m-thick blocky surface layer, showed seasonal deviations between a zero energy balance and the calculated sum of the energy balance components of around 5.5 $\mathrm{Wm^{-2}}$ in fall/winter, −0.9 $\mathrm{Wm^{-2}}$ in winter/spring and around −9.4 $\mathrm{Wm^{-2}}$ in summer. The calculations integrate heat exchange processes including thermal radiation between adjacent blocks, turbulent heat flux and energy storage change in the blocky surface layer. Finally, it is hypothesized that these deviations approximately equal unmeasured freezing and thawing processes within the blocky surface layer.

1 Introduction

Permafrost in high mountain environments is a common phenomenon at altitudes above 2400 m a.s.l in the European Alps. It can be found in areas with various subsurface characteristics such as solid rock, weathered rock with a fine grained surface cover, or talus slopes consisting of coarse debris. At debris-covered sites in high mountains relatively cold permafrost can be found at lower altitudes than would be expected from the prevailing mean annual air temperatures. One explanation for this are the thermal properties of blocky surfaces, which may lead to the existence of permafrost at sites where without such ground characteristics permafrost would not develop (e.g., Harris and Pedersen, 1998; Delaloye and Lambiel, 2005; Hanson and Hoelzle, 2004). One special permafrost form in talus slopes are rock glaciers. This type of permafrost is characterized by ice-supersaturated sediments covered by large blocks (Arenson et al., 2002). Rock glaciers often show lobes of tens of meters in wavelength and a few meters in amplitude at the surface (Kääb et al., 1998). Besides the subsurface material, the ice and water content of the ground, the energy balance at the surface is the most important factor for the existence of permafrost. Due to the coarse surface layer of a rock glacier with boulders of up to 10 m in diameter the determination of an energy balance at the surface is a complex problem. The surface in this case is rather to be seen as a blocky surface layer of several meters in thickness comprising a large part of the blocky surface layer with voids and the air column above (Herz et al., 2003). In surface energy balance measurements even under less complicated circumstances, e.g., arctic plains with sparse vegetation, there are usually deviation terms of up to 20 Wm^{-2} to a zero energy balance reported due to method-related errors and parametrizations (Westermann et al., 2009). A study by Mittaz et al. (2000) at the Murtèl rock glacier found deviations from a zero energy balance of up to 78 Wm^{-2} in winter and −130 Wm^{-2} in summer, which were explained by advective heat transfer processes through voids within the debris.

Several processes of advective and convective air circulation in blocky material have been described in the literature. A convective process is the Balch effect (Balch, 1900), which describes the replacement of warm air with cold air within the voids of the blocky material due to density differences. An additional convective/advective process on inclined blocky slopes called the chimney effect was first described in a study by Wakonigg (1996). In winter relatively warm air within the blocky layer ascends beneath the snow cover, creating melting holes in the upper part of the slope, which facilitates the aspiration of cold air inside the talus slope (Delaloye et al., 2003). Discharge of cold air in summer driven by gravity may lead to advective heat transfer by air circulation (Delaloye et al., 2003). Further studies stressing the importance of air circulation within the active layer of coarse blocky scree slopes for the Alps have been presented by, e.g., Vonder Mühll et al. (2003). Tanaka et al. (2000) describe these effects in a modeling study for mountainous regions in Korea and Japan. Similar effects are also described for anthropogenic structures, i.e., crushed rock highway embankments (e.g., Binxiang et al., 2007). Harris and Pedersen (1998) suggested an advective process that is characterized by continuous air exchange between the voids and the atmosphere. Air exchange with the atmosphere will result in almost instantaneous warming and cooling of the blocky debris to a considerable depth in response to changes in air temperature. Heat transfer by water flow from precipitation between the blocks is expected to lead to a reduction in the temperature gradients and in the blocky surface layer, which will be reflected in the thermal conduction term and the storage change (integrated over the entire surface layer).

In model studies aiming at the simulation of the hydrothermal regime and the response of permafrost to climate change, three-dimensional energy exchange processes in the active layer are often approximated by very low effective thermal conductivities of the coarse blocky material, i.e., the surface layer is treated as a "thermal semi-conductor" (see, e.g., Cheng et al., 2007 or Gruber and Hoelzle, 2008).

The aim of this study was to compare the energy balance of a calibrated one-dimensional heat and mass transfer model, which was used in an earlier study to simulate the thermal regime of the Murtèl rock glacier under the influence of climate change scenarios (see Scherler et al., 2013), to a measured energy balance.

This approach has been chosen as existing energy balance formulations do not account for the complex surface of block materials, which is addressed here by developing a volumetric energy balance. Because existing models, such as the Coup Model used in this study, do not account for all the energy exchange processes within such surface materials, a method to account for these by adding a sink/source component is examined here. The results of both methods and the relative strengths/weaknesses of the approaches with respect to different applications are discussed.

This approach also allowed for the indirect quantification of the total energy exchange by three-dimensional heat transfer processes. In the measured energy balance the use of additional terms for radiative heat transfer between the blocks (see, e.g., Kunii and Smith, 1960; Fillion et al., 2011), heat storage change and turbulent heat flux in the blocky surface layer allowed for the approximation of seasonal freezing and thawing processes within the active layer and the permafrost.

1.1 Coup Model

The model used in this study is a one-dimensional heat and mass transfer model for the soil–snow–atmosphere system (Coup Model; Jansson and Karlberg, 2011). The model was chosen for a sensitivity study involving transient hydrothermal simulations using RCM derived climate scenarios of the 21st century (see Scherler et al., 2013). The empirical parametrization used in the calibration of the model, as described below, was a part of this project. In this study, a detailed comparison of the simulated and the measured energy balance is presented.

Two coupled partial differential equations for water and heat flow are the core of the Coup Model. These equations are solved with an explicit forward difference method. A detailed description of the model including all its equations and parameters is given in Jansson and Karlberg (2011). Applications of the model are detailed in a number of studies (e.g., Johnsson and Lundin, 1991; Stähli et al., 1996; Bayard et al., 2005; Scherler et al., 2010, 2013). Processes that are important for permafrost, such as freezing and thawing of the soil (Lundin, 1990) as well as the accumulation, metamorphosis, and melt of a snow cover (Gustafsson et al., 2001), are included in the model. The model is driven by hourly averages of air temperature, relative humidity, wind speed, global radiation, incoming longwave radiation and precipitation. The number of iterations per day used in the simulation is 1440. The upper boundary condition is given by a surface energy balance at the soil–snow–atmosphere boundary layer. The lower boundary condition at the bottom of the soil column at a depth of 70 m is given as a zero heat flux and a seepage flow of percolating water. The model is initialized with an ice content of 85 % in the permafrost at depths of 3.4 to 22.4 m below the surface and a starting temperature of −1.5 °C.

To account for three-dimensional heat transfer by longwave radiation and air circulation between the blocks, which

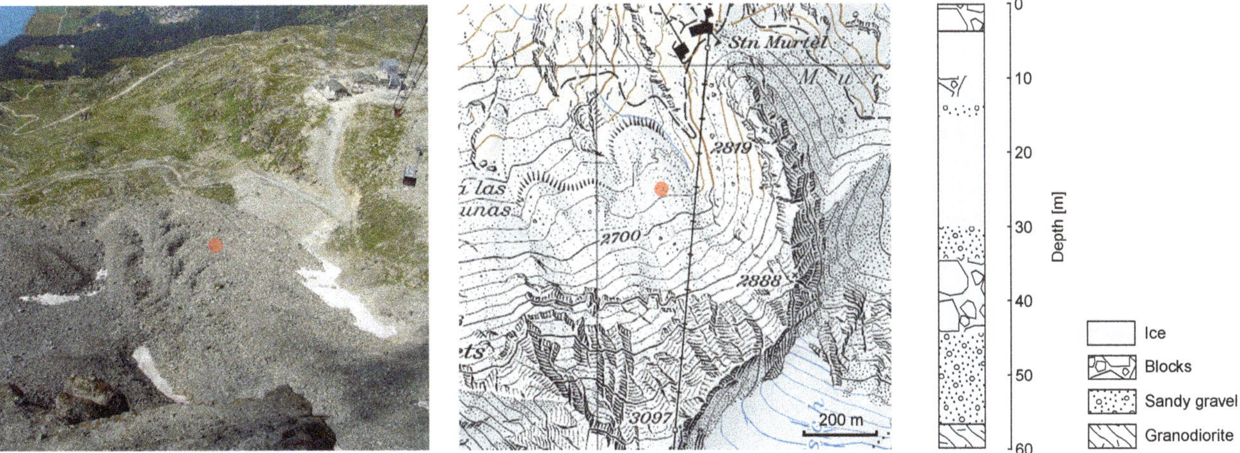

Figure 1. Field site photograph, situation map (reproduced by permission of swisstopo (BA14029)), and approximative stratigraphy (according to Arenson et al., 2002) indicating the depth of the permafrost table. The red dot shows the location of the borehole and the meteorological station.

cannot be simulated directly in a one-dimensional model, but are supposed to have a significant impact on the thermal regime of the active layer in coarse debris-covered permafrost (Delaloye and Lambiel, 2005; Mittaz et al., 2000; Hanson and Hoelzle, 2004), a layer that serves as a heat source/sink is introduced in the model. It adds energy to the soil system in the summer season (June–mid September) and extracts energy in winter (mid September–mid January). The layer is 1 m thick and is situated between 0.2 and 1.2 m depth. The thickness is chosen large enough to approximate the natural situation (i.e., 40 % porosity in the active layer) and thin enough not to cause numerical problems. This energy source/sink layer is parametrized based partly on knowledge taken from an observational study done by Mittaz et al. (2000), who found significant deviations to a zero energy balance in summer and winter in measurements at the Murtèl rock glacier site. The values for the parametrization were then adjusted experimentally during the calibration phase of the model. Heat source and heat sink are treated as constant in the respective seasons due to simplicity. Parametrization is considered as successful when measured borehole temperatures and simulated temperatures at two depths within the permafrost show the best fit. To reach near thermodynamic equilibrium conditions the model was run for four 11-year cycles in the case with a heat source and sink parametrization and for eleven 11-year cycles in the case without a heat pump. This discrepancy is due to the 85 % ice content in the respective layers (5.5 and 10.5 m), which has to be melted in the case of no additional heat sink/source in the model before reaching an approximate thermodynamic equilibrium.

2 Site description

The site of this study is the Murtèl–Corvatsch rock glacier in the Upper Engadine, Switzerland (see Fig. 1). The rock glacier reaches from 2850 to 2620 m a.s.l. and is 400 m long and 200 m wide, facing north-northwest. At the site a 60 m-deep borehole was drilled and equipped with thermistors in 1987 that have been manually logged in 1-month intervals until 1992, and since then data has been stored automatically by a logger collecting temperature data in 6 h intervals (Hoelzle et al., 2002). A micrometeorological station established in 1997 at 2700 m a.s.l. next to the borehole measures short-wave and long-wave incoming and outgoing radiation, air temperature, surface temperature, relative humidity, wind speed and direction (Mittaz et al., 2000; Hoelzle and Gruber, 2008). The site is characterized by a coarse blocky surface layer of approximately 3–3.5 m in thickness above a massive ice core down to 28 m and a frozen blocky layer underneath, reaching from 28 to 50 m probably adjacent to the bedrock (Arenson et al., 2002). The ice core has a temperature of −2 °C at 10 m depth and −1.4 °C at 25 m depth. The active layer has a thickness of 3.2 m on average. The diameters of the boulders forming the surface layer are in the range of decimeters up to several meters. The comparison of the stratigraphy of the studied borehole with the stratigraphies of two boreholes located within a distance of 30 m shows significant small-scale heterogeneities in the rock glacier (Vonder Mühll et al., 2001; Arenson et al., 2010). In direct proximity of the rock glacier, areas with fine grained subsurface material/soil as well as solid rock exist that show no permafrost conditions (Schneider et al., 2012). The rocks at the site mainly consist of metamorphic granodiorite and basalt (Schneider et al., 2012). Annual precipitation at the site is about 900 mm (982 mm St Moritz 1951–1980; 856 mm Piz Corvatsch 1984–1997). Typical maximum snow cover

Table 1. Meteorological station equipment and accuracy.

Variable	Sensor	Sensor type	Accuracy
	Logger (Campbell)	CR10X data logger; SDM-INT8 interval timer; AM416 multiplexer	
Radiation (short- and longwave)	CNR1 net radiometer (Kipp & Zonen)	2 pyranometer CM3; 2 pyrgeometer CG3; Pt-100 temp. sensor	±10 %; ±2 K
Air temperature/ humidity	MP-100A ventilated hygrometer (Rotronic)	RTD Pt-100; C94 hygrometers	±10 %
Wind speed	05103-5 model (Young)	Potentiometer	±0.3 m s^{-1}
Snow height	SR50 (Campbell)	Ultrasonic electrostatic transducer	±0.01 m
Surface temperature	Infrared thermometer	Irt/c.5	±1.5 °C
Borehole temperatures	YSI 44006 (Yellow Springs Instruments)	NTC thermistors	±0.02 °C
Precipitation	MeteoSwiss (Corvatsch summit)	Rain gauge	±30 %

thickness is between 1 m and 2 m. Mean annual air temperature is −1.7 °C for the observation period of March 1997 to March 2008. The study site has been described in more detail by Haeberli et al. (1988); Hoelzle et al. (2002); Vonder Mühll et al. (2003); Schneider et al. (2012).

2.1 Methods

2.2 Meteorological measurements

The meteorological parameters air temperature, surface temperature, relative humidity, incoming and outgoing shortwave and longwave radiation, wind speed and snow height, are measured by a micrometeorological station directly at the study site (Mittaz et al., 2000; Hoelzle and Gruber, 2008). Data for this study have been measured at the station for the period from January 1997 to March 2008 in a 10 min interval and were stored as 30 min averages by the logger (see Table 1). Precipitation data were taken from a nearby station of MeteoSwiss, located at the summit of Piz Corvatsch. Data gaps in the on-site measurements, which are caused by lightning, avalanches, or hoarfrost, were reconstructed with measurements from the summit station, corrected by the use of correlation coefficients determined between the two stations. This completed meteorological data set consisting of incoming shortwave and longwave radiation, air temperature, wind speed, and relative humidity, and precipitation is used as input in the Coup Model (see Sect. 1.1). For the energy balance calculations the original data are left unchanged and only seasons with sufficient measurements are included in the analysis (see Results section).

2.3 Energy balance

Generally, the energy balance at seasonally snow-covered sites refers to a unit area and includes the net short-wave and long-wave radiation components, turbulent fluxes composed of sensible heat and latent heat, ground heat flux, melt energy of the snow, and heat flux through the snow cover. The corresponding energy balance (Williams and Smith, 1989) at such a site is given as

$$Q_{\mathrm{rad}} + Q_{\mathrm{h}} + Q_{\mathrm{le}} + Q_{\mathrm{g}} + \Delta Q_{\mathrm{m}} + Q_{\mathrm{s}} = 0, \quad (1)$$

where Q_{rad} [W m^{-2}] is the net radiation, Q_{h} [W m^{-2}] is the sensible heat flux, Q_{le} [W m^{-2}] is the latent heat flux, Q_{g} [W m^{-2}] is the ground heat flux, ΔQ_{m} [W m^{-2}] is the melt energy at the snow surface, and Q_{s} [W m^{-2}] is the heat flux through the snow cover. Following convention (Oke, 1988), heat fluxes towards the surface are denoted as positive and heat fluxes away from the surface are denoted as negative (see Fig. 2).

Due to the blocky surface layer at the Murtèl–Corvatsch rock glacier, in this study the energy balance within a volumetric blocky surface layer was studied. This contrasts with the approach of other energy balance studies, which refer to the energy balance of a two-dimensional unit area (see, e.g., Stocker-Mittaz, 2002; Westermann et al., 2009).

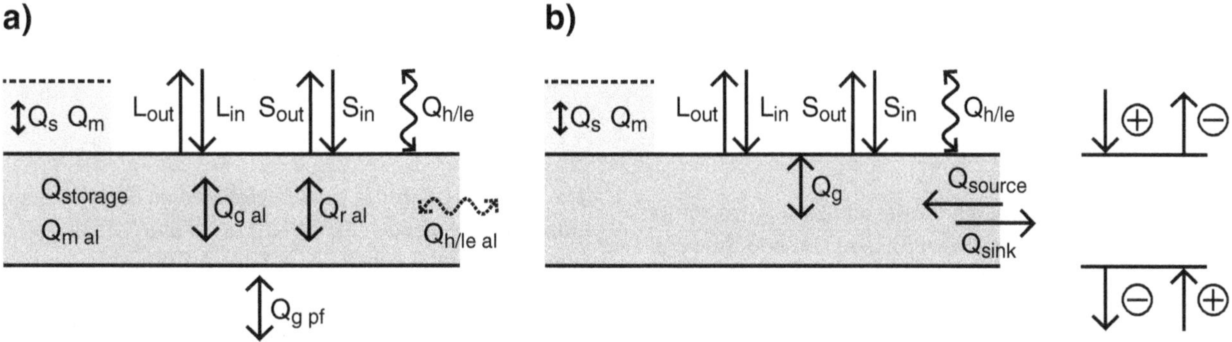

Figure 2. Illustration of the different energy exchange processes in **(a)** the energy balance calculations according to Eqs. (1–3) and **(b)** in the Coup Model. The scheme on the right-hand side shows the convention of positive and negative fluxes.

Processes within the blocky layer, added to the purely conductive ground heat flux usually applied, are convective or advective heat transfer by air flow in the voids between the blocks, net longwave radiation between adjacent blocks, melt and freezing energy within the active layer and at the permafrost table, and the heat storage change. The formulation of the energy balance term Q_g of Eq. (1) then becomes

$$Q_g = Q_{g_{al}} + Q_{g_{pf}} + Q_{le_{al}} + Q_{h_{al}} + Q_{rad_{al}} \\ -\Delta Q_{storage} - \Delta Q_{m_{al}} - \Delta Q_{m_{pf}}, \qquad (2)$$

where $Q_{g_{al}}$ [W m^{-2}] is the conductive heat flux through the blocks of the active layer, $Q_{g_{pf}}$ [W m^{-2}] is the ground heat flux through the permafrost table, $Q_{le_{al}}$ [W m^{-2}] and $Q_{h_{al}}$ [W m^{-2}] are the latent heat flux and the sensible heat flux in the voids between the blocks, $Q_{rad_{al}}$ [W m^{-2}] is the net radiative heat flux between the blocks, $\Delta Q_{storage}$ [W m^{-2}] is a source or sink term for heat energy in the blocks, and $\Delta Q_{m_{al}}$ [W m^{-2}] and $\Delta Q_{m_{pf}}$ [W m^{-2}] are the melt/freezing energy used in the active layer and at the permafrost table.

Energy balance components were calculated on an hourly time step (except melt energy, for which 24 h intervals are used) and were then averaged to monthly and seasonal values (June–September; October–January; February–May) as shown in Fig. 3. The seasonality chosen reflects the periods in which the different three-dimensional energy exchange processes in the active layer are assumed to act in the same direction (e.g., cooling processes due to air circulation in October–January, enhanced heat flux through the active layer due to longwave radiation between the blocks from June–September). In the following the individual terms of Eqs. (1) and (2) are explained in detail.

2.3.1 Radiative heat flux

Q_{rad} [W m^{-2}] is the net radiation at the surface and is calculated from direct radiation measurements at the micrometeorological station. The radiation measurements comprise incoming and outgoing short-wave and long-wave radiation:

$$Q_{rad} = L_{\downarrow} + S_{\downarrow} + L_{\uparrow} + S_{\uparrow} \qquad (3)$$

where L [W m^{-2}] denotes longwave radiation and S [W m^{-2}] denotes shortwave radiation.

The slope angle at the site was approximated by 10°, which reduces the radiation density on the surface. A further reduction in radiation density is expected due to surface roughness and shadow effects caused by the shape of the blocks. Therefore this value is corrected by a geometrical factor of 0.9. This factor is taken from a US patent 7305 983 B1, which gives insolation information on inclined roofs. This information is gained by calculating the insolation depending on roof orientation and inclination of buildings in a GIS. The reduction found by these authors ranges from 95 to 50 %. We use a value of 0.9, which represents a roof inclination of 35° to 45°, depending on the orientation of the roof.

2.3.2 Turbulent fluxes

The turbulent heat fluxes within the blocks and at the surface blocky layer are calculated following the bulk method (Oke, 1988). The sensible heat flux, Q_h, from the surface to the air is

$$Q_h = -C_a \kappa^2 z^2 \left(\frac{\Delta \overline{u}}{\Delta z} \frac{\Delta \overline{T}}{\Delta z} \right) (\Phi_M \Phi_H)^{-1}, \qquad (4)$$

where C_a [J kg^{-1} K^{-1}] is the specific heat capacity of air, κ is the von Karman constant, $\Delta \overline{u}$ [m s^{-1}] is the wind speed gradient between sensor and ground surface, Δz [m] is the height, $\Delta \overline{T}$ [K] is the temperature gradient between sensor and ground surface, Φ_H is a dimensionless stability function for heat and Φ_M is a dimensionless stability function to account for the curvature of the logarithmic wind profile due to buoyancy effects. z is the log mean height calculated after Brock et al. (2010) as

$$z = \ln\frac{\Delta h}{z_0} \times \ln\frac{\Delta h}{z_0}, \tag{5}$$

with Δh (2 m) being the height of the meteorological station and z_0 being the roughness length (0.18 m for snow-free conditions and 0.07 m for snow-covered conditions as found by Stocker-Mittaz, 2002).

The latent heat flux at the ground surface is given by

$$Q_{\mathrm{le}} = -\rho_{\mathrm{a}} L_{\mathrm{v}} \kappa^2 z^2 \left(\frac{\Delta\overline{u}}{\Delta z}\frac{\Delta\overline{\rho_{\mathrm{v}}}}{\Delta z}\right)(\Phi_M \Phi_V)^{-1}, \tag{6}$$

where ρ_{a} [kg m^3] is the density of air, L_{v} [kJ kg^{-1}] is the latent heat of evaporation, $\Delta\overline{\rho_{\mathrm{v}}}$ [kg kg^{-1}] is the gradient in specific humidity between the ground surface and the humidity sensor at 2 m height, and Φ_V is a dimensionless stability function for vapor.

The stability functions in Eqs. (4) and (6) are calculated as:

in the stable case (R_i positive)

$$(\Phi_M \Phi_x)^{-1} = (1 - 5R_i)^2 \tag{7}$$

in the unstable case (R_i negative)

$$(\Phi_M \Phi_x)^{-1} = (1 - 16R_i)^{3/4}, \tag{8}$$

where Φ_x is the respective stability function (Φ_H or Φ_V), R_i is the bulk Richardson number for categorizing atmospheric stability and the state of turbulence in the lower atmosphere calculated as

$$R_i = \frac{g}{\overline{T}}\frac{(\Delta\overline{T}/\Delta z)}{(\Delta\overline{u}/\Delta z)^2}, \tag{9}$$

where g [ms^{-2}] is the acceleration due to gravity and $\overline{T}$ [K] is the mean temperature in the Δz [m] layer.

2.3.3 Melt energy

The melt energy at the surface of the snow cover is calculated according to the difference in snow height in 24 h intervals, as measured by an ultrasonic sensor at the micrometeorological station (see Table 1). The threshold temperature for snowmelt is set to −3 °C; below this temperature no snowmelt is calculated even if a decrease in snow height is measured. In addition, a measurement error is expected if the snow height decreases by more than 0.2 m in a 24 h interval. In this case, no snowmelt is calculated. Snow density has not been measured. Instead, a constant value of 300 kg m^{-3} was chosen according to Keller (1994), who found snow densities ranging from 250–400 kg m^{-3} at the same site. The melt energy is thus given as

$$\Delta Q_{\mathrm{m}} = \frac{\Delta h \rho_{\mathrm{s}} L_{\mathrm{f}}}{\Delta t}, \tag{10}$$

where Δh [m] is the difference in snow height, ρ_{s} [kg m^3] is the density of snow, L_{f} [kJ kg^{-1}] is the specific latent heat of fusion of water and Δt [s] is the time interval.

2.3.4 Ground heat flux

The ground heat flux is calculated based on borehole temperature measurements at 0.55, 1.55, 2.55 and 3.55 m depth assuming a thermal conductivity k_r of 2.5 Wm^{-2} in the Fourier heat conduction equation (see Eq. 11), which is considered to be appropriate for the solid metamorphic rocks found within the blocky layer. The values were then multiplied by a correction factor of 0.6 to account for the reduction in conductive heat fluxes within the air-filled pores between the blocks, corresponding to a porosity of 40 % in the active layer. In contrast to the Coup Model simulations (see Sect. 1.1), changes in thermal conductivity due to water and ice content as well as latent heat processes are not accounted for here. Considering the low water retention capacity of the voids between the blocks, these parameters are supposed to be of minor importance for the thermal conductivity.

$$Q_g = 0.6 k_r \sum_{i=1}^{3} \frac{\Delta T_i}{\Delta z_i}, \tag{11}$$

where Δz_i are the respective layer thicknesses (here, 1 m). The heat flux within the permafrost layer, $Q_{g_{\mathrm{pf}}}$, is calculated likewise using the thermal conductivity of ice and temperatures measured at 3.55 and 4.55 m depth during the winter period (October–January) and with a 3.55 m temperature fixed at 0 °C during the spring and summer period (February–September). This is an assumption based on the concept that the lower boundary of the respective layer represents the permafrost table where thawing processes are supposed to keep the temperature at 0 °C during the summer period.

2.3.5 Net radiation between adjacent blocks

Due to the studied volumetric layer, a term accounting for radiative processes between the blocks due to temperature differences has been included in the energy balance (Eq. 2) (Kunii and Smith, 1960). The temperature gradient from the surface to the permafrost table leads to immediate heat flow from the warmer upper blocks to colder lower blocks in the active layer. This process is based on the emission and the absorption of thermal/longwave radiation between adjacent blocks of different temperature. The net flow of longwave radiation between two blocks with surface temperatures T_1 and T_2, which in this case are approximated by parallel plates, is given as

$$q_{net} = \varepsilon_{\mathrm{eff}} \sigma (T_1^4 - T_2^4), \tag{12}$$

where q_{net} is the net longwave radiation, ε is the emmisivity of the block (0.96) as determined by surface temperature and longwave radiation measurements, T [K] is the absolute temperature, and σ [W m^{-2} K^{-4}] is the Stefan–Boltzmann constant.

In the case of two opposite blocks with an emissivity of $\varepsilon < 1$, the reflection of radiation has to be considered following McAdams (1954):

$$\varepsilon_{1,2} = \frac{1}{\frac{1}{\varepsilon_1} + \frac{1}{\varepsilon_2} - 1} = \frac{\varepsilon_1 \varepsilon_2}{\varepsilon_1 + \varepsilon_2 - \varepsilon_1 \varepsilon_2}. \tag{13}$$

In the case where ε_1 equals ε_2, which is assumed for the blocks of the rock glacier, Eq. (13) becomes

$$\varepsilon_{\text{eff}} = \frac{\varepsilon^2}{2\varepsilon - \varepsilon^2}. \tag{14}$$

The calculation is based on the borehole temperatures at 0.55, 1.55, 2.55, and 3.55 m. Errors might arise from too high gradients due to measurement depth intervals of 1 m. Voids between individual blocks are assumed to be no larger than 0.33 m on average. The reduction in radiative heat flux between the blocks by a factor of 3 was chosen because of the temperature gradient within separate blocks, i.e., the block has a different temperature at its surface than what is measured by the thermistor within the block. Given a linear temperature gradient and equally spaced parallel plates, reduction by a factor of 1/3 results.

2.3.6 Snow heat flux

The snow heat flux is the heat flux within the snow layer. It is calculated following the Fourier heat conduction equation

$$Q_{\text{s}} = k_{\text{s}} \frac{T_{\text{s}} - T_{0.55}}{0.55 + h_{\text{s}}}, \tag{15}$$

where T_{s} [°C] is the snow surface temperature, $T_{0.55}$ [°C] is the temperature of the sensor at 0.55 m depth, h_{s} is the snow height and the thermal conductivity of snow k_{s} [W m^{-1} K^{-1}] is calculated following Devaux (1933) (see also Keller, 1994; Stocker-Mittaz, 2002):

$$k_{\text{s}} = 2.93 \left(\frac{\rho_{\text{s}}^2}{1\,000\,000} + 0.1 \right), \tag{16}$$

where ρ_{s} (300 kg m^{-3}) is the density of snow.

2.3.7 Energy storage

When looking at the energy balance of a volumetric layer the storage of energy has to be accounted for. The heating of blocks during the summer period will produce an energy sink term in the energy balance equation. The release of heat due to cooling of the blocks acts as an energy source in the balance equation (see Eq. 2). The storage change term is calculated as

$$\Delta Q_{\text{storage}} = \frac{c_{\text{r}} \Delta T m_{\text{r}}}{\Delta t}, \tag{17}$$

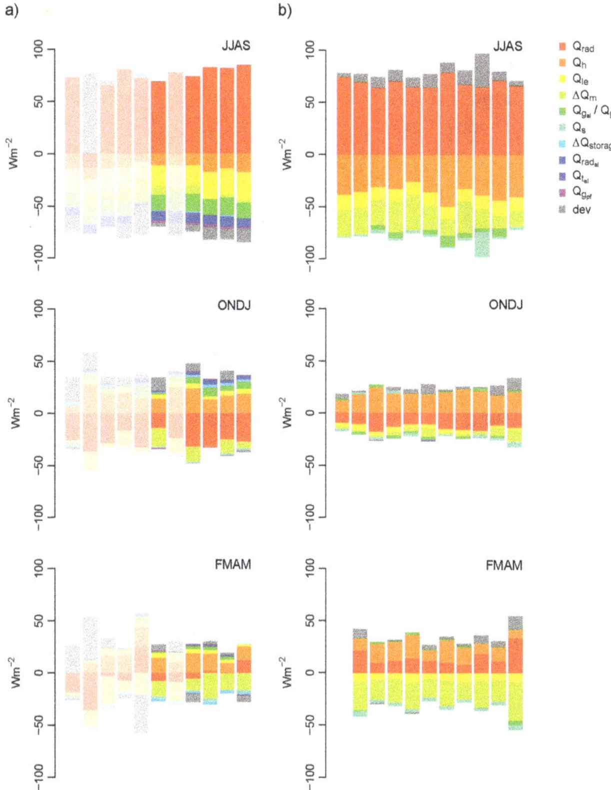

Figure 3. Seasonal energy balance components at the Murtèl–Corvatsch rock glacier. **(a)** shows the observations and **(b)** shows the modeled components. Faded colors indicate years with incomplete measurements. Q_{rad}: net radiation; Q_{h}: sensible heat flux at the surface; Q_{le}: latent heat flux at the surface; ΔQ_{m}: melt energy in the snow cover; Q_{s}: conductive heat flux through the snow; $Q_{\text{g}_{\text{al}}}$ / Q_{g}: conductive heat flux within the active layer **(a)**/ through the surface **(b)**; $Q_{\text{rad}_{\text{al}}}$: net radiation in the active layer; $\Delta Q_{\text{storage}}$: change in heat content in the active layer; $Q_{\text{t}_{\text{al}}}$: turbulent heat flux in the active layer; $Q_{\text{g}_{\text{pf}}}$: conductive heat flux through the permafrost table; dev: energy balance closure. The seasons are divided into: June–September (JJAS), October–January (ONDJ) and February–May (FMAM).

where c_{r} [J kg^{-1} K^{-1}] is the specific heat capacity of rock, ΔT [K] is the daily temperature difference and m_{r} [kg] is the rock mass. The porosity of the blocky layer is assumed to be 40 %.

Results

2.4 Energy balance

Figure 3 and tables 2, 3, and 4 show the seasonal energy balances at the study site from 1997 to 2007. Figure 3a shows the measured energy balance components, Fig. 3b shows the modeled energy balance components. In the measured energy balance, the following criteria were used to select seasons to be excluded in the energy balance calculations: (1) seasons with too many missing values overall (> 30 %), (2) seasons missing important variables (i.e., sur-

Table 2. Seasonal (June–September) averages of the energy balance components with deviations (in Wm^{-2}) and corresponding ice thickness equivalents (in m). Years marked with an asterik were not considered for the calculation of the average and the standard deviation. Q_{rad}: Net radiation; Q_h: Sensible heat flux at the surface; Q_{le}: Latent heat flux at the surface; ΔQ_m: Melt energy in the snow cover; Q_s: Conductive heat flux through the snow; $Q_{g_{al}}$: Conductive heat flux within the active layer; $Q_{rad_{al}}$: Net radiation in the active layer; $\Delta Q_{storage}$: Change in heat content in the active layer; $Q_{t_{al}}$: Turbulent heat flux in the active layer; $Q_{g_{pf}}$: Conductive heat flux through the permafrost table; dev: Energy balance closure; ice: Ice melt equivalent of the energy balance closure.

Year	Q_{rad}	Q_h	Q_{le}	ΔQ_m	Q_s	$Q_{g_{al}}$	$Q_{rad_{al}}$	$\Delta Q_{storage}$	$Q_{t_{al}}$	$Q_{g_{pf}}$	dev	ice
1997*	73.7	−15.2	−22.2	–	–	−12.6	−6.9	−1.5	−0.5	−1.1	−13.7	−0.47
1998*	–	−11.7	−12.5	−15.0	0.0	−14.4	−7.9	−0.3	−0.5	−1.2	63.6	2.19
1999*	65.9	−14.1	−22.6	−5.5	0.0	−15.4	−8.5	−2.3	−0.5	−1.6	4.6	0.16
2000*	81.1	−13.8	−20.8	−11.6	0.0	−12.8	−7.1	0.0	−0.5	−1.3	−13.0	−0.45
2001*	72.8	−7.9	−16.0	−21.8	0.2	−1.8	−1.0	4.5	−0.5	−0.6	−28.0	−0.96
2002	69.4	−11.2	−19.5	−8.7	0.1	−15.0	−8.2	−0.8	−0.5	−1.8	−3.9	−0.13
2003*	77.8	−11.3	−20.5	−2.7	0.0	−20.4	−11.2	−0.3	−0.5	−1.0	−10.1	−0.35
2004	74.1	−11.0	−18.8	−8.6	0.0	−16.7	−9.2	−1.3	−0.5	−1.3	−6.6	−0.23
2005	82.4	−17.7	−25.8	–	–	−15.1	−8.3	−0.3	−0.5	−1.7	−13.1	−0.45
2006	81.9	−14.3	−24.3	−3.3	–	−17.0	−9.3	−1.4	−0.5	−1.8	−10.1	−0.35
2007	84.9	−17.6	−26.5	−2.8	0.0	−14.8	−8.1	−0.1	−0.5	−1.1	−13.5	−0.46
Average	78.5	−14.3	−23.0	−5.9	0.0	−15.7	−8.6	−0.8	−0.5	−1.4	−9.4	−0.32
Stdev	5.8	2.9	3.2	2.8	0.1	0.9	0.5	0.5	0.0	0.3	3.7	0.13

face temperature T_{ir} or long- and shortwave radiation), and (3) complete years with two missing seasons following the above criteria. The long-term seasonal average energy balance is shown in Fig. 4. Measured and simulated energy balances differ significantly in most of the terms. Radiative heat flux at the surface is smaller in the model than in the measurements in the summer and winter seasons, whereas sensible heat flux is larger in the respective seasons. Latent heat is larger by a factor of 2 in the measurements compared to the simulation. In the model latent heat is always negative, i.e., flowing away from the surface. Melt energy is larger in the model during summer and equal from February to May.

Both measured and simulated energy balances have deviation terms to close the energy balance to zero. The deviations may arise from various sources that can differ between model and measurements; see also the corresponding Sect. 3.1.9 in the Discussion.

Seasonal deviation terms of the measurements (only complete measurement years considered) range from 16.9 Wm^{-2} in October–February (Table 3) to −13.3 Wm^{-2} in February–May (Table 4). Deviation terms in the model range from 31.9 Wm^{-2} in June–September to −1.0 Wm^{-2} in February–May (see Fig. 3). The sum of the average seasonal deviations of the measurements (see tables 2, 3, and 4) is equal to a net melt/refreezing rate of −0.10 m a^{-1} of ice in the blocky layer. This is comparable to the net melt rate of −0.05 m a^{-1} found by Kääb et al. (1998) for the same site.

2.5 Simulation of the thermal regime

Figure 6 shows the measured and the simulated temperatures at two depths for the simulated period. The green lines in Fig. 6 show the results for the simulation with only meteorological measurement input and the red lines the results with measured meteorological input as well as an additional seasonal heat source and heat sink in the active layer representing advective and radiative heat fluxes. It can be seen that in the case where no additional heat source or sink is active, thermal conditions do not favor the development of permafrost if local meteorological data is used to drive the model. Temperatures are well above 0 °C in summer down to 11 m below the surface, indicating that permafrost is not present in this simulation. In the other case with an additional heat source and sink, permafrost is present at the respective depths. The values found by experimental calibration are about 28.9 Wm^{-2} for the heat sink during the period from mid September to mid January. The heat source in the period from June to mid September amounts to 26 Wm^{-2}.

3 Discussion

3.1 Uncertainties in the energy balance measurements

Regarding the energy balance measurements, there are some general points that need to be addressed. First, the categorization of seasons may be based on prevailing meteorological conditions, processes in the active layer or a combination of both (Westermann et al., 2009; Langer et al., 2011; Schneider et al., 2012). Here, three seasons, approximately based on the heat source and heat sink seasons in the model, were differentiated. This may lead to problems in so far as processes may occur in multiple seasons to different proportions depending on meteorological conditions on the one hand and may counteract each other on the other hand. Thus, an interpretation

Table 3. Seasonal (October–January) averages of the energy balance components with deviations (in Wm^{-2}) and corresponding ice thickness equivalents (in m). Years marked with an asterik were not considered for the calculation of the average and the standard deviation. Q_{rad}: Net radiation; Q_h: Sensible heat flux at the surface; Q_{le}: Latent heat flux at the surface; ΔQ_m: Melt energy in the snow cover; Q_s: Conductive heat flux through the snow; $Q_{g_{al}}$: Conductive heat flux within the active layer; $Q_{rad_{al}}$: Net radiation in the active layer; $\Delta Q_{storage}$: Change in heat content in the active layer; $Q_{t_{al}}$: Turbulent heat flux in the active layer; $Q_{g_{pf}}$: Conductive heat flux through the permafrost table; dev: Energy balance closure; ice: Ice melt equivalent of the energy balance closure.

Year	Q_{rad}	Q_h	Q_{le}	ΔQ_m	Q_s	$Q_{g_{al}}$	$Q_{rad_{al}}$	$\Delta Q_{storage}$	$Q_{t_{al}}$	$Q_{g_{pf}}$	dev	ice
1997*	−25.9	6.6	0.6	−1.3	−1.6	−3.3	−1.9	3.8	–	−0.6	23.6	0.82
1998*	−37.0	26.8	7.6	−17.7	−4.2	3.7	2.2	1.7	–	−0.1	16.7	0.58
1999*	−29.0	18.9	4.4	−4.1	−2.2	1.5	1.1	2.1	–	−0.4	7.7	0.27
2000*	−16.9	19.2	6.1	−15.2	−1.5	1.4	1.1	1.1	–	−0.6	5.2	0.18
2001*	−33.0	14.1	3.4	−4.3	0.0	8.9	5.0	2.8	–	0.7	2.4	0.08
2002	−14.2	14.0	4.1	−18.2	−0.8	1.3	1.1	0.9	–	−0.9	12.7	0.44
2003*	−24.0	25.0	6.7	−14.1	−2.7	4.1	2.6	2.2	–	−0.1	0.2	0.01
2004	−32.1	23.5	5.4	−14.3	−1.6	5.4	3.2	3.0	–	0.1	7.5	0.26
2005	−32.1	13.3	3.3	−0.3	–	7.8	5.0	3.1	–	0.1	−0.3	−0.01
2006	−24.8	16.7	5.2	−13.6	−1.8	4.3	2.8	2.9	–	−0.5	8.9	0.31
2007	−26.9	18.8	4.8	−6.6	−1.6	6.5	4.0	2.4	–	0.2	−1.5	−0.05
Average	−26.0	17.2	4.5	−10.6	−1.5	5.1	3.2	2.4	–	−0.2	5.5	0.19
Stdev	6.6	3.7	0.7	6.4	0.4	2.2	1.3	0.8	–	0.4	5.5	0.23

of typical processes within a season is difficult. Nevertheless, some characteristics in the magnitude and the direction of individual energy balance components are obvious. Also, the deviations show seasonal similarities and may even be interpreted as freezing and thawing processes due to their directions. Finally, it also has to be considered that data gaps, random measurement errors and parametrizations may have a significant influence on the results presented herein.

In the following, potential sources of error calculation of the individual energy balance components are discussed in detail.

3.1.1 Net radiation

The factor by which measured incoming radiation is reduced is cos 10° and a correction factor of 0.9, which is assumed to account for both slope and surface geometry. As radiative heat flux at the surface is a very important term in the energy balance, small errors in the geometrical correction factor may lead to uncertainties. A further source of error is a possible underestimation of radiation density during the snow-covered season due to a snow cover on the upward looking sensor.

3.1.2 Turbulent fluxes at the surface

Turbulent fluxes, as calculated following Eqs. (4) and (6), are strongly dependent on wind speed, which is generally very low at the site. In the model an enhancing parameter is used that avoids effects of extreme stable stratification during periods of low wind speeds. This may lead to an overestimation of turbulent fluxes in the model. Furthermore, saturated conditions at the ground surface were assumed for the calculation of the latent heat flux, which is probably a reasonable choice for the depressions between the blocks of a rock glacier, but may lead to an overestimation for the dry conditions at the top of the blocks. Eddy covariance measurements, which were not available at the study site, would certainly improve the calculations.

3.1.3 Melt energy

The calculation of melt energy based solely on snow height measurements and assuming a constant snow density, as it was done in this study, may lead to errors. Snow density will certainly vary over the winter period in nature, reaching a peak in spring with the beginning of snowmelt. So, with the assumption of a constant snow density, melt rates will be overestimated in the beginning of winter, when snow is less dense, and underestimated in late winter and spring when snow is probably denser than 300 kg m^{-3}. Refreezing of melted snow within the snow cover will lead to more melt than would be expected by measurements of the difference in snow heights. Settlement of snow might be mistaken for melt when occurring above the threshold temperature of −3 °C. During a 24 h period snowmelt and snowfall may occur (snowfall in the morning, snowmelt in the afternoon), which is not considered in the calculations. This situation is most likely to occur in the melting period from April to July and during the summer season. Thus the values calculated for the respective period are likely to be too low. It has to be considered that snow density estimation above permafrost is complicated, because of low ground temperatures that lead to a different snow densification pattern in spring than would

Table 4. Seasonal (February–May) averages of the energy balance components with deviations (in Wm^{-2}) and corresponding ice thickness equivalents (in m). Years marked with an asterik were not considered for the calculation of the average and the standard deviation. Q_{rad}: Net radiation; Q_h: Sensible heat flux at the surface; Q_{le}: Latent heat flux at the surface; ΔQ_m: Melt energy in the snow cover; Q_s: Conductive heat flux through the snow; $Q_{g_{al}}$: Conductive heat flux within the active layer; $Q_{rad_{al}}$: Net radiation in the active layer; $\Delta Q_{storage}$: Change in heat content in the active layer; $Q_{t_{al}}$: Turbulent heat flux in the active layer; $Q_{g_{pf}}$: Conductive heat flux through the permafrost table; dev: Energy balance closure; ice: Ice melt equivalent of the energy balance closure.

Year	Q_{rad}	Q_h	Q_{le}	ΔQ_m	Q_s	$Q_{g_{al}}$	$Q_{rad_{al}}$	$\Delta Q_{storage}$	$Q_{t_{al}}$	$Q_{g_{pf}}$	dev	ice
1997*	−18.4	−1.2	−1.5	–	–	−3.2	−1.7	–	–	0.1	25.9	0.88
1998*	−36.1	8.8	2.0	−14.9	−1.8	1.9	1.0	−1.0	–	0.2	40.1	1.36
1999*	−3.4	16.5	6.0	−26.4	−2.3	2.0	1.0	−1.8	–	0.5	7.8	0.26
2000*	−7.3	16.7	5.4	−10.7	−2.1	1.4	0.8	−0.4	–	−0.1	−3.7	−0.12
2001*	25.7	18.7	6.6	−20.7	−1.2	3.5	2.0	0.6	–	0.3	−35.3	−1.20
2002	−7.8	14.3	4.9	−14.7	−2.4	1.5	0.5	−2.3	–	0.9	5.1	0.17
2003*	−9.0	10.2	7.7	−18.1	−1.1	2.2	1.3	−2.1	–	0.0	8.9	0.30
2004	−5.5	18.6	4.7	−11.0	−2.1	2.8	1.5	−1.2	–	0.5	−8.2	−0.28
2005	2.1	16.2	4.2	−24.7	−2.9	2.9	1.4	−2.6	–	0.6	2.7	0.09
2006	−0.3	9.5	3.0	−16.4	–	3.1	1.6	−2.6	–	0.4	1.9	0.06
2007	12.2	13.4	1.8	−16.7	−2.2	0.2	−0.2	−2.0	–	−0.3	−6.3	−0.21
Average	0.1	14.4	3.7	−16.7	−2.4	2.1	1.0	2.1	–	0.3	−0.9	−0.03
Stdev	7.0	3.0	1.2	4.5	0.3	1.1	0.7	0.5	–	0.3	5.3	0.18

be expected for non-permafrost sites. Keller (1994) showed that snow with lower densities than the one used herein can be found above mountain permafrost. Thus the value chosen is considered a good approximation for the average density over the entire snow-covered period.

3.1.4 Ground heat flux

The strong influence of the thermal conductivity on the conductive heat flux may lead to significant uncertainties. Further uncertainties are added due to the unknown porosity (fraction of blocks to air-filled voids) of the blocky layer, the position of the thermistors in the borehole and the reduction factor chosen in this study.

3.1.5 Net radiation between adjacent blocks

The calculation of the thermal radiation between adjacent blocks in this study is based on the assumption of three equally large quadratic blocks with an area 1 m^2 parallel stacked with a spacing of 0.33 m. The surface temperatures are assumed to be equal to the temperatures measured at 0.55 m, 1.55, 2.55, and 3.55 m. The reduction by a factor of 1/3, as described in the Methods section, accounts for this rough estimation, which may lead to significant uncertainties, especially in seasons with large thermal gradients.

3.1.6 Snow heat flux

Using the temperature at 0.55 m depth instead of the ground surface temperature in Eq. (15) may lead to errors, which can be considered as small due to nearly isothermal conditions in the active layer during the snow-covered period.

3.1.7 Energy storage

Errors in the calculation of the energy storage change may be due to assumption of the rock mass, varying rock density and heat capacity as well as borehole temperature measurements.

3.1.8 Turbulent fluxes between blocks

Measurements for the calculations were only available for the time period of mid June to mid July 2006, thus values available for the respective period have been taken for the complete summer season of June to September. Values for the other two seasons were not available and are missing in the energy balance. This may produce significant errors in the energy balance in the fall period, where these processes have been shown to be large due to the advection of cold air (Panz, 2006).

3.1.9 Energy balance closure

In the model the energy balance is not supposed to be closed due to convective heat transfer by precipitation and snow as well as surface runoff. In the measurements there are other sources of error. Besides the effects of radiative, convective and advective heat transfer, which are the subject of this study and are thus expected to cause deviations, there are other sources of error, such as direct measurement errors at the meteorological station, i.e., icing and snow at radiation sensors (see Table 1). Finally, unmeasured freezing and thawing processes within the blocky surface layer can add significant uncertainties to the energy balance.

When summing up all energy balance components following Eq. (1) and considering the energy exchange processes

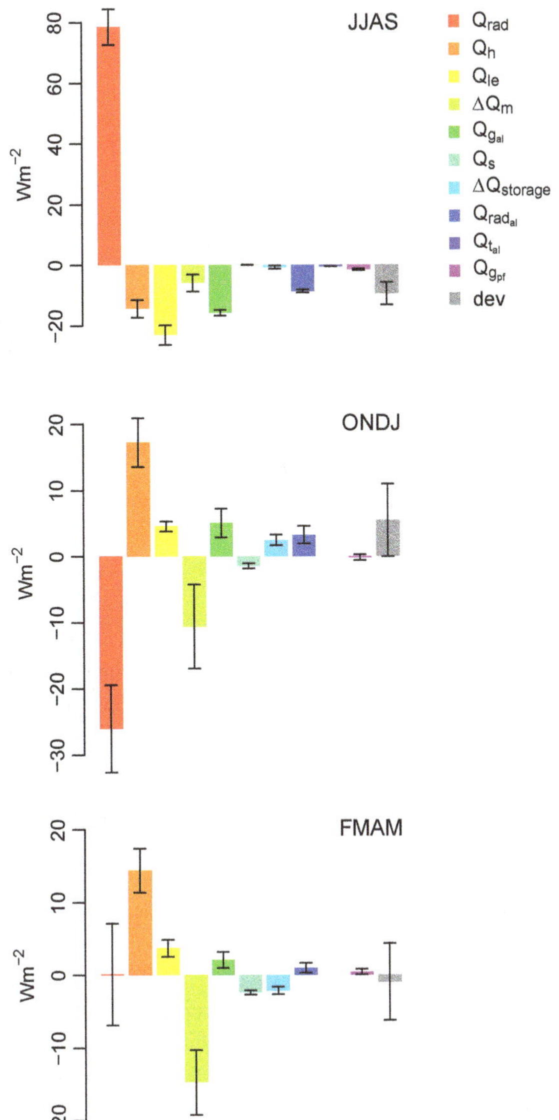

Figure 4. 5-year averages (with corresponding standard deviations) of the seasonal energy balance components at the Murtèl–Corvatsch rock glacier. Q_{rad}: net radiation; Q_{h}: sensible heat flux at the surface; Q_{le}: latent heat flux at the surface; ΔQ_{m}: melt energy in the snow cover; Q_{s}: conductive heat flux through the snow; $Q_{\mathrm{g_{al}}}$: conductive heat flux within the active layer; $Q_{\mathrm{rad_{al}}}$: net radiation in the active layer; $\Delta Q_{\mathrm{storage}}$: change in heat content in the active layer; $Q_{\mathrm{t_{al}}}$: turbulent heat flux in the active layer; $Q_{\mathrm{g_{pf}}}$: conductive heat flux through the permafrost table; dev: energy balance closure. The seasons are divided into: June–September (JJAS), October–January (ONDJ) and February–May (FMAM).

in the blocky layer following Eq. (2) we assume that the result should be zero. As the two terms $\Delta Q_{\mathrm{m_{al}}}$ and $\Delta Q_{\mathrm{m_{pf}}}$ in Eq. (2) as well as the magnitude of the lateral turbulent fluxes in the active layer are unknown, Eq. (1) will have a deviation term to a zero energy balance. Random measurement errors and uncertainties in the parametrization of the energy balance component calculations will also add up to the total deviation term. Assuming that the random measurement errors and the parametrization uncertainties will even out over the studied time period, the deviation term could be interpreted as an indirect measure of the magnitude of the unmeasured processes in the blocky surface layer. If it is hypothesized that the missing processes are mainly associated with freezing and thawing of water in the active layer ($\Delta Q_{\mathrm{m_{al}}}$) and at the permafrost table ($\Delta Q_{\mathrm{m_{pf}}}$), then the sum of the deviation terms would be an indirect measure of net melt or net refreezing rates. This result can then be compared to the net melt rate found in a study by Kääb et al. (1998) based on geodetic measurements as well as calculations of the vertical deformation of the rock glacier mass over the period from 1987 to 1996. The value −0.05 m per year found by Kääb et al. (1998) is smaller than the one of −0.17 m found in this study. This discrepancy may be due to measurement and parametrization errors as well as missing lateral fluxes on the one hand and/or a real increase in the net melt rate caused by a warmer climate in recent years on the other. The IPCC (IPCC, 2013) reports an average global surplus in anthropogenic radiative forcing of 2.29 $\mathrm{W m^{-2}}$ over the industrial era, which would correspond to a net melt rate of 0.24 m per year.

3.2 Model

The values found for the heat source and sink layer by calibrating the model to match observed borehole temperatures have to be treated with care because the source/sink layer is 1 m thick and placed close to the surface. This means that the heat extracted or added is transferred to larger depths (i.e., the depths shown in Fig. 6) by conduction and percolating water. This transfer will not act immediately on the temperatures at depth, but will take some time. In nature however the heat transfer by thermal radiation as well as convection and advection of air in the voids between the blocks may act more directly on the thermal regime in the permafrost below. Because of this phase shift in heat transfer it is possible that the timing of the heat source/sink in the layer, located near the surface, is not identical to the timing of the processes in nature. Furthermore it is not likely that the three-dimensional heat transfer within the blocky layer will be constant over periods of four consecutive months, as it is assumed for the parametrization of the model. The approach chosen to calibrate a process-based soil model in this study differs from similar model studies on sites with coarse debris cover (e.g., Gruber and Hoelzle, 2008). The presented solution with a heat source and sink layer is considered to be useful as an additional instrument for both the calibration of the model and for an approximate quantification of three-dimensional heat transfer within the active layer.

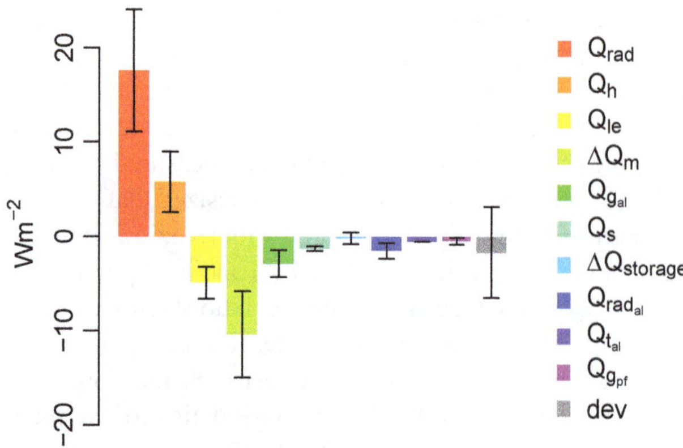

Figure 5. 5-year averages over all three seasons (with corresponding standard deviations) of the energy balance components at the Murtèl–Corvatsch rock glacier. Q_{rad}: net radiation; Q_h: sensible heat flux at the surface; Q_{le}: latent heat flux at the surface; ΔQ_m: melt energy in the snow cover; Q_s: conductive heat flux through the snow; $Q_{g_{al}}$: conductive heat flux within the active layer; $Q_{rad_{al}}$: net radiation in the active layer; $\Delta Q_{storage}$: change in heat content in the active layer; $Q_{t_{al}}$: turbulent heat flux in the active layer; $Q_{g_{pf}}$: conductive heat flux through the permafrost table; dev: energy balance closure.

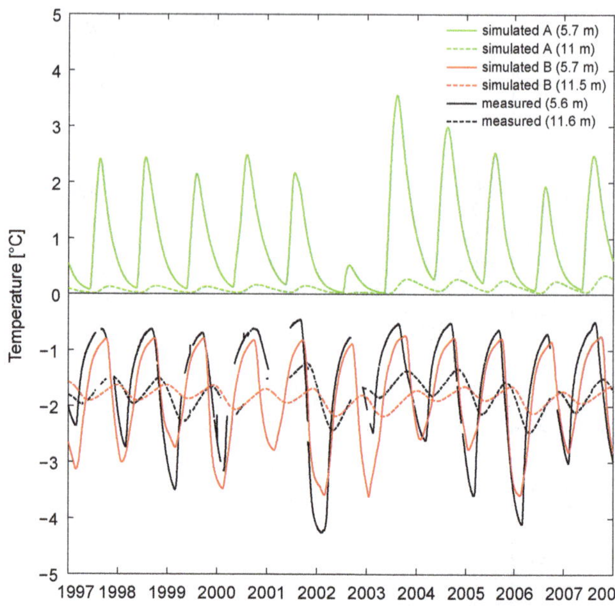

Figure 6. Simulated and measured temperatures at the Murtèl–Corvatsch rock glacier showing the calibration without (simulated A) and with thee source/sink term (simulated B) at depths of 5.6 and 11.5 m below the surface, within the permafrost layer.

3.3 Synopsis

Besides energy balance studies as presented herein, the indirect approach for the quantification of three-dimensional heat transfer by air circulation and longwave radiation by applying a heat source and sink layer in a permafrost model can serve as an additional instrument in the investigation of such processes. The direct comparison of model parameters for the heat sink and source with the deviations found in energy balance measurements is difficult to interpret because of possible measurement errors on the one hand and process simplification in the model on the other. The comparison of the measured and the simulated energy balance reveals large differences for some of the components, especially the sensible heat flux during the summer season. This can be attributed to the different reference units (unit area versus volumetric surface layer) in the model and the measurements and to a correction parameter in the Coup Model, which enhances turbulent exchange during periods with low wind speeds. Measured energy fluxes are studied in a volumetric surface layer that includes processes such as radiative heat transfer between blocks in the nature. As such processes are not integrated in the model; they are likely to be compensated by other heat transfer mechanisms. A surplus of energy at the surface in the model will not completely be transferred to the ground, but will rather be emitted to the atmosphere by turbulent fluxes and longwave radiation. This difference will be most significant in the summer season, as typically measured wind speeds tend to be low during this period. In the model turbulent fluxes are enhanced by a correction parameter during such conditions.

However, similarities can be found in the direction of the heat flow processes found in the energy balance measurements, i.e., the radiative heat transfer between the blocks of the active layer and the deviation term, and the parametrization of the heat source/sink layer. The value found for the heat source found by the calibration was 26 Wm^{-2}, and the respective energy flow in the measurements (radiative heat flux and deviation) is −18.1 Wm^{-2}, which means that energy used to melt ice in the active layer in the measurements and in the model energy has to be added to the system to account for this additional melt. During the heat sink period in the model, 28.9 Wm^{-2} are extracted from the system, which corresponds to 8.7 Wm^{-2} surplus in the measurements. The differences between the amount of energy in the two approaches could be explained by an excess of heat flow from the surface during the summer season in the model and by missing lateral energy fluxes in the measurements during summer and winter.

4 Conclusions

In this study we applied a numerical soil model integrating freezing and thawing processes and a dynamical snow cover to simulate a 13-year period of the active layer and the permafrost at the Murtèl–Corvatsch rock glacier in the Upper Engadine, Swiss Alps. Other than considering the blocky layer with voids as a thermal semi-conductor, a different approach is presented, which integrates a heat source and sink into the active layer. A measured energy balance over a volumetric surface layer including terms for radiative heat transfer between adjacent blocks, turbulent fluxes in the active layer and energy storage change is presented and compared to the modeled energy balance at the site. The deviations in the measured energy balance were used indirectly to quantify unmeasured latent heat processes involving freezing and thawing processes in the active layer and at the permafrost table.

- The approach chosen to calibrate a process-based soil model differs from similar model studies on sites with coarse debris cover. The presented solution with a heat source and sink layer is considered to be useful for both the calibration of the model and the approximate quantification of three-dimensional heat transfer processes within the active layer, which so far cannot be modeled explicitly.

- The unmeasured heat transfer processes within the blocky active layer could be approximated in the model to act as a heat sink of 28.9 $\mathrm{Wm^{-2}}$ during a period from mid September to mid January and as a heat source in the period from June to mid September of 26 $\mathrm{Wm^{-2}}$.

- Measured and simulated energy balances differ significantly. The differences can partly be attributed to the different reference units (i.e., a unit area in the model and a volumetric surface layer in the measurements) and thus different energy exchange processes.

- The integration of additional energy balance components reduced the deviations to a zero energy balance significantly compared to earlier studies at the same site (Mittaz et al., 2002).

- The remaining deviations in the measured energy balance are hypothesized to be due to latent heat effects (i.e., freezing and thawing) in the active layer and at the permafrost table. This hypothesis is supported by the results of an earlier study by Kääb et al. (1998), which showed a net melt rate on the same order of magnitude at the same site.

In future studies the emphasis should be on both, more detailed measurements of energy balance components including the blocky layer (i.e., heat transfer by longwave radiation and turbulent fluxes due to convective and advective air circulation), and more physically based modeling of the heat source and sink terms, i.e., the representation of radiation coupled with thermal gradients in the active layer and air flow coupled to meteorological conditions.

Acknowledgements. This study was part of the German "Sensitivity of mountain Permafrost to Climate Change" Bündel project (SPCC, http://www.spcc-project.de/) funded by the German National Science Foundation (DFG, HA3475/3-1)), and the Swiss National Science Foundation "The evolution of mountain permafrost in Switzerland" project (TEMPS, SNF, CRSII2_136279). We would like to thank the Swiss Permafrost Monitoring Network (PERMOS) for providing the measurement data. We also thank Corvatsch Bergbahn AG for transport and accommodation.

Edited by: M. Koppes

References

Arenson, L., Hoelzle, M., and Springman, S.: Borehole deformation measurements and internal structure of some rock glaciers in Switzerland, Permafrost Periglac., 13, 117–135, 2002.

Arenson, L., Hauck, C., Hilbich, C., Seward, L., Yamamoto, Y., and Springman, S.: Sub-surface heterogeneities in the Murtèl-Corvatsch rock glacier, Switzerland, in: Proceedings of the 6th Canadian Permafrost Conference, 1494–1500, 2010.

Balch, E. S.: Glaciers or freezing caverns, Allen, Lane and Scott, Philadelphia, PA, 1900.

Bayard, D., Staehli, M., Parriaux, A., and Fluehler, H.: The influence of seasonally frozen soil on the snowmelt runoff at two alpine sites in southern Switzerland, J. Hydrol., 309, 66–84, 2005.

Binxiang, S., Lijun, Y., and Xuezu, X.: Onset and evaluation on winter-time natural convection cooling effectiveness of crushed-rock highway embankment, Cold Reg. Sci. Technol., 48, 218–231, 2007.

Brock, B. W., Mihalcea, C., Kirkbride, M. P., Diolaiuti, G., Cutler, M. E. J., Smiraglia, C.: Meteorology and surface energy fluxes in the 2005–2007 ablation seasons at the Miage debris-covered glacier, Mont Blanc Massif, Italian Alps, J. Geophys. Res.: Atmospheres, 115, D9, doi:10.1029/2009JD013224, 2010.

Cheng, G. D., Lai, Y. M., San, Z. Z., and Jiang, F.: The "thermal semi-conductor" effect of crushed rocks, Permafrost Periglac., 18, 151–160, 2007.

Delaloye, R. and Lambiel, C.: Evidence of winter ascending air circulation throughout talus slopes and rock glaciers situated in the lower belt of alpine discontinuous permafrost (Swiss Alps), Norsk Geogr. Tidssk., 59, 194–203, 2005.

Delaloye, R., Reynard, E., Lambiel, C., Marescot, L., and Monnet, R.: Thermal anomaly in a cold scree slope (Creux du Van, Switzerland), in: Proceedings of the 8th International Conference on Permafrost, Zurich, Switzerland, edited by: Phillips, M., Springman, S. M., and Arenson, L., 175–180, Balkema Publishers, Lisse, 2003.

Devaux, J.: Radiothermic economy of fields of snow and glaciers, Science Abstr., Serias A, 36, 980–981, 1933.

Fillion, M.-H., Côté, J., and Konrad, J.-M.: Thermal radiation and conduction properties of materials ranging from sand to rock-fill, Can. Geotech. J., 48, 532–542, 2011.

Gruber, S. and Hoelzle, M.: The cooling effect of coarse blocks revisited: a modeling study of a purely conductive mechanism, in: 9th Int. Conf. on Permafrost, Institute of Northern Engineering, Fairbanks, University of Alaska, 29 June–3 July 2008, edited by: Kane, D. and Hinkel, K., 557–561, 2008.

Gustafsson, D., Stähli, M., and Jansson, P.-E.: The surface energy balance of a snow cover: comparing measurements to two differentsimulation models, Theor. Appl. Climatol., 70, 81–96, 2001.

Haeberli, W., Huder, J., Keusen, H., Pika, J., and Röthlisberger, H.: Core drilling through rock glacier-permafrost, in: Proceedings of the Fifth International Conference on Permafrost, Vol. 2, 937–942, by Senneset-Kaare, 1988.

Hanson, S. and Hoelzle, M.: The thermal regime of the active layer at the Murtèl rock glacier based on data from 2002, Permafrost Periglac., 15, 273–282, 2004.

Harris, S. A. and Pedersen, D. E.: Thermal regimes beneath coarse blocky materials, Permafrost Periglac., 9, 107–120, 1998.

Herz, T., King, L., and Gubler, H.: Microclimate within coarse debris of talus slopes in the alpine periglacial belt and its effect on permafrost, in: Proceedings of the 8th International Conference on Permafrost, Zurich, Switzerland, edited by: Phillips, M., Springman, S. M., and Arenson, L., 383–387, Balkema Publishers, Lisse, 2003.

Hoelzle, M. and Gruber, S.: Borehole and ground surface temperatures and their relationship to meteorological conditions in the Swiss Alps, in: 9th Int. Conf. on Permafrost, Institute of Northern Engineering, Fairbanks, University of Alaska, 29 June–3 July 2008, edited by: Kane, D. L. and Hinkel, K. M., 723–728, 2008.

Hoelzle, M., Vonder Mühll, D., and Haeberli, W.: Thirty years of permafrost research in the Corvatsch - Furtschellas area, Eastern Swiss Alps: A review, Norsk Geogr. Tidsskr., 56, 137–145, 2002.

IPCC: Working Group I Contribution to the IPCC Fifth Assessment Report (AR5), Climate Change 2013: The Physical Science Basis, 2013.

Jansson, P.-E. and Karlberg, L.: Coupled heat and mass transfer model for soil-plant-atmosphere systems, Royal Institute of Technology, Dept of Civil and Environmental Engineering, Stockholm, http://www.lwr.kth.se/Vara%20Datorprogram/CoupModel/index.htm, 2011.

Johnsson, H. and Lundin, L.-C.: Surface runoff and soil water percolation as affected by snow and soil frost, J. Hydrol., 122, 141–159, 1991.

Kääb, A., Gudmundsson, G. H., and Hoelzle, M.: Surface deformation of creeping mountain permafrost. Photogrammetric investigations on rock glacier Murtèl, Swiss Alps, in: Proceedings of the 7th International Conference on Permafrost, 531–537, 1998.

Keller, F. U.: Interaktionen zwischen Schnee und Permafrost: Eine Grundlagenstudie im Oberengadin, Ph.D. thesis, Dissertation an der Abteilung Naturwissenschaften der Eidgenössisch Technischen Hochschule Zürich, 1994.

Kunii, D. and Smith, J.: Heat transfer characteristics of porous rocks, AIChE J., 6, 71–78, 1960.

Langer, M., Westermann, S., Muster, S., Piel, K., and Boike, J.: The surface energy balance of a polygonal tundra site in northern Siberia – Part 2: Winter, The Cryosphere, 5, 509–524, 2011, http://www.the-cryosphere-discuss.net/5/509/2011/.

Lundin, L. C.: Hydraulic properties in an operational model of frozen soil, J. Hydrol., 118, 289–310, 1990.

McAdams, W. H.: Heat transmission, McGraw-Hill Book Co., Inc., New York, NY, 1954.

Meder, S. E., Pennetier, O. A., Ansberry, D. M., Brunner, I. M. I. M.: Assessment of solar energy potential on existing buildings in a region, U.S. Patent 7305 983 B1, December 11, 2007.

Mittaz, C., Hoelzle, M., and Haeberli, W.: First results and interpretation of energy-flux measurements over Alpine permafrost, Ann. Glaciol., 31, 275–280, 2000.

Mittaz, C., Imhof, M., Hoelzle, M., and Haeberli, W.: Snowmelt Evolution Mapping Using an Energy Balance Approach over an Alpine Terrain, Arct. Antarct. Alp. Res., 34, 274–281, 2002.

Oke, T. R.: Boundary layer climates, Routledge, 1988.

Panz, M.: Analyse von Austauschprozessen in der Auftauschicht des Blockgletschers Murtèl-Corvatsch, Oberengadin, Master's thesis, Ruhr Universität (unpublished), 2006.

Scherler, M., Hauck, C., Hoelzle, M., Stähli, M., and Völksch, I.: Meltwater infiltration into the frozen active layer at an alpine permafrost site, Permafrost Periglac., 21, 325–334, 2010.

Scherler, M., Hauck, C., Hoelzle, M., and Salzmann, N.: Modeled sensitivity of two alpine permafrost sites to RCM-based climate scenarios, J. Geophys. Res., 118, 780–794, doi:10.1002/jgrf.20069, 2013.

Schneider, S., Hoelzle, M., and Hauck, C.: Influence of surface and subsurface heterogeneity on observed borehole temperatures at a mountain permafrost site in the Upper Engadine, Swiss Alps, The Cryosphere, 6, 517–531, doi:10.5194/tc-6-517-2012, 2012.

Stähli, M., Jansson, P.-E., and Lundin, L.-C.: Preferential water flow in a frozen soil – a two-domain model approach, Hydrol. Process., 10, 1305–1316, 1996.

Stocker-Mittaz, C.: Permafrost distribution modeling based on energy balance data, Ph.D. thesis, Univ. of Zurich, Zurich, Switzerland, 2002.

Tanaka, H., Nohara, D., and Yokoi, M.: Numerical simulation of wind hole circulation and summertime ice formation at Ice Valley in Korea and Nakayama in Fukushima, Japan, Journal-Meteorological Society of Japan, Series 2, 78, 611–630, 2000.

Vonder Mühll, D., Arenson, L., and Springman, S.: Two new boreholes through the Murtèl–Corvatsch rock glacier, Upper Engadin, Switzerland, in: 1st European Permafrost Conference, Abstract, p. 83, 2001.

Vonder Mühll, D., Arenson, L., and Springman, S.: Temperature conditions in two Alpine rock glaciers, in: 8th International Conference on Permafrost, Zürich, Swets & Zeitlinger, Lisse, 1195–1200, 2003.

Wakonigg, H.: Unterkühlte schutthalden, Arbeiten aus dem Institut für Geographie der Karl-Franzens-Universität Graz, 33, 209–223, 1996.

Westermann, S., Lüers, J., Langer, M., Piel, K., and Boike, J.: The annual surface energy budget of a high-arctic permafrost site on Svalbard, Norway, The Cryosphere, 3, 245–263, 2009, http://www.the-cryosphere-discuss.net/3/245/2009/.

Williams, P. J. and Smith, M. W.: The frozen earth: fundamentals of geocryology, Vol. 306, Cambridge University Press, Cambridge, 1989.

8

Impact of change in erosion rate and landscape steepness on hillslope and fluvial sediments grain size in the Feather River basin (Sierra Nevada, California)

M. Attal[1], S. M. Mudd[1], M. D. Hurst[1,*], B. Weinman[2], K. Yoo[2], and M. Naylor[1]

[1]School of GeoSciences, Univ. of Edinburgh, Drummond Street, Edinburgh, EH8 9XP, UK
[2]Department of Soil, Water, and Climate, University of Minnesota, 439 Borlaug Hall, 1991 Upper Buford Circle, St. Paul, MN 55108-6028, USA
[*]now at: British Geological Survey, Keyworth, Nottingham, NG12 5GG, UK

Correspondence to: M. Attal (mikael.attal@ed.ac.uk)

Abstract. The characteristics of the sediment transported by rivers (e.g. sediment flux, grain size distribution – GSD) dictate whether rivers aggrade or erode their substrate. They also condition the architecture and properties of sedimentary successions in basins. In this study, we investigate the relationship between landscape steepness and the grain size of hillslope and fluvial sediments. The study area is located within the Feather River basin in northern California, and studied basins are underlain exclusively by tonalite lithology. Erosion rates in the study area vary over an order of magnitude, from $> 250\,\mathrm{mm\,ka^{-1}}$ in the Feather River canyon to $< 15\,\mathrm{mm\,ka^{-1}}$ on an adjacent low-relief plateau. We find that the coarseness of hillslope sediment increases with increasing hillslope steepness and erosion rates. We hypothesise that, in our soil samples, the measured 10-fold increase in D_{50} and doubling of the amount of fragments larger than 1 mm when slope increases from 0.38 to $0.83\,\mathrm{m\,m^{-1}}$ is due to a decrease in the residence time of rock fragments, causing particles to be exposed for shorter periods of time to processes that can reduce grain size. For slopes in excess of $0.7\,\mathrm{m\,m^{-1}}$, landslides and scree cones supply much coarser sediment to rivers, with D_{50} and D_{84} more than one order of magnitude larger than in soils. In the tributary basins of the Feather River, a prominent break in slope developed in response to the rapid incision of the Feather River. Downstream of the break in slope, fluvial sediment grain size increases, due to an increase in flow competence (mostly driven by channel steepening) as well as a change in sediment source and in sediment dynamics: on the plateau upstream of the break in slope, rivers transport easily mobilised fine-grained sediment derived exclusively from soils. Downstream of the break in slope, mass wasting processes supply a wide range of grain sizes that rivers entrain selectively, depending on the competence of their flow. Our results also suggest that, in this study site, hillslopes respond rapidly to an increase in the rate of base-level lowering compared to rivers.

1 Introduction

In the rock cycle, clastic sediment is produced in upland mountainous areas. The type of sediment delivered from hillslopes to the fluvial system conditions the characteristics of the sediment that is transported by rivers and ultimately exported from mountain ranges to sedimentary basins (Knighton, 1982; Parker, 1991; Heller et al., 2001; Attal and Lavé, 2006; Sklar et al., 2006; Chatanantavet et al., 2010; Whittaker et al., 2010; Bennett et al., 2014; Michael et al., 2014). The grain size distribution (GSD) within hillslope soils and weathering profiles exerts a strong control on hillslope hydrology (e.g. Lohse and Dietrich, 2005) and chemical weathering rates by modulating particle surface areas (e.g. White and Brantley, 2003; Yoo and Mudd, 2008) and water residence time (Maher, 2010). In bedrock rivers, sediment flux and GSD affect the ability of rivers to erode their substrate in two ways: they control (i) the availability and effectiveness of tools for bedrock erosion and (ii) the extent of the protective alluvial cover that the rivers need to mobilise during floods for erosion to happen (e.g. Gilbert, 1877; Sklar and Dietrich, 2004; Cowie et al., 2008; Hobley et al., 2011). They also control the architecture and properties of the stratigraphic successions in sedimentary basins, because the distance travelled by sediment particles before being deposited is dictated primarily by their grain size (e.g. Duller et al., 2010; Whittaker et al., 2010; Armitage et al., 2011). In the short term, fluvial sediment flux and GSD condition whether a river aggrades or incises, both in upland areas and throughout sedimentary basins (e.g. Lane et al., 2007; Duller et al., 2010). This point is of particular relevance when considering the impact of climate change and land use on river dynamics and human infrastructures within river basins, since both changing climate and land use modify sediment and water fluxes from hillslopes to rivers, with a potentially negative impact on the capacity of rivers to hold water within their channels (Lane et al., 2007).

The GSD of the sediment supplied to rivers is one of the main controls on the characteristics of the sediment transported by rivers (i.e. GSD, bedload-to-total-load ratio and lithologic content), the other main controls being abrasion, selective transport and sediment fluxes from hillslopes (Wolcott, 1988; Attal and Lavé, 2006; Sklar et al., 2006; Whittaker et al., 2010). Numerical models suggest that, in areas where rivers are actively incising into bedrock and net deposition is negligible, the continuous supply of fresh material from hillslopes in uniformly eroded landscapes may offset the reduction in grain size by abrasion and prevent downstream fining (Attal and Lavé, 2006; Sklar et al., 2006). Models have also shown that spatial variations in the grain size of the sediment supplied to rivers could have a significant impact on the GSD of the sediment in the river: whereas the effect of a coarser point source would vanish a few kilometres downstream of the location of the point source (Sklar et al., 2006), a general coarsening or fining of the sediment supplied to the river over a given area would lead to significant and potentially abrupt coarsening or fining of the fluvial sediment, which could persist downstream for kilometres (Attal and Lavé, 2006; Sklar et al., 2006). These model results have been corroborated by field observations in rivers in the Himalayas and in the Apennines (Attal and Lavé, 2006; Whittaker et al., 2010). However, whereas sediment fluxes from hillslopes have been quantified in many places over a range of time scales (e.g. Brown et al., 1995; Bierman and Steig, 1996; Granger et al., 1996; Hovius et al., 1997; West et al., 2005), little is known about the GSD of the sediment being delivered to rivers and about the controls upon it (Wolcott, 1988; Casagli et al., 2003; Attal and Lavé, 2006; Whittaker et al., 2010).

In non-glaciated areas, previous studies have shown that differences in hillslope steepness are associated with differences in hillslope processes: as gradient increases, shallow hillslope erosion processes, e.g. ravelling and creeping, are replaced by "steep-slope" erosion processes, e.g. landslides, rock fall and formation of large scree cones. Such observations have been made in varied landscapes and contrasting climatic settings (e.g. San Gabriel Mountains, California – Lavé and Burbank, 2004; Nanga Parbat massif, Himalayas – Burbank et al., 1996; Oregon Coast Range, Oregon – Roering et al., 1999) and are consistent with the results of experimental studies of hillslope sediment transport (Roering et al., 2001). Furthermore, initial data from one catchment in the Apennines (Whittaker et al., 2007, 2010) suggest that erosion processes operating on steep hillslopes provide coarser material to the fluvial system than erosion processes operating on gentle hillslopes. Lavé and Burbank (2004) made similar qualitative observations in California. In addition, Attal and Lavé (2006) showed that lithology exerts a major control on the GSD of the sediment supplied by landslides to the Marsyandi River (Nepal, Himalayas). However, most of these observations are qualitative, and the few studies that produced detailed GSD of the sources of sediment for rivers focused either on landslide deposits (Casagli et al., 2003; Attal and Lavé, 2006; Whittaker et al., 2010) or soils (Marshall and Sklar, 2012).

This study proposes to bridge this gap by assessing the impact of increased erosion rates and associated slope steepening on sediment characteristics, both on hillslopes and in rivers. The study area is the Feather River basin (California), which comprises both low- and high-relief areas with erosion rates varying over an order of magnitude, from $> 250\,\text{mm ka}^{-1}$ in the steepest parts of the landscape to $< 15\,\text{mm ka}^{-1}$ on the low-relief plateau (Riebe et al., 2000; Hurst et al., 2012). This morphological contrast results from the rapid incision of the Feather River in response to a relative drop in base level, causing the formation of a deep gorge (Fig. 1). Tributary basins are still responding to the relative drop in base level and typically exhibit a topographic break in slope separating a low-relief relict topography (plateau) from a steepened landscape (Figs. 1 and 2). Hillslope and river sediment characteristics were measured both on the plateau and downstream of the main topographic break in slope in a series of tributary basins to identify potential changes in sediment sources and to assess the impact of changes in source and variations in channel slope on the characteristics of the sediment transported by rivers.

After a description of the study area and methods, we present the GSD data for the hillslope sites (sources) and for the fluvial sites. In light of these data, we analyse the relationships between source and fluvial sediment characteristics and discuss the potential links between tectonics, slope steepness, and sediment delivery and transport in mountain rivers.

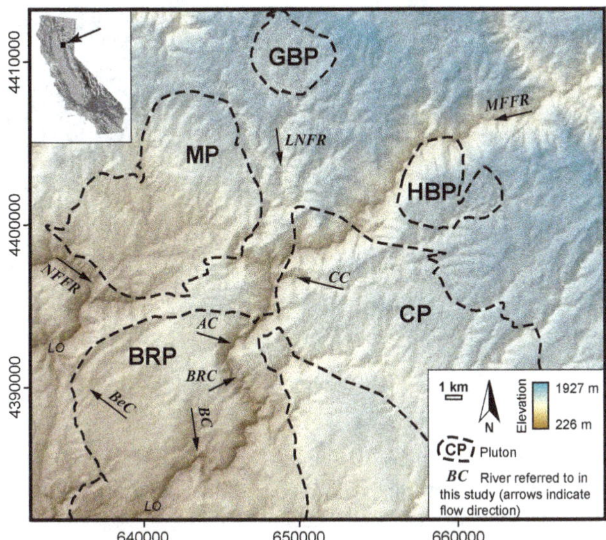

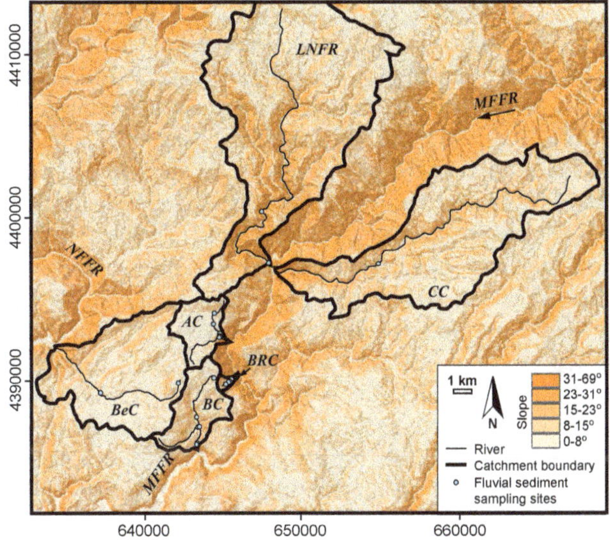

Figure 1. Overview of study area. Topographic data from USGS (National Elevation Dataset). The spatial reference system is UTM Zone 10N with units in metres. Top panel: shaded relief of the study area showing the distribution of the Mesozoic plutons (from Geological map of the Chico Quadrangle; Saucedo and Wagner, 1992) and the studied rivers. Inset shows location of study area in California. Plutons (bold): BRP – Bald Rock; CP – Cascade; GBP – Granite Basin; HBP – Hartman Bar; MP – Merrimac. Rivers (bold italic): AC – Adams Creek BC – Bean Creek; BeC – Berry Creek; BRC – Bald Rock Creek; CC– Cascade Creek; LNFR – Little North Fork River; MFFR – Middle Fork Feather River; NFFR – North Fork Feather River. LO – Lake Oroville. Bottom panel: slope map of the study area draped on shaded relief, showing boundaries of studied basins and sampling sites.

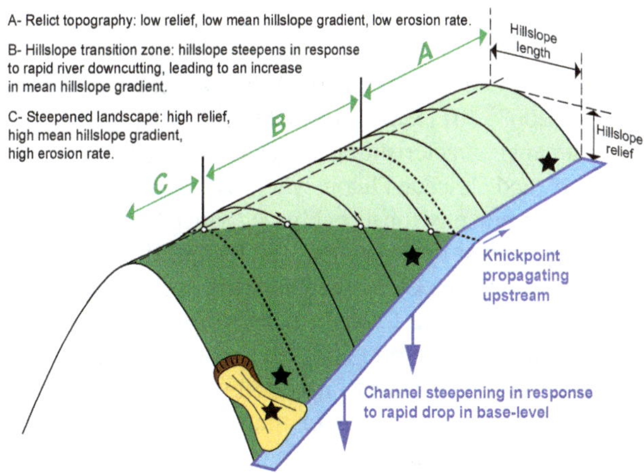

Figure 2. Schematic illustrating the typical morphology of the Feather River's tributary basins (adapted from Hurst et al., 2012). In response to a rapid drop in base level, a knickpoint propagates upstream along the channel, separating the steepened landscape from the relict topography. A break in slope also propagates up the hillslopes (dots with arrows) in response to the increase in channel downcutting rate. Stars schematically represent the location of sampling sites with respect to morphological domains: on the relict topography (domain A, site identifier POMD), in the transition zone where the hillslopes have not completely adjusted (domain B, site identifier FTA) and in the steepened area below the break in slope (domain C, site identifier BRC and BRB for soils and LD for landslides). Note that the width of domain B is a function of the response time of the hillslopes (the shorter the response time, the narrower the domain B). The mean hillslope gradient S_h used in this study is the ratio of hillslope relief to hillslope length (shown on figure).

2 Study area and methods

This study focuses on basins draining an area where Mesozoic plutons have intruded a metamorphic basement, east of Lake Oroville, in the Sierra Nevada of California (Fig. 1). In response to an increase in the rate of base-level lowering, the origin and timing of which are still debated (Wakabayashi and Sawyer, 2001; Stock et al., 2004; Gabet, 2014), the North Fork and Middle Fork Feather rivers have formed gorges up to 600 m deep (Fig. 1). These gorges dissect a relict low-relief landscape (Fig. 1) that has erosion rates an order of magnitude lower than the gorges: cosmogenic radionuclide-derived erosion rates in basins draining the Bald Rock and Cascade plutons vary from $14.4 \pm 1.6\,\mathrm{mm\,ka^{-1}}$ on the plateau to rates in excess of $250\,\mathrm{mm\,ka^{-1}}$ in the steepest parts of the landscape (Riebe et al., 2000 – see samples within their "Fall River" area; Hurst et al., 2012). Many of the tributary basins which drain from the relict surface to the North and Middle Fork Feather rivers have been left hanging (Figs. 1 and 2): these basins typically exhibit a prominent convexity on their hillslopes and river profiles, marking the boundary between the lower basin which has steepened in response to the increase in the rate of base-level drop and the upper basin which has not yet detected the change in base-level lowering rate (Figs. 1 and 2) (e.g. Whipple and Tucker, 2002; Crosby and Whipple, 2006; Whittaker et al., 2008; Attal et al., 2008, 2011). We have measured hillslope and fluvial sediment characteristics in tributary basins that drain the Bald Rock and Cascade Pluton, where the source rock lithology is predominantly tonalite (Fig. 1) (Saucedo and Wagner, 1992). The fluvial data set was complemented with sites in two large tributaries of the Feather River: Cascade Creek, which incises into the Cascade Pluton in the lower half of its course, and Little North Fork River, which mostly drains the metamorphic basement intruded by the Mesozoic plutons. Both basins show signs of transience (break in slope on hillslopes and along the river) and were investigated to assess whether their behaviour was consistent with the one exhibited by the smaller basins in response to the increase in the rate of base-level lowering.

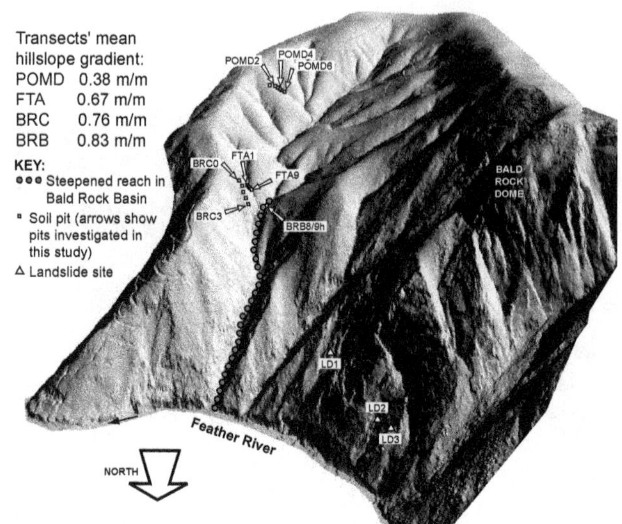

Figure 3. Shaded relief of Bald Rock basin derived from lidar (1 m resolution) data, showing the location of the soil and landslide sites. Horizontal length of steepened reach is ~550 m.

2.1 Sources of sediment for the rivers

Sources of sediment in the study area comprise soils from soil-mantled hillslopes on the low-relief plateau and patchy soils, landslide deposits, scree cones and debris-flow deposits in the steep, incised valleys near the Feather River. Evidence for rock failures of various dimensions, from individual fragments to the release of hundreds of cubic metres of debris, is widespread on slopes above $0.7\,\mathrm{m\,m^{-1}}$. No recent debris flows were documented in the study area, but evidence of past debris flows was found along rivers in the steepened landscape below the prominent topographic break in slope. However, the GSD of the debris-flow deposits found in the field could not be characterised because these had undergone substantial reworking after their emplacement. All source sites were chosen on the Bald Rock tonalite pluton, identified by both field observations and geological map (Saucedo and Wagner, 1992). Sampling and measurement methods are similar to the ones used by Attal and Lavé (2006) (see below).

All soil sampling sites are located in the Bald Rock Creek basin (Figs. 1 and 3). Soil pits were dug along hillslope transects in three morphologically distinct areas (Figs. 2 and 3): on the relict topography above the break in slope (POMD, mean hillslope gradient $S_h = 0.38\,\mathrm{m\,m^{-1}}$), in the transition zone where the hillslopes have not completely adjusted to the base-level fall (FTA, $S_h = 0.67\,\mathrm{m\,m^{-1}}$) and below the break in slope (BRC and BRB, $S_h = 0.75$ and $0.84\,\mathrm{m\,m^{-1}}$, respectively); the mean hillslope gradient (S_h) represents the ratio of hillslope relief over the horizontal length of the hillslope (Fig. 2), which is a reliable proxy for erosion rate in this area (Hurst et al., 2012). At or below the break in slope (FTA, BRC and BRB), we found that the soils lack distinct illuvial B horizons. In contrast, redder soils with slightly clay-enriched B horizons were present above the break in slope (POMD). The soils in the area belong to either sandy-skeletal, mixed, mesic Lithic Xerorthents (Waterman Series) or coarse-loamy, mixed, superactive, mesic Typic Dystroxerepts (Chaix Series) (Soil Survey Staff, accessed 13 February 2015). At each site, soil pits were excavated to the depth of 20–30 cm below the soil–saprolite boundary. The material extracted from the pits was sieved in the field using 10, 20 and 40 mm square mesh sieves (Fig. 4a and b). Each fraction was weighed using a portable scale (accuracy = 20 g), and fragments larger than 80 mm were weighed individually. Large fragments were found to be very lightly weathered; the size of the fragments larger than 80 mm was thus calculated assuming that they were spheres with a density of $2650\,\mathrm{kg\,m^{-3}}$. Approximately 1 kg of the fraction finer than 10 mm was sampled, and its GSD was determined in the lab using 8, 5.6, 4, 2.8, 2, 1.4 and 1 mm square mesh sieves. The GSD of the fraction finer than 1 mm was determined using a Malvern laser grain size analyser. At the soil sampling sites, the mass of sediment sieved and weighed ranged between 122 and 550 kg per pit, except at one site, where the soil was thin compared to the other sites and the soil–saprolite boundary was reached quickly: 63 kg of sediment was dug out and sieved at the steepest Bald Rock site (BRB, $S_h = 0.84\,\mathrm{m\,m^{-1}}$) (Fig. 3).

Landslide deposits and scree cones were investigated exclusively east of the Bald Rock Dome, immediately north of the Bald Rock Basin, where vegetation is scarce and debris are being actively accumulated below the rocky face (Figs. 3 and 4c). This was the only location where an active landsliding area could be accessed safely in the study area. The three sampling points are located in places where the debris accumulation has been cut by gullies or by the path, providing a clear cut through the deposit. LD1 is located at mid-height in a debris fan, whereas LD2 and LD3 are situated near the top of landslide fans. The surface material was removed down to

Figure 4. Photographs of hillslope sites (see Fig. 3 for location of sites). Panels **(a)** and **(b)**: sediment at soil sites and 10, 20 and 40 mm square mesh sieves. **(a)** Sediment coarser than 10 mm at site FTA9 (transition zone); clasts are fresh and angular and show little evidence of chemical weathering. **(b)** Sediment at site POMD6 (plateau), including 540 kg of sediment finer than 10 mm (heap to the right), and 7.5 and 3.0 kg of clasts in the fractions 10–20 and 20–40 mm, respectively (on the tarpaulin). **(c)** Sediment at the landslide site LD2. Upper panel: overview with close-up of pit (inset). Lower panel: different sediment fractions extracted from the pit. Hammer is 300 mm long. Swiss army knife is 90 mm long.

the depth of the largest clast exposed in the vicinity of the site to avoid bias caused by winnowing of the surface or kinetic sieving during landsliding. Eighty-five to 115 kg of sediment was dug out, sieved and weighed (Fig. 4c). The procedure for determining the GSD of the landslide sediment in the field and in the lab is identical to the one applied to soil material (see above).

Additionally, photographs of the surface of the scree cones and landslide deposits were taken at various locations below Bald Rock Dome. The field of each photograph was typically 1 to 2 m wide and high. A scale was placed at the centre of the field before each photograph was taken. These images were then used to determine GSD: following Kellerhals and Bray (1971), a regular square grid with 100 line intersections was placed on each photograph and the smallest axis of the clast at each intersection was measured. Clasts within the landslide deposits have no preferential orientation, which means that the length of the smallest axis measured on the photograph is a minimum estimate of the intermediate axis of the clasts. However, tonalite clasts were typically found to be neither elongated nor platy (Fig. 4c), thus limiting the deviation between the length measured on the photograph and the actual length of the clast's intermediate axis. Following Kellerhals and Bray's (1971) recommendation based on the voidless cube model, clasts covering *n* grid intersections were counted *n* times. According to this model, the GSD by number obtained from the photographs is directly comparable to the GSD by mass derived from the volumetric samples. The limitations associated with this model are discussed further (Sect. 2.3).

2.2 Fluvial sediment

The methods used to determine the GSD of fluvial sediment are similar to the ones used by Attal and Lavé (2006). Gravel bars were identified along the studied rivers, including the river basin where soil sediment was investigated (Figs. 1, 5 and 6). We focused on material that had been unambiguously transported by fluvial processes and avoided lag deposits found where old debris-flow or landslide deposits had been reworked (these latter deposits are characterised by extremely coarse sediment with locked, moss-covered particles that are indicative of low mobility). Surface and subsurface were distinguished to account for the armouring that typically characterises fluvial deposits (Bunte and Abt, 2001). Surface GSD was determined by photo analysis (see previous paragraph); because pebbles tend to lie preferentially with their small axis perpendicular to the surface of the gravel bar, the smallest visible axis on the photograph was considered as the intermediate axis of the pebble. Subsurface sediment was excavated from a pit after removing the surface material over an area of approximately 0.5 by 0.5 m. This subsurface material was subjected to the same sieving and weighing procedure as the soil samples (see Sect. 2.1). We maximised the amount of sediment sieved with respect to the size of the largest clast at each site, but our efforts were restricted by the size of the gravel bars in these mountainous settings: some of the bars were small ($< 2\,m^2$) and bounded by bedrock, which reduced the volume of sediment available for sieving. We were also unable to dig deep below the water level. The mass of sediment sieved and weighed at each site typically ranged between 23 and 154 kg.

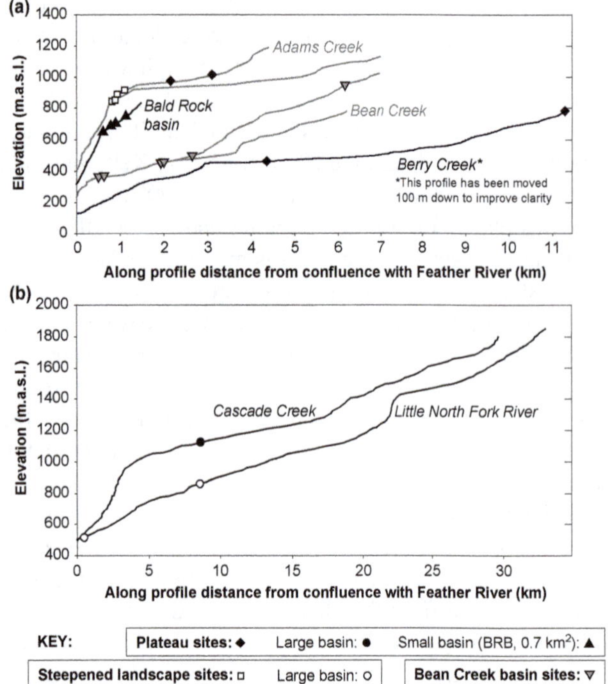

Figure 5. Long profiles of the studied rivers with location of the sampling sites for fluvial sediment. **(a)** Small tributary basins of the Feather River (length < 12 km). **(b)** Two large tributaries of the Feather River (note change in scale on the *x* axis). We classify sites based on their position with respect to the topographic break in slope (see key); Bean Creek sites are treated separately due to the lack of a clear morphological distinction between relict surface and steepened landscape.

Most of the rivers investigated have a large convexity on their long profile which marks the transition from the relict topography to the steepened landscape downstream (Figs. 1, 2 and 5). Along-stream variations in fluvial sediment GSD were determined in three basins draining the tonalite pluton: measurements were performed at four sites in Bald Rock basin and at six sites in Bean Creek and Adams Creek basins (Figs. 1 and 5a). Additional measurements in adjacent basins on the pluton were carried out on the plateau (Berry Creek basin) (Figs. 1 and 5a). Measurements in large rivers draining multiple lithologies were carried out on the plateau (Cascade Creek) and in the gorge (Little North Fork River) (Figs. 1 and 5b).

In the following analysis and discussion, we will distinguish "plateau" sites from "steepened landscape" sites (Fig. 5). Bean Creek sites will be treated separately due to the absence of a clear morphological boundary between plateau and incised landscape.

Figure 6. Example of gravel bar investigated in this study (site BC1 on Bean Creek). **(a)** Overview and **(b)** sediment after sieving.

2.3 Sampling method bias and precision of measurements

Many sample-size recommendations have been made for representatively sampling granular material (see extensive review in Bunte and Abt, 2001). For material typically coarser than 128 mm, Church et al. (1987) recommend that the largest particle in a sample represents no more than 5 % of the total mass of the sample to avoid unrepresentative positive skewness of the grain size distributions due to a few large clasts representing a large proportion of the total sample. Due to logistical and geomorphological constraints, this recommendation was not fulfilled at one of the soil sites, at all landslide sites and at more than half of the fluvial sites (see Tables A1 and A2 in Appendix).

To assess the impact of the largest clast representing an excessively large fraction of the volumetric sample on the determination of characteristic sediment grain sizes (i.e. median grain size D_{50} and 84th percentile D_{84}), the following procedure was applied. In the following example, the mass m of the largest clast represents n % of the total mass of the sample. Firstly, the largest clast was removed from the distribution to estimate D_{50} and D_{84}, had this large clast not been sampled; secondly, a large clast was added to the distribution, its mass calculated so that it represents n % of the mass of the new volumetric sample (we observe that in all cases, this calculated mass is 1 to 1.1 times the mass m of the largest clast in the actual sample). This procedure gives a robust estimate of the potential variation in D_{50} and D_{84} induced by the inclusion or omission of large clasts in the sample (see Tables A1 and A2 in Appendix). In the following, error bars on grain sizes in figures represent the range of values between the two scenarios mentioned above rather than uncertainty, which cannot be calculated without a priori knowledge of the true grain size distribution or applying an inevitably imperfect model to represent this distribution.

Grid counts were performed on the surface of landslide and gravel bar sediment using photographic methods (see Sects. 2.1 and 2.2). The number of clasts counted on each photograph typically ranged between 65 and 100, due to the image being obscured by leaves, water or shadows in places (Table A2). Ideally, the size of the grid applied to the photographs should be chosen so that no more than one grid intersection falls on one sediment clast. Unfortunately, such a requirement is nearly impossible to fulfil at all sites in this mountainous setting where boulders larger than 0.5 m are present and gravel bars can sometimes be less than 2 m long. As mentioned above, clasts covering n grid intersections were counted n times, following Kellerhals and Bray's method (1971) based on the voidless cube model. Whereas Bunte and Abt (2001) agree that the voidless cube model may be applicable to armoured coarse gravel or cobble beds, thus allowing a direct comparison of grid-by-number and volume-by-mass samples (Kellerhals and Bray, 1971; Bunte and Abt, 2001), they highlight that multiple counting of particles over-represents large particles and produces GSDs that are too coarse in their coarse part. The effect of multiple counting is minimal on D_{50} but can be substantial on D_{84} estimates. To assess the impact of large clasts covering $n > 1$ grid nodes on the photographs on the determination of D_{50} and D_{84}, we applied a similar procedure to the one used for the volumetric sample. Firstly, the largest clast was removed from the distribution to estimate D_{50} and D_{84}, had this large clast not been sampled; secondly, a large clast similar in size to the largest clast in the actual sample was added to the distribution, covering the same number of grid nodes than this largest clast (Table A2). This procedure does not account for multiple clasts covering multiple nodes, but it gives a rough estimate of the variation in grain size potentially induced by the largest clast on the distribution, which is particularly significant for D_{84} (Table A2) (Bunte and Abt, 2001). As with the volumetric samples, error bars on grain size in figures will represent the range of values between the two scenarios mentioned above.

2.4 Flow competence and sediment grain size

Flow competence dictates the maximum size of grains transported by a river for a given discharge. Competence is commonly expressed as a function of fluvial shear stress (e.g. Buffington and Montgomery, 1997), but this quantity is difficult to estimate in mountain rivers. Instead, an alternative approach involves the use of water discharge per unit flow width. According to theory and flume experiments, a power relationship (with an exponent 2/3 in the case of uniform grain size) is expected between the grain size of the sediment entrained by a given water discharge Q and the quantity $\omega_m = Q\,S^M / W$, where S is channel slope, W is channel width and M is an exponent ranging between 1.12 and 1.17 (Schoklitsch, 1962; Bathurst et al., 1987; Whitaker and Potts, 2007; Bathurst, 2013) (note that ω_m would be proportional to specific stream power if M were equal to unity). Measurements of the maximum grain size entrained in a series of natural rivers also show a power relationship with Q/W for a given slope, thus supporting the theory (measurements were made at a given site over a range of discharges for each river: Whitaker and Potts, 2007; Bathurst, 2013).

Large variations in ω_m are expected along the rivers in the study area, in particular at the main topographic break in slope where both discharge and slope will increase downstream. In a situation where all grain sizes are potentially available for transport in the river, river sediment is expected to become coarser as ω_m increases, which we will assess in the following. For simplicity, we assume that (1) sediment in gravel bars is representative of the sediment that is typically transported during floods, (2) sediment in all the gravel bars investigated has been mobilised during an event of similar magnitude, and (3) fluvial sediment transport and subsequent deposition in gravel bars has occurred during floods resulting from storm events with no spatial variation in intensity across the entire study area. To maximise the validity of these assumptions, we consistently chose gravel bars that contained sediment that had been unambiguously transported by fluvial processes and that showed evidence of recent transport (i.e. we avoided bars with significant vegetation and/or moss cover). It is worth noticing that the climate in the study area is characterised by high seasonality, with 90 % of the precipitation falling between October and April during storms lasting from a few hours to up to 10 days (see data for Brush Creek hydrologic station (BRS) located in the headwaters of the Adams Creek basin at latitude 39.692 and longitude −121.339; data accessed on 9 February 2015 on the California Data Exchange Center website at http://cdec.water.ca.gov/cgi-progs/stationInfo?station_id=BRS; maximum daily precipitation recorded since 1986 was 292 mm on 1 January 1997). This implies that sediment transport in the study catchments is likely to happen suddenly and synchronously during storms. We thus consider that discharge scales with drainage area A (e.g. Snyder et al., 2003) and therefore assume that flow competence can be expressed as a function of $\omega'_m = A\,S^M / W$; we use a value of $M = 1.15$ as representative of the range of values published in the literature (between 1.12 and 1.17; e.g. Whitaker and Potts, 2007; Bathurst, 2013). Topographic metrics and river profiles were extracted from a 1 m resolution lidar-derived digital elevation model (DEM) obtained via the National Center for Airborne Laser Mapping (NCALM). The data were complemented by 10 m resolution topographic data from USGS (National Elevation Dataset) in three basins that were not entirely covered by the lidar data: Berry Creek, Cascade Creek and Little North Fork River. For each site, drainage area and channel slope were ex-

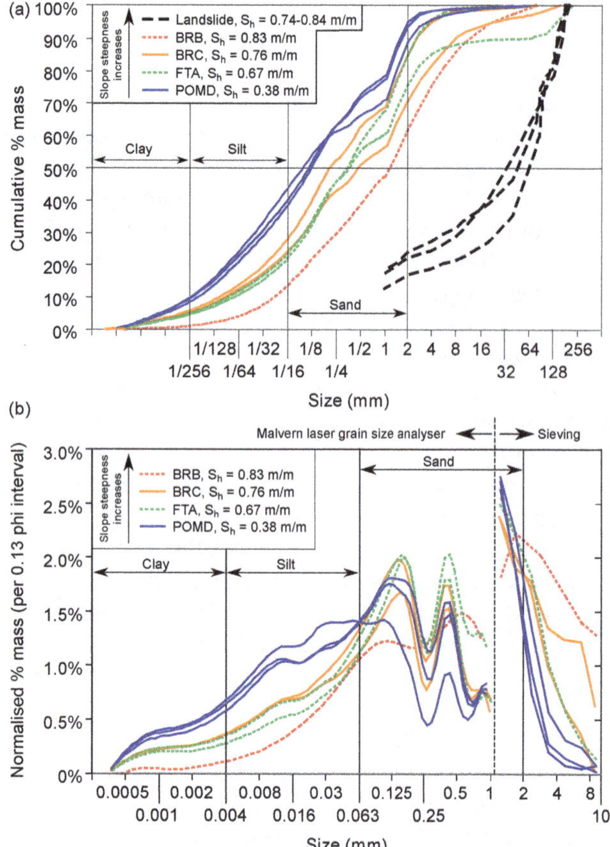

Figure 7. **(a)** Cumulative grain size distributions measured for the sources of sediment in Bald Rock basin (see sites location in Fig. 3). Line patterns reflect hillslope steepness at the sites (S_h is mean hillslope gradient) and type of source (soil or landslide). Note the $\log_2$ scale on the x axis. **(b)** Non-cumulative grain size distribution of the fraction finer than 10 mm of soil samples. Line patterns are the same as in **(a)**. Grain size distributions of fraction coarser and finer than 1 mm were determined using sieves and a Malvern laser grain size analyser, respectively. For both methods, the percent mass has been normalised to represent the value per 0.13ϕ interval. Lines connecting the curves produced with the two methods (at 1 mm) have been removed for clarity; the peak at the transition is real: sediment in the fraction 1–2 mm is significantly more abundant than sediment in the fraction 0.5–1 mm. Note the $\log_2$ scale on the x axis.

tracted from the 1 m resolution DEM, except for the Cascade Creek and lowermost Berry Creek sites, where the USGS DEM data were used instead (see Table A3 in Appendix). Slope was estimated over a 100-m distance; based on field observations, this distance, which represents between 5 and 50 channel widths, is deemed to reflect reach-scale geometry rather than the local pool and riffle morphology. Similarly, minimum and maximum channel widths were measured for each site over a 100-m long stretch on the lidar-derived DEM and Google Maps images: the mean width was used for the calculation of ω'_m and the difference between mean and extrema was used as deviation for width.

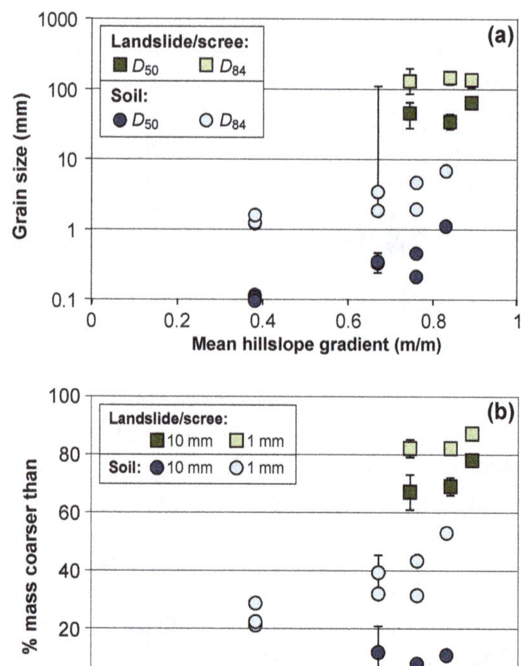

Figure 8. Grain size data for the sources of sediment: soils (circles) and landslides (squares). **(a)** Median grain size D_{50} and 84th percentile D_{84} as a function of hillslope steepness. **(b)** Percent mass of the total sample finer than 1 and 10 mm as a function of hillslope steepness. Error bars represent plus or minus values calculated according to procedure described in Sect. 2.3.

3 Results

3.1 Sources of sediment for the rivers

Our results show that sediments from landslides and scree cones are significantly coarser than those from hillslope soils with no evidence of mass wasting (Figs. 7 and 8). For slopes steeper than $0.7\,\mathrm{m\,m^{-1}}$, mass wasting such as landsliding and formation of scree cones delivers sediment with grain sizes more than one order of magnitude larger than soils, as shown in median grain size D_{50} and 84th percentile of the distribution D_{84} (Figs. 7 and 8a). Soils typically contain less than 12 % mass of grains larger than 10 mm, whereas fragments larger than 10 mm represent ~70 % mass of the landslide deposits investigated (Fig. 8b). When considering the cut-off size of 1 mm that separates material that can potentially be transported as suspended load from grains that will be transported as bedload, the difference is less accentuated but still substantial: landslide deposits contain around twice as much material coarser than 1 mm than soils do (Fig. 8b).

Furthermore, the type of source seems to influence the size of the largest particle to be supplied to the river: the size of the largest particle found in the soil pits is on average 89 ± 64 mm (± standard deviation), compared to

Table 1. Statistical results for regression of $D = k(\omega'_m)^b$ for Adams Creek data. D is taken as D_{50}, D_{84} and D_{100}, both for subsurface and surface samples. Results are highly significant for the exponent b (p value < 0.01 and high test statistic value t, except for D_{100} for the surface samples), but k is poorly constrained.

Grain size	b	Standard error	t value	p value	log k	Standard error	t value	p value	Multiple R^2
D_{50} sub.	0.55	0.10	5.5	0.005	−1.19	0.48	−2.5	0.070	0.88
D_{50} surface	0.61	0.10	6.3	< 0.001	−1.31	0.46	−2.8	0.023	0.83
D_{84} sub.	0.43	0.06	7.0	0.002	−0.16	0.29	−0.6	0.611	0.92
D_{84} surface	0.53	0.08	6.8	< 0.001	−0.55	0.37	−1.5	0.176	0.85
D_{100} sub.	0.20	0.06	3.6	0.024	1.14	0.27	4.3	0.013	0.76
D_{100} surface	0.40	0.08	5.1	< 0.001	0.24	0.38	0.6	0.549	0.76

191 ± 15 mm in the landslide deposits (Fig. 8a). In addition, GSD derived from 18 photos of the surface of landslide deposits in the Bald Rock Dome area yielded D_{50} and D_{84} values of 81 ± 84 and 187 ± 126 mm, respectively; this indicates that surface landslide GSD is spatially highly variable and that boulder-size fragments are widespread within the landslide deposits, despite them not being found in the pits we dug. The GSD of the measured landslide deposits falls within the range of GSD measured by Casagli et al. (2003) in areas underlain by turbidites and shales in the Apennines and by Attal and Lavé (2006) in areas underlain by quartzites, gneiss and schists along the Marsyandi River (Himalayas).

Within the soils, data from the Bald Rock basin seem to indicate an increase in D_{50}, D_{84} and fractions coarser than 1 and 10 mm with increasing transect slope steepness (Fig. 8). The difference in D_{50} between soils on slopes with gradients of 0.38 and 0.83 m m^{-1} is an order of magnitude, whereas D_{84} is larger in the steepest soils by a factor of 4 (Fig. 8a). The fraction coarser than 10 mm increases from 2 to 11 % with increasing slope from 0.38 to 0.83 m m^{-1}, while the fraction coarser than 1 mm doubles, from around 25 % to more than 50 % of the sample (Fig. 8b).

3.2 Sediment transported by rivers

Sediment characteristics have been measured along the river in three basins: Adams Creek basin, Bean Creek basin and Bald Rock basin (Fig. 9). These basins have a drainage area of 10.1, 14.7 and 0.7 km^2, respectively. First, we observe that most gravel bars show an armouring of the surface, with surface sediment coarser than subsurface sediment (squares and circles in Fig. 9, respectively). The Adams Creek basin exhibits the most prominent break in slope (Figs. 1, 5a and 9a). It is also the basin in which grain size changes the most dramatically across the main profile convexity: both surface and subsurface grain size (D_{50} and D_{84}) increase substantially downstream of the break in slope separating the plateau from the steepened landscape. In the Bean Creek basin, the landscape is generally steeper than in the Adams Creek basin and the transition from steepened landscape to upper catchment is more subdued (Figs. 1, 5a and 9b). Sediment tends to be coarser in this catchment than in the Adams Creek basin, except for the two lowermost sites, which have a GSD similar to the GSD at the two lowermost sites in the Adams Creek basin. Upstream of these two sites, data seem to show an overall downstream fining, with the uppermost site having the coarsest subsurface sediment in the entire Bean Creek basin. The Bald Rock Basin is the smallest of the three basins (Fig. 9c). Sediment in the channel is fine-grained compared to the basins discussed above, with D_{50} and D_{84} not exceeding 13 and 54 mm, respectively. The data seem to show a slight downstream coarsening of the sediment, with the uppermost site showing the finest GSD and the lowermost site exhibiting the coarsest GSD (see GSD on right panel in Fig. 9c). The amount of sediment within this channel is low compared to the other studied basins, as demonstrated by substantial bedrock exposure in the channel, in particular downstream of the break in slope where no sediment was found.

As mentioned in Sect. 2.4, a power relationship between the grain size of the grains entrained by the river and ω'_m would be expected: $D = k(\omega'_m)^b$, with k a constant and b an exponent equal to 2/3 in the case of uniform grain sizes (e.g. Bathurst, 2013). The whole data set collected in this study is noisy (Fig. 10), but it can be seen that sites with the higher flow competence tend to have the coarsest sediment and vice versa. In Adams Creek, where ω'_m spans over 2 orders of magnitude, D_{50} and D_{84} data show a good agreement with a power relationship, demonstrating an increase in both flow competence and grain size past the main topographic break in slope (Table 1, Fig. 10). However, the maximum grain size D_{100} is not as well correlated with ω'_m, in particular in the subsurface, in contradiction with theory, flume and field studies (Whitaker and Potts, 2007; Bathurst, 2013). The exponent b tends to be higher for D_{50} than for D_{84}, both for surface and subsurface samples. The range of ω'_m in Bean Creek and Bald Rock basin is too small to produce meaningful regressions.

When considering the morphological divisions in the studied landscape, it is noticeable that steepened landscape sites have systematically higher flow competence than plateau

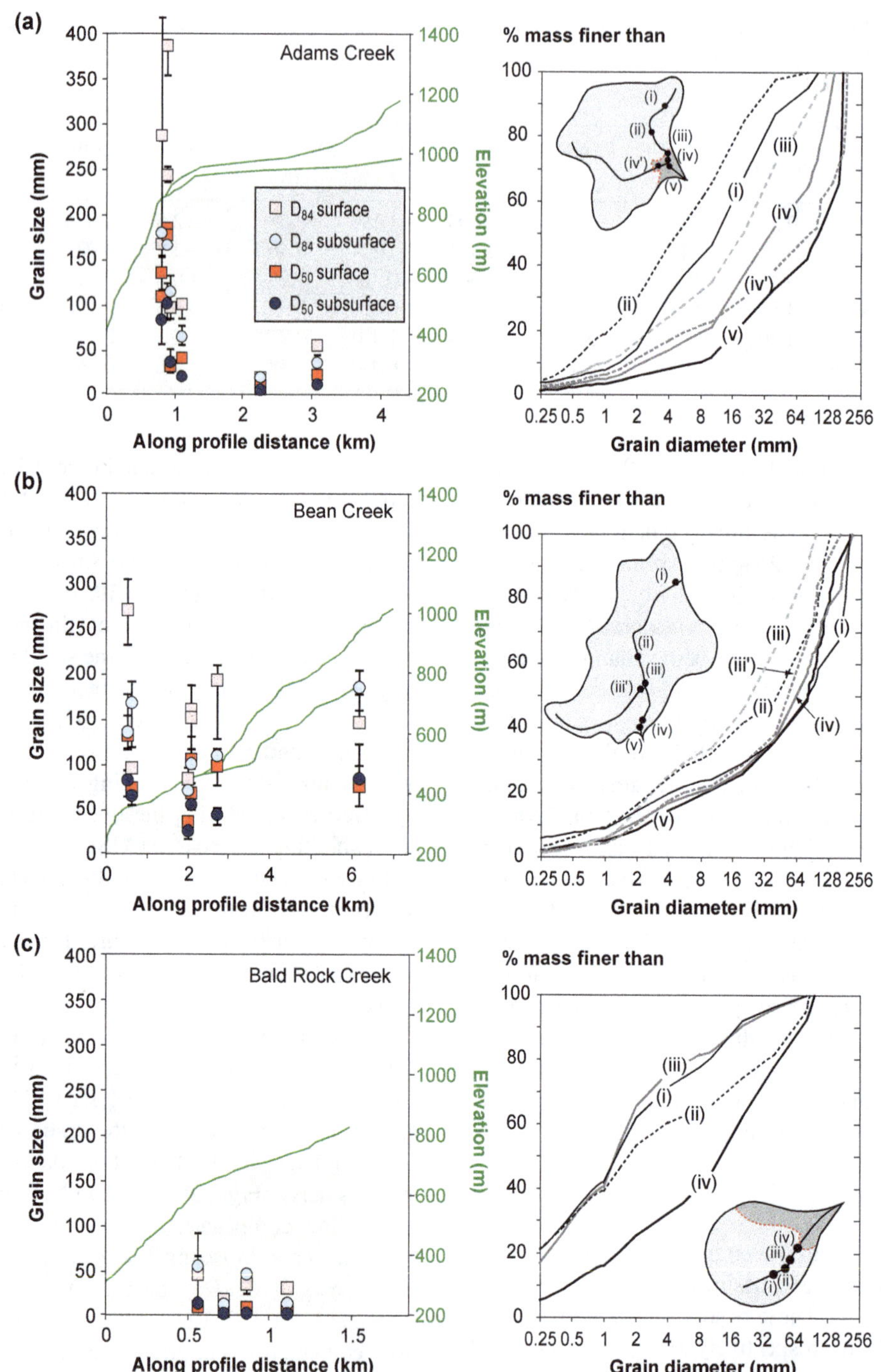

Figure 9. Grain size of the fluvial sediment along **(a)** Adams Creek, **(b)** Bean Creek and **(c)** Bald Rock Creek. Left panels: D_{50} and D_{84} measured along the rivers (subsurface and surface). Error bars represent plus or minus values calculated according to procedure described in Sect. 2.3. River profiles are shown in green; note change in scale on the *x* axis. Right panels: cumulative grain size distribution of subsurface sediment (note the $\log_2$ scale on the *x* axis). Inset grey shapes are schematic map representations of the basins showing the distribution of the samples – (i) being the site the further upstream; transition from plateau to steepened landscape is highlighted by shading and the red dotted line, except in Bean Creek, where there is no clear transition.

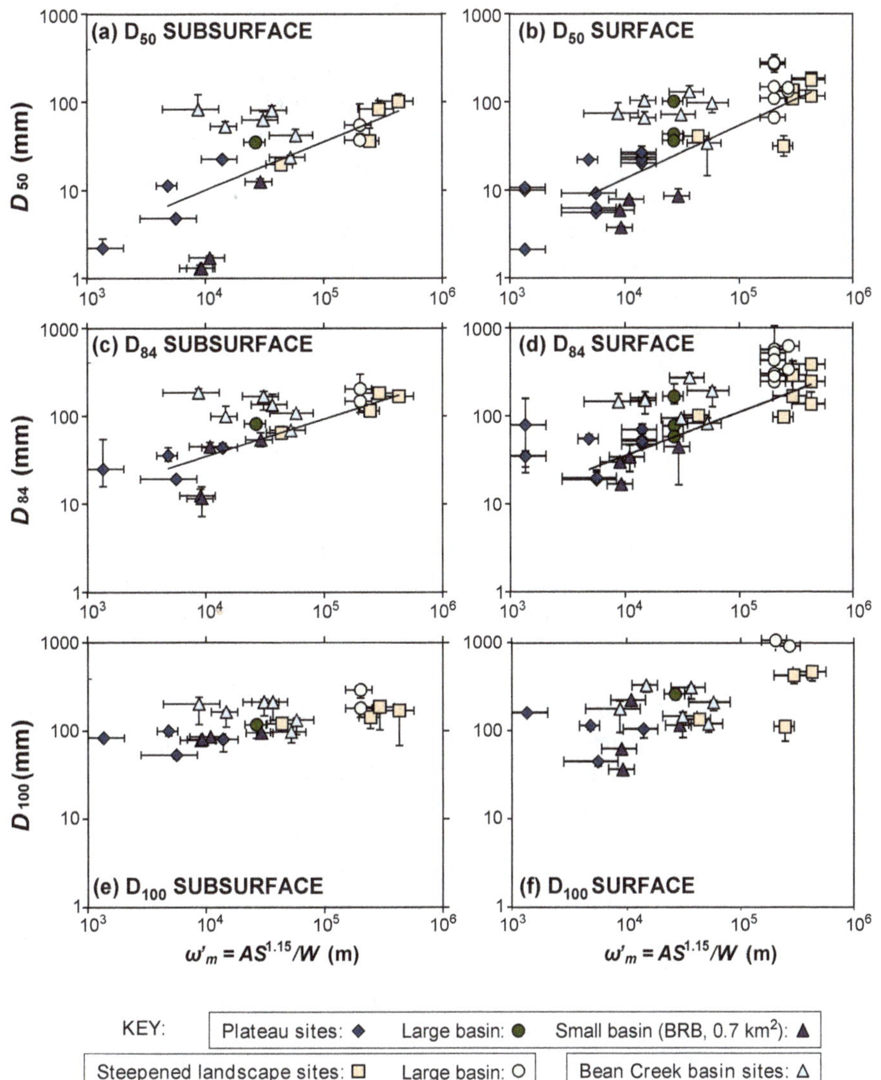

Figure 10. Fluvial sediment grain size as a function of ω'_m. Grain sizes are shown for both subsurface (left panels) and surface (right panels): D_{50} (top panels), D_{84} (middle panels) and D_{100} (bottom panels). Note the log–log scale. Vertical error bars represent plus or minus values calculated according to procedure described in Sect. 2.3. Horizontal bars reflect variability of channel width at the scale of the 100 m reaches considered (Sect. 2.3). We classify sites based on their position with respect to the topographic break in slope (see key and Fig. 5); Bean Creek sites are treated separately due to the lack of a clear morphological distinction between relict surface and steepened landscape. Lines in **(a)** to **(d)** represent best-fit power regression for Adams Creek sites (Table 1) and are shown as a reference for comparison with other sites.

sites (Fig. 10). Importantly, plateau sites tend to have smaller grain sizes than steepened landscape sites; this is exemplified by the subsurface samples which have not experienced armouring and are therefore more likely to be representative of the sediment transported by the river (Fig. 10): D_{50}, D_{84} and D_{100} on the plateau do not exceed 35, 82 and 118 mm, respectively; in the steepened landscape, D_{50}, D_{84} and D_{100} are in the ranges 20–101, 65–202 and 120–290 mm, respectively. These observations stand irrespective of basin size: the plateau site in the large Cascade Creek basin (solid circles in Fig. 10) has the highest ω'_m and the coarsest sediment of all plateau sites but lower ω'_m and finer sediment than the steepened landscape sites. Similarly, the data points from the sites along the incised Little North Fork River (open circles in Fig. 10) fall within the grain size and ω'_m domains delineated by the steepened landscape data points. The Bean Creek sites sit at an intermediate position between the steepened landscape and plateau sites in terms of flow competence but share the range of grain size with steepened landscape sites.

4 Discussion

4.1 Landscape steepness and the characteristics of the sources of sediment

Our results indicate that hillslope steepness partly controls the grain size of the sediment supplied to rivers by controlling the type of process by which sediment is supplied (Figs. 7 and 8). Slope failures and scree cones are observed on slopes steeper than $0.7\,\mathrm{m\,m^{-1}}$. They provide much coarser sediment to river systems than the erosion of soils does. Within soils, grain size seems to generally increase with increasing slope steepness (Figs. 7 and 8).

The hillslope relief in a landscape is related to both erosion rate and the efficiency of sediment transport processes (e.g. Roering et al., 2007). The soils we sampled developed on a similar parent material and have been subjected to a similar climate with similar vegetation (Chaparral, Oak, Pine). Our samples are only separated by several hundred metres laterally and less than 150 m vertically. We thus assume that sediment transport efficiency is similar at all of these sites. Consequently, differences in hillslope relief or mean hillslope gradient S_h (the ratio of hillslope relief over the horizontal length of the hillslope) in our field area should be driven by differences in erosion rates (Roering et al., 2007; Hurst et al., 2012). These quantities can serve as a proxy for erosion rates as long as slope gradients remain gentler than a threshold slope beyond which landsliding processes begin to dominate; this threshold slope typically varies between 0.8 and $1.2\,\mathrm{m\,m^{-1}}$ (e.g. Roering et al., 1999; Binnie et al., 2007; DiBiase et al., 2010; Matsushi and Matsuzaki, 2010) and was estimated $\sim 0.8\,\mathrm{m\,m^{-1}}$ in the study area (Hurst et al., 2012). In the Bald Rock basin we see that both D_{50} and D_{84} seem to generally increase with increasing S_h (Fig. 8) and therefore with erosion rate (Hurst et al., 2012). It is notable that sediment flux is directly related to erosion rates: a doubling of erosion rate should lead to a doubling of sediment flux to the river. An increase in erosion rate and hillslope steepness will therefore result in rivers being supplied with larger amounts of coarser sediment, making an increase in erosion rate more likely to influence fluvial sediment GSD than a simple change in source GSD.

Erosion rates and soil thicknesses combine to control how long particles spend in the soil (e.g. Small et al., 1999; Mudd and Furbish, 2006; Brantley and White, 2009; Mudd and Yoo, 2010; Yoo et al., 2011). A greater time spent in the soil gives particles longer exposure to processes that can reduce grain sizes, such as exposure to salt weathering (e.g. Wells et al., 2008), fracturing of rock due to root growth and tree throw (e.g. Gabet and Mudd, 2010; Roering et al., 2010), and/or clay and secondary mineral formation due to chemical weathering (e.g. Yoo and Mudd, 2008; Maher, 2010; Sweeney et al., 2012). We infer that the time particles spend within the weathering zone is significantly shorter in the steeper transects, giving chemical weathering processes less time to weaken parent material and resulting in coarser sediment. It can be seen in the fraction finer than 10 mm that the lower the mean hillslope gradient (and thus, we infer, the lower the erosion rate), the higher the clay and silt content and the lower the content in the fraction 1–10 mm (Fig. 7b). The particle size distributions for this fraction tend to be bimodal, exhibiting a low at 0.5–1 mm; this is consistent with previous observations that rocks which weather to sand (e.g. granite, sandstone) will produce a distinct bimodal distribution compared to rocks which weather to clays (Wolcott, 1988; Marshall and Sklar, 2012).

One metric to describe how long particles remain in the soil is the turnover time, which is the ratio of soil thickness to erosion rate multiplied by the ratio of soil density to rock density (Almond et al., 2007; Mudd and Yoo, 2010). In a steadily eroding soil, the turnover time is equivalent to the mean residence time of the particles (Mudd and Yoo, 2010). We quantified turnover time in the two "equilibrated" transects above and below the break in slope (POMD and BRC, respectively; Figs. 2 and 3). In our study area, erosion rates can be estimated as a function of hilltop curvature (Hurst et al., 2012): we calculate erosion rates of 0.06 and $0.1\,\mathrm{mm\,ka^{-1}}$ for POMD and BRC, respectively. Soil thickness is 0.51 ± 0.09 m and 0.45 ± 0.12 m for POMD and BRC, respectively (Yoo et al., 2011). Assuming a soil to rock density ratio of 1/2, a ratio common in granitic landscapes (Heimsath et al., 2001; Riggins et al., 2011), we calculate a turnover time of ~ 4.3 and 2.3 ka for the plateau and steepened landscape transects, respectively, showing that landscape steepening causes a halving of the turnover time in our study area.

Geochemical analysis of these soils shows that chemical weathering is most pronounced in the plateau transect POMD ($S_h = 0.38\,\mathrm{m\,m^{-1}}$, Figs. 2 and 3); it has the highest pedogenic crystalline iron oxide concentrations and is also the most enriched in Zr and Ti, indicating a greater extent of chemical weathering (Yoo et al., 2011). Thus the difference in grain size amongst the hillslope samples can be at least partially explained through a chemical weathering mechanism: weathering of primary silicate minerals results in clay production and so one would expect more chemically weathered soils to be enriched in clays, as is the case in our field area (Fig. 7). Chemical weathering does not break up coarse clasts directly, but it can make clasts more susceptible to physical breakdown by weakening the clasts. We found that in the steep FTA, BRC and BRB transects ($S_h = 0.67$–$0.84\,\mathrm{m\,m^{-1}}$, Fig. 3), coarse clasts appeared to be nearly pristine in terms of chemical weathering: there was little iron oxide staining, and these clasts would ring when hit with a rock hammer (Fig. 4a). Clasts within the POMD transect tended to be stained with iron oxide (as supported by increased pedogenic crystalline Fe content; Yoo et al., 2011) and could be easily broken with a rock hammer. While these are admittedly qualitative observations, they are supported by the geochemical data which show enhanced weathering in POMD relative to FTA, BRC and BRB (Yoo et al., 2011).

4.2 Landscape steepness and fluvial sediment GSD in mountainous landscapes

Our data show that fluvial sediment grain size seems to generally increase with increasing flow competence (Fig. 10). The data are noisy but the trends are significant in the Adams Creek basin, where there is a clear increase in both flow competence and sediment grain size over the prominent break in slope that separates the steepened landscape from the relict topography (Table 1, Figs. 9a and 10). The exponents in the power relationship between grain size and ω'_m are lower than the value of 2/3 expected from theory and flume experiments with uniform grain size (Table 1), which may reflect significant hiding/exposure effects in sediment composed of such a wide range of grain sizes (up to boulder size) (Whitaker and Potts, 2007; Bathurst, 2013). The exponent on D_{50} tends to be higher than on D_{84}, indicating that the bulk of the sediment coarsens faster than the coarse tail of the distribution, though the exponents are not statistically discernible (Table 1). Depletion in fines could result in sediment coarsening, but it cannot be the sole cause for coarsening in our case: the plateau sites would still be significantly finer than the steepened landscape sites in Adams Creek even after complete removal of their fraction finer than 1, 2 or even 10 mm (Fig. 9a).

In a situation where all grain sizes are available for fluvial transport, increasing flow competence should lead to an increase in grain size through selective entrainment of larger grains. In our study area, field observations and inspection of the 1 m resolution lidar data suggest that a change in sediment source is also responsible for the increase in fluvial sediment grain size. On the plateau, hillslope gradient rarely exceeds $0.7\,\mathrm{m\,m^{-1}}$; hillslopes are soil-mantled and we find no evidence of landslides. In addition, we find no coarse sediment available for transport along the studied rivers on the plateau: clasts larger than cobble size are very rare on the plateau, whereas boulders are widespread on the steepened landscape (e.g. see D_{100} data in Fig. 10 and Table A2). This suggests that the fine-grained nature of the fluvial sediment on the plateau is primarily due to the scarcity of coarse sediment supply (Fig. 11a). Below the break in slope, landslides, scree cones and debris flows supply coarse sediment to the channels: evidence of reworked debris-flow deposits and selective mobilisation of sediment emplaced by mass wasting processes is widespread along the rivers below the break in slope (Fig. 11b). We therefore interpret the increase in sediment coarseness from the plateau to the steepened landscape as a result of an increase in both flow competence and the size of the sediment supplied from hillslopes to the channels. This observation is consistent with previous studies in tectonically or climatically perturbed landscapes. Whittaker et al. (2010) showed that the grain size of fluvial sediment along rivers in the Apennines increases at the transition from low-relief soil-mantled landscape to steep high-relief landslide-prone landscape (see also Whittaker et al., 2007; Attal et al., 2011). Attal and Lavé (2006) also showed that fluvial sediment grain size along the Marsyandi River (Himalayas) increases at the transition from previously glaciated till-covered landscape to steep high-relief landslide-prone landscape; Attal and Lavé's (2006) measurements further indicated that till was a source of finer sediment to the rivers than landslides.

Two basins depart noticeably from the general trend (Fig. 10). The Bald Rock basin sites (solid triangles in

Figure 11. Diagrams and photographs illustrating the contrast in sediment dynamics between sites on the plateau and sites across the steepened landscape. **(a)** On the plateau, rivers are fed with soil material and there is a clear lack of coarse material, even along large rivers; the drainage area of Cascade Creek where the photo was taken is $58\,\mathrm{km^2}$ and the largest pebble found in the area has a *b* axis of 260 mm. **(b)** Along the steepened reaches of the rivers and in the gorges, a wide range of grain size is available and "lag" deposits are widespread, that is, concentrations of very large clasts resulting from the reworking by fluvial processes of material deposited by mass wasting processes. The very large boulders are up to 10 m in size and are very unlikely to be mobilised by fluvial processes. Photographs show evidence of reworking of debris-flow deposits near the confluence of Cascade Creek with the Feather River (top panels), selective mobilisation of sediment supplied by landslides, and rock falls below Bald Rock Dome in the Feather River (bottom panels). Standing people are circled on the photographs for scale.

Fig. 10) may represent a situation where the grain size of the fluvial sediment is, at least temporarily, limited by sediment supply. In this small basin (0.7 km^2), fluvial sediment is fine-grained and scarce, with D_{50} and D_{84} not exceeding 13 and 54 mm, respectively (Fig. 9c). The channel has abundant bedrock exposure; no sediment was found in the channel downstream of the break in slope. We interpret this situation as resulting from a shortage of sediment in the channel. The basin is soil-mantled and entirely vegetated, and we found no evidence of recent slope failure within the basin, even below the break in slope. This may represent a transient situation where the material supplied to the channel has been completely evacuated from the basin; sediment will be replenished in the channel when sediment flux from hillslopes is – at least temporarily – substantially increased, e.g. following forest fires (Gabet, 2003; Lamb et al., 2011, 2013; DiBiase and Lamb, 2013; Riley et al., 2013).

The Bean Creek basin appears to have undergone a different type of response compared to the other basins. It exhibits no obvious topographic break in slope delimiting the plateau from the steepened landscape (Figs. 1, 5a and 9b). The whole basin is steeper than the plateau basins but less steep than the steepened landscape in adjacent basins. It is steep enough to experience landslides and debris flows, both processes supplying coarse sediment to the river, as observed in the field. Flow competence at the Bean Creek sites tends to be higher than at the plateau sites and lower than at the steepened landscape sites (Fig. 10). Fluvial sediment grain size in Bean Creek is coarser than at the plateau sites and within the range of values measured at the steepened landscape sites, testifying again to the influence of source type on fluvial sediment GSD (Fig. 10).

A series of observations suggest a rapid response of the hillslopes (in terms of source characteristics) to river steepening. Firstly, we observe that, only a few hundreds of metres downstream of the main topographic break in slope, fluvial sediment is significantly coarser than on the plateau and includes boulders that are typically absent on the plateau, as exemplified by the Adams Creek data (Fig. 9a). As rivers steepen and increase their competence in response to the increase in incision rate along the main stem of the Feather River, the adjacent hillslopes must steepen and respond rapidly to provide rivers with coarse sediment. Secondly, we note the absence of inner gorges in the steepened landscape, suggesting a tight coupling between the channel and hillslopes and a rapid response of hillslopes to an increase in the rate of river downcutting. These observations are consistent with the topographic analysis of Hurst et al. (2012) in the study area, which suggests that the response time of hillslopes in this landscape is rapid relative to that of the stream network. This rapid response means that the increase in flow competence and change in sediment sources occur at a similar location along the rivers, making isolating the relative influences of these two controls on the grain size of the sediment transported by the rivers challenging.

5 Conclusions

We have quantified the grain size distribution of sediment in both source areas (hillslope soils and landslide deposits) and channels in a mountainous landscape where the underlying lithology is exclusively tonalite and where erosion rates vary over an order of magnitude (Riebe et al., 2000; Hurst et al., 2012). We find that the coarseness of hillslope sediment increases with increasing mean hillslope gradient (where mean hillslope gradient represents the ratio of hillslope relief over the horizontal length of the hillslope) and erosion rate. We hypothesise that, in our soil samples, this is due to a decrease in residence time of rock fragments, causing particles to be exposed for shorter periods of time to processes that can reduce grain sizes, such as exposure to salt weathering, fracturing of rock due to root growth and tree throw, and/or clay formation due to chemical weathering. For slopes in excess of 0.7 $m\,m^{-1}$, mass wasting processes (e.g. landsliding) and scree cones supply much coarser sediment to rivers, with D_{50} and D_{84} more than 1 order of magnitude larger than in soils. Rapidly eroding landscapes also contribute more sediment to rivers than slowly eroding slopes per unit area; thus for basins of equal size a rapidly eroding basin will contribute a much larger amount of coarse sediment to the river network than a slowly eroding basin.

Changes in grain size and sediment fluxes from hillslopes are shown to impact the grain size of the sediment transported by the rivers. Fluvial sediment in the tributary basins hanging above the rapidly incising Feather River exhibits a significant downstream coarsening. The locus of the increase in grain size coincides with the prominent break in slope that developed along the river profiles in response to an increase in incision rate along the main stem of the Feather River. This increase in grain size is caused by an increase in flow competence (mostly driven by channel steepening) as well as a change in sediment source and in sediment dynamics: on the plateau upstream of the break in slope, rivers transport easily mobilised fine-grained sediment derived exclusively from soils. Downstream of the break in slope, mass wasting processes supply a wide range of grain sizes (up to bus-sized boulders) that rivers entrain selectively, depending on the competence of their flow. The absence of evidence, below the break in slope, of river reaches where the grain size of the fluvial sediment is limited by the grain size of the sediment supplied from hillslopes suggests that the response time of hillslopes to an increase in the rate of base-level lowering is rapid relative to that of the stream network in this landscape.

Data availability

Cumulative grain size distributions for all volumetric samples are available in two Excel spreadsheets in the supplementary material. The spreadsheet "Data_SoilLandslidesGSD_FeatherRiver_Attal2015" contains the grain size distributions of the 11 hillslope sites (one sheet per site): 8 in soils and 3 in landslides. The spreadsheet "Data_RiverSedimentGSD_FeatherRiver_Attal2015" contains the grain size distributions of the 21 fluvial sediment sites (one sheet per site).

Appendix A: Tables with description of sites and data

Table A1. Description of hillslope sites data. D_{100} is maximum grain size. Plus or minus values are calculated according to procedure described in Sect. 2.3.

Site ID	D_{50} (mm)	−	+	D_{84} (mm)	−	+	D_{100} (mm)	Total sample mass (kg)	% mass largest clast	% mass coarser than 1 mm	±	% mass coarser than 10 mm	±	Mean hillslope gradient S_h (m m^{-1})
Soils														
POMD2	0.12	< 0.01	< 0.01	1.25	< 0.01	< 0.01	58.3	227.0	0.1 %	21	< 0.5	1	< 0.5	0.38
POMD4	0.11	< 0.01	< 0.01	1.30	< 0.01	< 0.01	63.4	163.3	0.2 %	22	< 0.5	1	< 0.5	0.38
POMD6	0.10	< 0.01	< 0.01	1.65	< 0.01	< 0.01	40.0	550.4	0.02 %	29	< 0.5	2	< 0.5	0.38
FTA1	0.34	< 0.01	< 0.01	1.92	< 0.01	< 0.01	26.9	122.3	0.02 %	32	< 0.5	1	< 0.5	0.67
FTA9	0.36	0.11	0.12	3.55	1.43	109.20	220.0	173.0	8.5 %	39	6	12	9	0.67
BRC3	0.47	0.04	0.04	4.79	0.65	0.76	148.8	211.7	2.2 %	43	1	8	2	0.76
BRC0	0.22	< 0.01	< 0.01	2.00	< 0.01	< 0.01	79.1	189.0	0.4 %	31	< 0.5	2	< 0.5	0.76
BRB8-9h	1.15	< 0.01	< 0.01	7.02	< 0.01	< 0.01	77.3	63.8	1.0 %	53	< 0.5	11	< 0.5	0.83
Landslides														
LD1	34.95	7.62	9.86	148.35	28.98	23.97	176.8	84.8	9.0 %	82	2	69	3	0.84
LD2	66.16	8.56	10.79	138.57	30.73	9.34	189.2	113.3	8.3 %	87	1	78	2	0.89
LD3	46.41	18.21	20.18	133.01	45.48	68.29	207.2	81.4	15.2 %	82	3	67	6	0.74

Table A2. Description of fluvial sites data. Plus or minus values are calculated according to procedure described in Sect. 2.3.

Site ID	Subsurface									Surface									
	D_{50} (mm)			D_{84} (mm)			D_{100}	Total	%	D_{50} (mm)			D_{84} (mm)			D_{100}	Number	Nodes	% nodes
		−	=		−	+	(mm)	sample mass (kg)	mass largest clast		−	+		−	+	(mm)	of nodes sampled	covered by largest clast	covered by largest clast
Adams Creek																			
BAC5	11.3	0.9	0.9	35.7	4.8	8.2	99.4	28.1	5 %	22.1	0.2	1.3	54.8	2.0	5.7	114.7	74	2	3 %
BAC6	4.8	0.2	0.3	19.3	1.2	1.6	53.3	42.8	2 %	5.6	0.4	0.1	19.4	1.9	3.6	38.2	92	3	3 %
										9.2	0.1	0.1	18.9	0.1	0.7	44.3	89	1	1 %
										6.3	0.1	0.2	19.7	1.1	4.3	45.0	69	1	1 %
BAC1	19.7	2.0	1.8	64.5	9.5	11.8	120.1	49.4	5 %	40.7	2.2	3.4	100.5	16.1	4.5	135.0	68	4	6 %
BAC2	36.2	8.0	14.0	114.6	14.4	17.6	142.1	36.6	11 %	31.4	7.4	9.6	96.9	12.7	13.8	112.3	94	13	14 %
BAC4	83.2	27.5	20.4	180.3	15.8	4.8	185.8	71.6	12 %	135.4	23.7	18.9	289.1	6.0	131.5	424.9	87	12	14 %
										108.7	4.1	7.9	167.6	14.9	125.1	295.2	82	10	12 %
BAC3	101.4	18.6	22.3	166.7	3.9	1.7	169.1	57.2	12 %	185.6	1.6	3.2	389.4	33.2	6.4	466.6	66	2	3 %
										178.0	6.9	0.7	244.7	8.0	9.4	259.6	86	9	10 %
										116.6	2.6	1.6	136.2	5.0	49.4	209.2	59	5	8 %
Bald Rock B.																			
BRB-f2	1.3	0.1	0.1	12.5	2.0	2.4	78.1	27.5	3 %	5.9	0.1	0.1	29.8	0.7	2.3	62.4	88	1	1 %
BRB-f1	1.7	0.1	0.1	45.1	4.5	5.1	85.3	34.8	2 %	7.8	0.4	0.2	34.0	10.8	12.3	223.5	77	5	6 %
BRB-10f	1.3	0.1	0.0	11.6	4.4	4.2	80.0	23.0	4 %	3.7	0.1	0.1	16.7	0.6	0.9	36.6	91	1	1 %
BRB-8/9f	12.5	1.1	1.1	53.7	8.8	11.1	94.9	26.6	4 %	8.6	1.0	1.7	44.3	27.9	46.0	115.5	66	5	8 %
Bean Creek																			
BC6	82.7	31.1	38.2	184.4	25.8	18.5	200.9	59.4	19 %	73.5	4.4	23.3	145.3	3.1	31.1	177.9	60	8	13 %
BC5	42.0	11.7	7.1	107.7	7.4	7.6	132.4	36.9	9 %	96.6	21.9	1.0	191.9	65.6	16.7	211.5	84	10	12 %
BC3	23.6	1.1	1.3	68.9	4.4	4.9	96.9	44.6	3 %	33.8	19.4	3.2	81.9	7.8	12.3	120.2	74	5	7 %
BC4	52.8	5.7	7.1	98.9	5.8	30.0	163.3	61.5	10 %	103.4	7.7	11.4	159.1	3.7	2.4	328.5	55	4	7 %
										66.0	7.6	10.9	149.9	45.0	36.1	192.5	83	10	12 %
BC2	62.9	10.7	14.9	166.7	49.7	23.3	211.0	118.1	11 %	71.5	0.3	0.3	93.8	1.3	1.1	146.5	59	1	2 %
BC1	80.5	18.8	10.5	134.0	19.6	41.9	211.4	114.2	11 %	130.1	14.7	22.2	270.4	39.2	33.9	309.3	93	11	12 %
Berry Creek																			
Baldf	2.2	0.4	0.6	25.1	9.2	29.4	83.2	11.4	7 %	2.1	0.1	0.1	78.7	56.2	78.6	161.2	89	10	11 %
										10.0	0.1	0.6	35.1	8.9	3.3	138.9	94	3	3 %
										10.7	0.2	0.0	34.8	0.5	5.1	133.8	98	2	2 %
BerC1f	22.5	0.5	0.5	43.9	3.4	5.1	80.0	28.6	3 %	23.4	0.8	1.8	53.3	5.7	13.4	92.2	77	5	6 %
										25.2	2.5	1.0	69.5	6.5	10.5	104.8	98	7	7 %
										20.1	0.1	0.7	47.4	3.9	3.7	77.3	84	3	4 %
										22.2	1.3	1.4	53.3	0.5	0.6	86.6	97	1	1 %
										26.7	4.6	4.5	51.5	7.7	17.7	75.1	98	13	13 %

Table A2. Continued.

Site ID	Subsurface									Surface									
	D_{50} (mm)			D_{84} (mm)			D_{100}	Total	%	D_{50} (mm)			D_{84} (mm)			D_{100}	Number	Nodes	% nodes
		−	=		−	+	(mm)	sample mass (kg)	mass largest clast		−	+		−	+	(mm)	of nodes sampled	covered by largest clast	covered by largest clast
Cascade Creek																			
CC	35.1	2.3	2.6	81.7	1.4	2.4	117.3	63.6	4 %	42.8	1.2	0.6	76.7	1.2	3.1	117.3	94	3	3 %
										36.4	0.6	0.3	57.7	1.7	0.9	87.5	96	2	2 %
										101.0	1.1	4.5	166.2	29.2	63.9	260.4	93	8	9 %
Little North Fork River																			
LNF3										131.7	7.5	4.3	338.9	10.1	5.3	448.0	96	5	5 %
										144.8	11.1	0.4	624.8	1.2	2.5	915.8	96	2	2 %
LNF1-2	55.1	20.8	39.7	202.1	92.7	93.6	289.6	153.8	22 %	266.7	50.0	1.1	570.7	3.0	1.2	599.2	99	5	5 %
	37.1	4.7	6.6	146.6	20.3	9.6	179.0	119.5	7 %	283.8	17.1	57.8	522.3	77.1	534.5	1073.5	85	9	11 %
										276.8	14.5	4.4	433.0	11.0	24.2	1072.1	92	2	2 %
										109.5	3.9	5.2	302.8	1.1	2.9	563.6	98	3	3 %
										149.0	4.6	4.2	243.7	1.7	1.6	795.9	97	4	4 %
										66.1	0.6	1.3	283.6	0.5	0.5	316.4	96	1	1 %

Table A3. Location and description of fluvial sites. Coordinates are in the UTM reference system (WGS1984). For each river, sites are ordered downstream (asterisk indicates sites on tributaries). Drainage area and channel slope (calculated over 100 m) were extracted from the 1 m resolution lidar-derived DEM, except for the Cascade Creek and lowermost Berry Creek sites, where the 10 m resolution USGS DEM data were used instead. Minimum and maximum channel widths were measured over a 100 m long stretch on lidar-derived DEM complemented with Google Maps images; mean width is given, with ± representing the difference between mean and extrema.

Site ID	Easting (m)	Northing (m)	Elevation (m)	Drainage area (km^2)	Slope ($m\,m^{-1}$)	Width (m)
			Adams Creek			
BAC5	644 550	4 394 120	1012	0.63	0.032	2.5 ± 0.5
BAC6	644 460	4 393 420	970	1.39	0.028	4 ± 2
BAC1	644 870	4 392 860	918	2.02	0.118	4 ± 1
BAC2	644 840	4 392 720	887	2.07	0.345	2.5 ± 0.5
BAC4*	644 880	4 392 640	849	7.79	0.149	3 ± 1
BAC3	644 910	4 392 640	843	9.87	0.169	3 ± 1
			Bald Rock Basin			
BRB-f2	645 245	4 389 820	739	0.17	0.206	3 ± 1
BRB-f1	645 420	4 389 900	703	0.32	0.136	3 ± 1
BRB-10f	645 485	4 389 940	692	0.37	0.134	4 ± 1
BRB-8/9f	645 575	4 390 100	643	0.52	0.271	4 ± 1
			Bean Creek			
BC6	644 500	4 390 200	947	0.15	0.154	2 ± 1
BC5	643 405	4 387 800	492	5.81	0.074	5 ± 2
BC3	643 535	4 387 220	454	10.04	0.038	4.5 ± 1.5
BC4*	643 500	4 387 200	456	3.15	0.032	4 ± 1
BC2	643 425	4 386 220	362	14.23	0.023	6 ± 2
BC1	643 400	4 386 140	360	14.33	0.026	6 ± 2
			Berry Creek			
Baldf	642 215	4 389 860	946	0.04	0.096	2 ± 1
BerC1f	637 324	4 389 271	610	23.16	0.006	4.5 ± 1.5
			Cascade Creek			
CC	654 814	4 397 091	1119	58.42	0.011	12.5 ± 2.5
			Little North Fork River			
LNF3	647 534	4 400 281	839	104.13	0.075	20 ± 5
LNF1-2	648 124	4 396 981	492	119.21	0.052	20 ± 5

Acknowledgements. This work was funded by the Carnegie Trust for the Universities of Scotland (grant awarded to M. Attal), the US National Science Foundation (grant EAR0819064 awarded to K. Yoo and S. Mudd) and the UK Natural Environment Research Council (grant NE/H001174/1 awarded to S. Mudd). We are extremely grateful to K. Maher and K. Mayer for sample handling and valuable assistance in the field. We would like to thank L. Sklar and J. Marshall for stimulating discussions and for allowing us to use some equipment in their sedimentology lab. We thank the four anonymous reviewers (including two reviewers of an earlier version of the manuscript that had been submitted to another journal) as well as the associate editor J. West for their very helpful and constructive comments, which allowed us to greatly improve the manuscript. We thank H. Parsons and M. Johnston for help collecting the landslide data.

Edited by: A. J. West

References

Almond, P., Roering, J., and Hales, T. C.: Using soil residence time to delineate spatial and temporal patterns of transient landscape response, J. Geophys. Res., 112, F03S17, doi:10.1029/2006JF000568, 2007.

Armitage, J. J., Duller, R. A., Whittaker, A. C., and Allen, P. A.: Transformation of tectonic and climatic signals from source to sedimentary archive, Nat. Geosci., 4, 231–235, doi:10.1038/ngeo1087, 2011.

Attal, M. and Lavé, J.: Changes of bedload characteristics along the Marsyandi River (central Nepal): Implications for understanding hillslope sediment supply, sediment load evolution along fluvial networks, and denudation in active orogenic belts, Geological Society of America Special Paper 398, Geological Society of America, Boulder, Colorado, 143–171, doi:10.1130/2006.2398(09), 2006.

Attal, M., Tucker, G. E., Whittaker, A. C., Cowie, P. A., and Roberts, G. P.: Modeling fluvial incision and transient landscape evolution: Influence of dynamic channel adjustment, J. Geophys. Res., 113, F03013, doi:10.1029/2007JF000893, 2008.

Attal, M., Cowie, P. A., Whittaker, A. C., Hobley, D., Tucker, G. E., and Roberts, G. P.: Testing fluvial erosion models using the transient response of bedrock rivers to tectonic forcing in the Apennines, Italy, J. Geophys. Res., 116, F02005, doi:10.1029/2010JF001875, 2011.

Bathurst, J. C., Graf, W. H., and Cao, H. H.: Bed load discharge equations for steep mountain rivers, in: Sediment Transport in Gravel-bed Rivers, edited by: Thorne, C. R., Bathurst, J. C., and Hey, R. D., John Wiley, Chichester, UK, 453–477, 1987.

Bathurst, J. C.: Critical conditions for particle motion in coarse bed materials of nonuniform size distribution, Geomorphology, 197, 170–184, doi:10.1016/j.geomorph.2013.05.008, 2013.

Bennett, G. L., Molnar, P., McArdell, B. W., and Burlando, P.: A probabilistic sediment cascade model of sediment transfer in the Illgraben, Water Resour. Res., 50, doi:10.1002/2013WR013806, 2014.

Bierman, P. and Steig, E. J.: Estimating rates of denudation using cosmogenic isotope abundances in sediment, Earth Surf. Proc. Land., 21, 125–139, doi:10.1002/(SICI)1096-9837(199602)21:2<125::AID-ESP511>3.0.CO;2-8, 1996.

Binnie, S. A., Phillips, W. M., Summerfield, M. A., and Fifield, L. K.: Tectonic uplift, threshold hillslopes, and denudation rates in a developing mountain range, Geology, 35, 743–746, 2007.

Brantley, S. L. and White, A. F.: Approaches to Modeling Weathered Regolith in Thermodynamics and Kinetics of Water-Rock Interaction, edited by: Oelkers, E. H. and Schott, J., Rev. Mineral. Geochem., 70, 435–484, 2009.

Brown, E. T., Stallard, R. F., Larsen, M. C., Raisbeck, G. M., and Yiou, F.: Denudation rates determined from the accumulation of in situ-produced ^{10}Be in the Luquillo experimental forest, Puerto Rico, Earth Planet. Sc. Lett., 129, 193–202, doi:10.1016/0012-821X(94)00249-X, 1995.

Buffington, J. M. and Montgomery, D. R.: A systematic analysis of eight decades of incipient motion studies, with special reference to gravel-bedded rivers, Water Resour. Res., 33, 1993–2029, 1997.

Bunte, K. and Abt, S. R.: Sampling surface and subsurface particle-size distributions in wadable gravel-and cobble-bed streams for analyses in sediment transport, hydraulics, and streambed monitoring, Gen. Tech. Rep. RMRS-GTR-74, US Department of Agriculture, Forest Service, Rocky Mountain Research Station, Fort Collins, CO, p. 428, 2001.

Burbank, D. W, Leland, J., Fielding, E., Anderson, R. S., Brozovic, N., Reid, M. R., and Duncan, C.: Bedrock incision, rock uplift and threshold hillslopes in the northwestern Himalayas, Nature, 379, 505–510, 1996.

Casagli, N., Ermini, L., and Rosati, G.: Determining grain size distribution of the material composing landslide dams in the Northern Apennines: sampling and processing methods, Eng. Geol., 69, 83–97, doi:10.1016/S0013-7952(02)00249-1, 2003.

Chatanantavet, P., Lajeunesse, E., Parker, G., Malverti, L., and Meunier, P.: Physically based model of downstream fining in bedrock streams with lateral input, Water Resour. Res., 46, W02518, doi:10.1029/2008WR007208, 2010.

Church, M., McLean, D. G., and Walcott, J. F.: River bed gravels: sampling and analysis, in: Sediment Transport in Gravel-Bed Rivers, edited by: Thorne, C. R., Bathurst, J. C., and Hey, R. D., John Wiley and Sons, Chichester, 43–88, 1987.

Cowie, P. A., Whittaker, A. C., Attal, M., Roberts, G. P., Tucker, G. E., and Ganas, A.: New constraints on sediment-flux-dependent river incision: Implications for extracting tectonic signals from river profiles, Geology, 36, 535–538, doi:10.1130/G24681A.1, 2008.

Crosby, B. T. and Whipple, K. X.: Knickpoint initiation and distribution within fluvial networks: 236 waterfalls in the Waipaoa River, North Island, New Zealand, Geomorphology, 82, 16–38, doi:10.1016/j.geomorph.2005.08.023, 2006.

DiBiase, R. A. and Lamb, M. P.: Vegetation and wildfire controls on sediment yield in bedrock landscapes, Geophys. Res. Lett., 40, 1093–1097, doi:10.1002/grl.50277, 2013.

DiBiase, R. A., Whipple, K. X., Heimsath, A. M., and Ouimet, W. B.: Landscape form and millennial erosion rates in the San Gabriel Mountains, CA, Earth Planet. Sc. Lett., 289, 134–144, doi:10.1016/j.epsl.2009.10.036, 2010.

Duller, R. A., Whittaker, A. C., Fedele, J. J., Whitchurch, A. L., Springett, J., Smithells, R., Fordyce, S., and Allen, P. A.: From grain size to tectonics, J. Geophys. Res., 115, F03022, doi:10.1029/2009JF001495, 2010.

Gabet, E. J.: Post-fire thin debris flows: field observations of sediment transport and numerical modeling, Earth Surf. Proc. Land., 28, 1341–1348, doi:10.1002/esp.590, 2003.

Gabet, E. J.: Late Cenozoic uplift of the Sierra Nevada, California? A critical analysis of the geomorphic evidence, Am. J. Sci., 314, 1224–1257, doi:10.2475/08.2014.03, 2014.

Gabet, E. J. and Mudd, S. M.: Bedrock erosion by root fracture and tree throw: A coupled biogeomorphic model to explore the humped soil production function and the persistence of hillslope soils, J. Geophys. Res., 115, F04005, doi:10.1029/2009JF001526, 2010.

Gilbert, G. K.: Report on the geology of the Henry Mountains: Geographical and geological survey of the rocky mountain region, US Government print-off, Washington, D.C., p. 160, 1877.

Granger, D. E., Kirchner, J. W., and Finkel, R.: Spatially averaged long-term erosion rates measured from in situ-produced cosmogenic nuclides in alluvial sediment, J. Geol., 104, 249–257, doi:10.1086/629823, 1996.

Heimsath, A. M., Chappell, J., Dietrich, W. E., Nishiizumi, K., and Finkel, R. C.: Late Quaternary erosion in southeastern Australia: A field example using cosmogenic nuclides, Quatern. Int., 83–85, 169–185, doi:10.1016/S1040-6182(01)00038-6, 2001.

Heller, P. L., Beland, P. E., Humphrey, N. F., Konrad, S. K., Lynds, R. M., McMillan, M. E., Valentine, K. E., Widman, Y. A., and Furbish, D. J.: Paradox of downstream fining and weathering-rind formation in the lower Hoh River, Olympic Peninsula, Washington, Geology, 29, 971–974, 2001.

Hobley, D. E. H., Sinclair, H. D., Mudd, S. M., and Cowie, P. A.: Field calibration of sediment flux dependent river incision, J. Geophys. Res., 115, F04017, doi:10.1029/2010JF001935, 2011.

Hovius, N., Stark, C. P., and Allen, P. A.: Sediment flux from a mountain belt derived by landslide mapping, Geology, 25, 231–234, doi:10.1130/0091-7613(1997)025<0231:SFFAMB>2.3.CO;2, 1997.

Hurst, M. D., Mudd, S. M., Walcott, R., Attal, M., and Yoo, K.: Using hilltop curvature to derive the spatial distribution of erosion rates, J. Geophys. Res., 117, F02017, doi:10.1029/2011JF002057, 2012.

Kellerhals, R. and Bray, D. I.: Sampling procedures for coarse fluvial sediments, J. Hydraul. Div.-ASCE, 97, 1165–1180, 1971.

Knighton, A. D.: Longitudinal changes in the size and shape of stream bed material: evidence of variable transport conditions, Catena, 9, 25–34, 1982.

Lamb, M. P., Scheingross, J. S., Amidon, W. H., Swanson, E., and Limaye, A.: A model for fire-induced sediment yield by dry ravel in steep landscapes, J. Geophys. Res., 116, F03006, doi:10.1029/2010JF001878, 2011.

Lamb, M. P., Levina, M., DiBiase, R. A., and Fuller, B. M.: Sediment storage by vegetation in steep bedrock landscapes: Theory, experiments, and implications for postfire sediment yield, J. Geophys. Res., 118, 1147–1160, doi:10.1002/jgrf.20058, 2013.

Lane, S. N., Tayefi, V., Reid, S. C., Yu, D., and Hardy, R. J.: Interactions between sediment delivery, channel change, climate change and flood risk in a temperate upland environment, Earth Surf. Proc. Land., 32, 429–446, 2007.

Lavé, J. and Burbank, D.: Denudation processes and rates in the Transverse Ranges, southern California: Erosional response of a transitional landscape to external and anthropogenic forcing, J. Geophys. Res., 109, F01006, doi:10.1029/2003JF000023, 2004.

Lohse, K. A. and Dietrich, W. E.: Contrasting effects of soil development on hydrological properties and flow paths, Water Resour. Res., 41, W12419, doi:10.1029/2004WR003403, 2005.

Maher, K.: The dependence of chemical weathering rates on fluid residence time, Earth Planet. Sc. Lett., 294, 101-110, doi:10.1016/j.epsl.2010.03.010, 2010.

Marshall, J. A. and Sklar, L. S.: Mining soil databases for landscape-scale patterns in the abundance and size distribution of hillslope rock fragments, Earth Surf. Proc. Land., 37, 287–300, doi:10.1002/esp.2241, 2012.

Matsushi, Y. and Matsuzaki, H.: Denudation rates and threshold slope in a granitic watershed, central Japan, Nucl. Instrum. Meth. Phys. Res. B, 268, 1201–1204, 2010.

Michael, N. A., Whittaker, A. C., Carter, A., and Allen, P. A.: Volumetric budget and grain-size fractionation of a geological sediment routing system: Eocene Escanilla Formation, south-central Pyrenees, Geol. Soc. Am. Bull., 126, 585–599, doi:10.1130/B30954.1, 2014.

Mudd, S. M. and Furbish, D. J.: Using chemical tracers in hillslope soils to estimate the importance of chemical denudation under conditions of downslope sediment transport, J. Geophys. Res., 111, F02021, doi:10.1029/2005JF000343, 2006.

Mudd, S. M. and Yoo, K.: Reservoir theory for studying the geochemical evolution of soils, J. Geophys. Res., 115, F03030, doi:10.1029/2009JF001591, 2010.

Parker, G.: Selective sorting and erosion of river gravel, II: Applications, J. Hydraul. Eng.-ASCE, 117, 150–171, 1991.

Riebe, C. S., Kirchner, J. W., Granger, D. E., and Finkel, R. C.: Erosional equilibrium and disequilibrium in the Sierra Nevada, inferred from cosmogenic Al-26 and Be-10 in alluvial sediment, Geology, 28, 803–806, 2000.

Riggins, S. G., Anderson, R. S., Anderson, S. P., and Tye, A. M.: Solving a conundrum of a steady-state hilltop with variable soil depths and production rates, Bodmin Moor, UK, Geomorphology, 128, 73–84, doi:10.1016/j.geomorph.2010.12.023, 2011.

Riley, K. L., Bendick, R., Hyde, K. D., and Gabet, E. J.: Frequency-magnitude distribution of debris flows compiled from global data, and comparison with post-fire debris flows in the western U.S., Geomorphology, 191, 118–128, doi:10.1016/j.geomorph.2013.03.008, 2013.

Roering, J. J., Kirchner, J. W., and Dietrich, W. E.: Evidence for nonlinear, diffusive sediment transport on hillslopes and implications for landscape morphology, Water Resour. Res., 35, 853–870, 1999.

Roering, J. J., Kirchner, J. W., Sklar, L. S., and Dietrich, W. E.: Hillslope evolution by nonlinear creep and landsliding: An experimental study, Geology, 29, 143–146, 2001.

Roering, J. J., Perron, J. T., and Kirchner, J. W.: Functional relationships between denudation and hillslope fonn and relief, Earth Planet. Sc. Lett., 264, 245–258, 2007.

Roering, J. J., Marshall, J., Booth, A. M., Mort, M., and Jin, Q. S.: Evidence for biotic controls on topography and soil production, Earth Planet. Sc. Lett., 298, 183–190, 2010.

Saucedo, G. J. and Wagner, D. L. Geologic map of the Chico quadrangle 1 : 250,000, Regional Geologic Map Series, Map no. 7A, California Department of Conservation, Division of Mines and Geology, http://ngmdb.usgs.gov/Prodesc/proddesc_63087.htm (last access: March 2015), 1992.

Schoklitsch, A.: Handbuch des Wasserbaues, 3rd Edn., Springer-Verlag, Vienna, 1962.

Sklar, L. S. and Dietrich, W. E.: A mechanistic model for river incision into bedrock by saltating bed load, Water Resour. Res., 40, W06301, doi:10.1029/2003WR002496, 2004.

Sklar, L. S., Dietrich, W. E., Foufoula-Georgiou, E., Lashermes, B., and Bellugi, D.: Do gravel bed river size distributions record channel network structure?, Water Resour. Res., 42, W06D18, doi:10.1029/2006WR005035, 2006.

Small, E., Anderson, R., and Hancock, G.: Estimates of the rate of regolith production using ^{10}Be and ^{26}Al from an Alpine hillslope, Geomorphology, 27, 131–150, 1999.

Snyder, N. P., Whipple, K. X., Tucker, G. E., and Merritts, D. J.: Channel response to tectonic forcing: field analysis of stream morphology and hydrology in the Mendocino triple junction region, northern California, Geomorphology, 53, 97–127, 2003.

Stock, G. M., Anderson, R. S., and Finkel, R. C.: Pace of landscape evolution in the Sierra Nevada, California, revealed by cosmogenic dating of cave sediments, Geology, 32, 193–196, 2004.

Sweeney, K. E., Roering, J. J., Almond, P., and Reckling, T.: How steady are steady-state landscapes? Using visible–near-infrared soil spectroscopy to quantify erosional variability, Geology, 40, 807–810, doi:10.1130/G33167.1, 2012.

Wakabayashi, J. and Sawyer, T. L.: Stream incision, tectonics, uplift, and evolution of topography of the Sierra Nevada, California, J. Geol., 109, 539–562, 2001.

Wells, T., Hancock, G., and Fryer, J.: Weathering rates of sandstone in a semi-arid environment (Hunter Valley, Australia), Environ. Geol., 54, 1047–1057, doi:10.1007/s00254-007-0871-y, 2008.

West, A. J., Galy, A., and Bickle, M.: Tectonic and climatic controls on silicate weathering, Earth Planet. Sc. Lett., 235, 211–228, doi:10.1016/j.epsl.2005.03.020, 2005.

Whipple, K. X. and Tucker, G. E.: Implications of sediment-flux-dependent river incision models for landscape evolution, J. Geophys. Res., 107, 2039, doi:10.1029/2000JB000044, 2002.

White, A. and Brantley, S.: The effect of time on the weathering of silicate minerals: why do weathering rates differ in the laboratory and field?, Chemical Geology, 202, 479-506, 2003.

Whitaker, A. C. and Potts, D. F.: Analysis of flow competence in an alluvial gravel bed stream, Dupuyer Creek, Montana, Water Resour. Res., 43, W07433, doi:10.1029/2006WR005289, 2007.

Whittaker, A. C., Cowie, P. A., Attal, M., Tucker, G. E., and Roberts, G. P.: Contrasting transient and steady-state rivers crossing active normal faults: new field observations from the Central Apennines, Italy, Basin Research, 19, 529–556, doi:10.1111/j.1365-2117.2007.00337.x, 2007.

Whittaker, A. C., Attal, M., Cowie, P. A., Tucker, G. E., and Roberts, G. P.: Decoding temporal and spatial patterns of fault uplift using transient river long profiles, Geomorphology, 100, 506–526, doi:10.1016/j.geomorph.2008.01.018, 2008.

Whittaker, A. C., Attal, M., and Allen, P. A.: Characterizing the origin, nature and fate of sediment exported from catchments perturbed by active tectonics, Basin Research, 22, 809–828, doi:10.1111/j.1365-2117.2009.00447.x, 2010.

Wolcott, J.: Nonfluvial control of bimodal grain-size distributions in river-bed gravels, J. Sediment. Petrol., 58, 979–984, 1988.

Yoo, K. and Mudd, S. M.: Discrepancy between Mineral Residence Time and Soil Age: Implications for the Interpretation of Chemical Weathering Rates, Geology, 36, 35–38, doi:10.1130/G24285A.1, 2008.

Yoo, K., Weinmann, B., Mudd, S. M., Hurst, M. D., Attal, M., and Maher, K.: Evolution of hillslope soils: The geomorphic theater and the geochemical play, Appl. Geochem., 26, S149-S153, doi:10.1016/j.apgeochem.2011.03.054, 2011.

9

Field investigation of preferential fissure flow paths with hydrochemical analysis of small-scale sprinkling experiments

D. M. Krzeminska[1], T. A. Bogaard[1], T.-H. Debieche[2,4], F. Cervi[2,5], V. Marc[2], and J.-P. Malet[3]

[1]Water Resources Section, Faculty of Civil Engineering and Geosciences, Delft University of Technology, Delft, Netherlands

[2]Université d'Avignon et des Pays de Vaucluse, EMMAH UMR1114 INRA-UAPV, 33 rue Louis Pasteur, 84000 Avignon, France

[3]Institut de Physique du Globe de Strasbourg, CNRS UMR7516, Université de Strasbourg, Ecole et Observatoire des Sciences de la Terre, 5 rue Descartes, 67084 Strasbourg, France

[4]Water and Environment Team, Geological Engineering Laboratory, Jijel University, P.O. Box 98, 18000 Jijel, Algeria

[5]Dipartimento di Ingegneria Civile, Chimica, Ambientale e dei Materiali (DICAM) Università di Bologna, Viale Risorgimento 2, 40136, Bologna, Italy

Correspondence to: D. M. Krzeminska (dkrzeminska@onet.eu)

Abstract. The unsaturated zone largely controls groundwater recharge by buffering precipitation while at the same time providing preferential flow paths for infiltration. The importance of preferential flow on landslide hydrology is recognised in the literature; however, its monitoring and quantification remain difficult.

This paper presents a combined hydrological and hydrochemical analysis of small-scale sprinkling experiments. It aims at showing the potential of such experiments for studying the spatial differences in dominant hydrological processes within a landslide. This methodology was tested in the highly heterogeneous black marls of the Super-Sauze landslide. The tests were performed in three areas characterised by different displacement rates, surface morphology and local hydrological conditions. Special attention was paid to testing the potential of small-scale sprinkling experiments for identifying and characterising preferential flow patterns and dominant hydrological processes.

1 Introduction

In the last two decades, the understanding of hydrological processes in hillslopes has advanced due to improved monitoring techniques (McDonnell, 1990; Kirchner, 2003; Tromp-van Meerveld and McDonnell, 2006) and, consequently, improved understanding of mass movement dynamics (Haneberg 1991; Uchida et al., 2001; Bogaard et al., 2004; Malet et al., 2005; de Montety et al, 2007; Wienhofer et al, 2011). Nevertheless, current knowledge is still incomplete, especially concerning infiltration and percolation processes, subsurface flow paths and residence time of groundwater (Bogaard et al., 2004). The main difficulties stem from strong heterogeneity of hillslope lithology and spatio-temporal variation of hydrological properties as well as dominant hydrological processes. This is particularly true when dealing with highly heterogeneous, unconsolidated, partly weathered silty-clay sediments, such as black marls. Additionally, in slow-moving clayey landslides, (constant) movement of sliding material results in the formation of fissures, due to compression or extension, in relation to the differential movement and deformation rate (Anderson, 2005; Schulson and Duval, 2009; Niethammer et al., 2012, Walter et al., 2012; Stumpf et al., 2012). Here, the term "fissures" refers to geo-mechanically induced cracks that are filled or partly filled with reworked material. Accordingly, the term "preferential flow" refers to rapid water flow bypassing the bulk of the matrix (Beven and Germann, 1982), which develops through the areas where water fluxes are favoured by the presence of fissures.

The presence of fissures creates so-called "dual permeability" or "multiple permeability" systems. Dual permeability theory (Gwo et al., 1995; Greco et al., 2002; Šimůnek et al., 2003; Gerke, 2006; Jarvis, 2007) considers the porous medium as two (or more) interacting and overlapping but yet distinct continua. The water flow occurs in both continua but it is mainly ruled by the fracture continuum (macropore or fissure), which generates preferential flow. This way, the presence of fissures may increase the rate of groundwater recharge (preferential vertical infiltration). On the other hand, it may increase the rate of drainage, which limits the building up of pore water pressure (preferential slope parallel drainage). However, when talking about dead-end fissures (disconnected fissure network, limited drainage capacity), they contribute to maintain high pore water pressures in the surrounding soils (McDonnell, 1990; Pierson, 1983; Van Beek and Van Asch, 1999; Uchida et al., 2001).

The quantification of groundwater recharge, especially by means of preferential flow, is an important point to be tackled for an advanced understanding of hydrological systems in hillslopes and landslides (Savage et al., 2003; Coe et al., 2004; Weiler and McDonnell 2007). However, the complexity of the processes and their high spatial variability make it very difficult to measure preferential flow in the field and to build up process models (Van Schaik, 2010). Various experimental techniques are currently used to gain insight into processes controlling preferential flow, e.g. dye tracing (Flury et al., 1994), tension infiltrometers (Angulo-Jaramillo et al., 1996) and continuous sampling of water drainage (e.g. multisampler Wicky lysimeter; Boll et al., 1992). Nevertheless, a consistent measurement method for evaluating preferential flow has not yet been formulated (Allaire et al., 2009).

Environmental tracing (Kabeya et al., 2007) and artificial tracing (Mali et al., 2007) in combination with hydrological survey are the most convenient investigation methods in field conditions. The experiments vary from laboratory tests (e.g. Allaire-Leung et al., 2000; Larsbo and Jarvis, 2006) to field experiments at different scales (e.g. Collins et al, 2002; Weiler and Naef, 2003; Mali et al., 2007; Kienzler and Naef, 2008). However, there are no plot-scale field measurements dedicated to monitoring and quantifying preferential fissure flow, being a special case of macropores with apertures up to tens of centimetres.

The main objective of this research is to test the potential of a small-scale (1×1 m^2) sprinkling experiment to identify, study and quantify the dominant hydrological processes within an active, highly heterogeneous landslide. The idea of using small-scale (1×1 m^2) sprinkling experiments arose after the successful performance of large-scale (approximately 100 m^2) sprinkling tests in summer 2007 at two landslide sites located in France: the Super-Sauze landslide and the Laval landslide (Debieche et al., 2012; Garel et al., 2012). These two experiments gave valuable insight into the preferential infiltration and preferential later drainage processes in those unstable clay-shale hillslopes. However, due to the size and long duration, these kinds of experiments are logistically and financially very demanding, and cannot be undertaken on a regular basis across the study area.

This paper presents the results obtained from three small-scale sprinkling tests performed on morphologically different areas of the persistently active Super-Sauze landslide (French Alps). The hydrological and hydrochemical observations were generalised into flow regimes and collated with current knowledge about the landslide.

2 Methodology

2.1 Experimental design

The sprinkling experiments were performed with the use of a sprinkling apparatus with one nozzle (1/4HH-10SQ), which was fixed at the top centre at around 2 m high. The apparatus was calibrated in order to provide a relatively homogeneous distribution of the sprinkling water over the 1×1 m^2 experimental plot. Water supply was pumped in with regulated constant pressure (1.1 bars). The sprinkling was carried out in blocks of 15 min sprinkling and 15 min break, with sprinkling intensity of approximately 20–30 mm 15 min^{-1}. This intensity is a trade-off between the feasibility of the sprinkling equipment (pump and nozzle) and a realistic sprinkling rate (applied intensity is comparable to the observed intensity of summer and autumn storms reaching 50 mm h^{-1} in 15 minutes; Malet et al., 2003). Moreover, it is optimised to establish a reasonably good spatial distribution and is high enough to ensure infiltration both in the matrix and the fissure compartments. To monitor the actual sprinkling volume, and determine its distribution within the sprinkling plots, rain gauges (five per plot) were installed. In order to protect the experiment from wind disturbances and to minimise evaporation, the experimental areas were covered with a tent. It is important to stress that the setup of the sprinkling experiment was designed to identify different patterns of the hydrological responses rather than being used for, as an example, infiltration capacity measurement.

The 1×1 m^2 sprinkling tests were carried out in two periods of 7–8 h sprinkling, composed of 14–17 sprinkling blocks (SB = 15 min rain + 15 min break), during two consecutive days. This way, the first day of each sprinkling test started with dry initial conditions, while the second one represented wet initial conditions. The water used for the sprinkling tests was first collected in water tanks and blended with chemical tracers. The artificial tracing was introduced in order to get insight into the subsurface water flow paths and event and pre-event water mixing proportions. Therefore, the tracing was realised with two tracers: bromide (Br^-) during the first day of experiment and chloride (Cl^-) during the second day.

Within each sprinkling plot 4–5 piezometers were installed: one in the middle of the plot, two in the direction of the expected (sub-)surface water movement (in the direction of the fissures, if they were visible at the surface), and one upslope with respect to the plot, as a reference (Fig. 1b). The piezometers were made of PVC tubes with 0.50 m filters, covered with standard filter protection, surrounded by filter sand and closed with granular bentonite. All three experimental setups were built up 2 days before the sprinkling experiment started.

Groundwater responses were monitored manually every 15 min and with the use of automatic recording water pressure devices (Diver and Keller devices with reported measurement accuracy of 5 mm) with a 3 min time resolution. The water for hydrochemical analyses was sampled every 1 h from all piezometers during the sprinkling experiment and one time per day for two consecutive days after the experiment. Additionally, the sprinkling plots were equipped with 1 m long access tubes for theta probes (PR1/6w-02, Delta-T Devices, with reported accuracy of $\pm 0.06\,m^3\,m^{-3}$; Van Bavel and Nichols, 2002) in order to monitor changes in the soil moisture profile at six depths (0.1, 0.2, 0.3, 0.4, 0.6 and 1.0 m). If the installation of theta probes was not possible (e.g. technical problems), the initial surface soil moisture (0–0.10 m depth) was measured with a manual field-operated time-domain reflectometry probe (FOM TDR). The reported accuracy of the FOM TDR is $\pm 0.02\,m^3\,m^{-3}$ (IA PAS, 2006).

2.2 Analysis methodology

The soil column for water balance and tracer mass balance calculation was bounded laterally by the $1 \times 1\,m^2$ sprinkling surface area and vertically by the maximum depth of the piezometer installed in the centre of each plot. The water balance of the sprinkling experiment for 7 or 8 h duration is

$$P + GW_{in} = GW_{out} + OF + E + \Delta S, \qquad (1)$$

where P is the precipitation (sprinkling), which represents the amount of sprinkling water; GW_{in} and GW_{out} are the groundwater inflow and outflow; OF is the overland flow; E is the evaporation; and ΔS is the change in storage over the duration of the sprinkling experiment. The groundwater outflow includes subsurface flow (SSF) and vertical "deep percolation" (Pe). Here we define the "deep percolation" flux as the water flowing down the lowest piezometric observation point in the experimental plot (see Fig. 4).

Moreover, a depletion curve analysis was applied with the analogy of hydrograph recession analysis by using the linear reservoir concept (Hornberger et al., 1991; Mikovari and Leibundgut, 1995; Sivapalan et al., 2002). Additionally, assuming that the groundwater level is a direct function of a change in drained volumes (therefore, a change in storage), it was possible to identify differences in types of storages based on the occurrence of inflexion points in the drawdown curves. The time for the depletion of the storages is indicated by a depletion factor (K) calculated for all segments of the drawdown curve using the empirical method explained by Linsley et al. (1982):

$$h_{t+\Delta t} = h_t \cdot e^{\frac{-\Delta t}{K}}, \qquad (2)$$

where h_t is the groundwater level at time t and Δt is the temporal resolution of groundwater level observations (min). In general, the steeper part of the curve represents fast drainage, assumed to be preferential flow, whereas the gentle slope part represents slower drainage, associated with matrix flow.

Besides a qualitative description of the infiltration process, the concentration of the conservative tracers (Br^- and Cl^-) was used to calculate the proportion of different water sources (event vs. pre-event water) using a two-component end-member mixing (EMMA) model. The EMMA model has been widely used for hydrological studies to separate the different contributions of streamflow (Christophersen and Hooper, 1992; Mulholland and Hill, 1997; Soulsby et al., 2003; James and Roulet, 2006; Cras et al., 2007). The end members are usually defined from the reservoir characteristics; therefore mixing diagrams inform about the variable source areas of runoff. At the same time, they could be used to understand the flow processes which take place during infiltration. The mixing proportions ($\alpha(t)$ and $\beta(t)$) are calculated by solving the following equations:

$$\begin{cases} \alpha_1(t) \cdot C_{Br^-,EW1} + \beta_1(t) \cdot C_{Br^-,PE} = C_{Br^-}(t) \\ \alpha_2(t) \cdot C_{Cl^-,EW2} + \beta_2(t) \cdot C_{Cl^-,PE} = C_{Cl^-}(t) \\ \alpha_{1,2}(t) + \beta_{1,2}(t) = 1, \end{cases} \qquad (3)$$

where $C_{Br/Cl,EW/PE}$ and $C_{Br/Cl}(t)$ are tracer concentrations [$mg\,L^{-1}$] measured during the sprinkling experiment at different times (sampled from piezometers). PE and EW indicate the pre-event and event water, and the numbered indexes are related to the first and second day of experiments, respectively.

Besides the added conservative tracers Br^- and Cl^-, the sulfate concentration in groundwater prior to the experiment was used as an independent variable to define a "pre-event" end member ($C_{SO4,PE}$). Sulphate is the major component of the groundwater chemistry and it can be used as a tracer as long as the impact of the difference between groundwater and rainwater concentrations remains far larger than that of the water–rock interaction. Since this was the case, the applied sprinkling sulfate content ($C_{SO4,EW}$) was considered as the second end member and Eq. (3) was formulated accordingly. The two estimated mixing proportions (for artificial and environmental tracers) for both experiments were plotted to analyse and validate the mixing assumption.

Furthermore, the simple mass balance equations were used (for Br^- and Cl^-) to estimate the most probable water (and

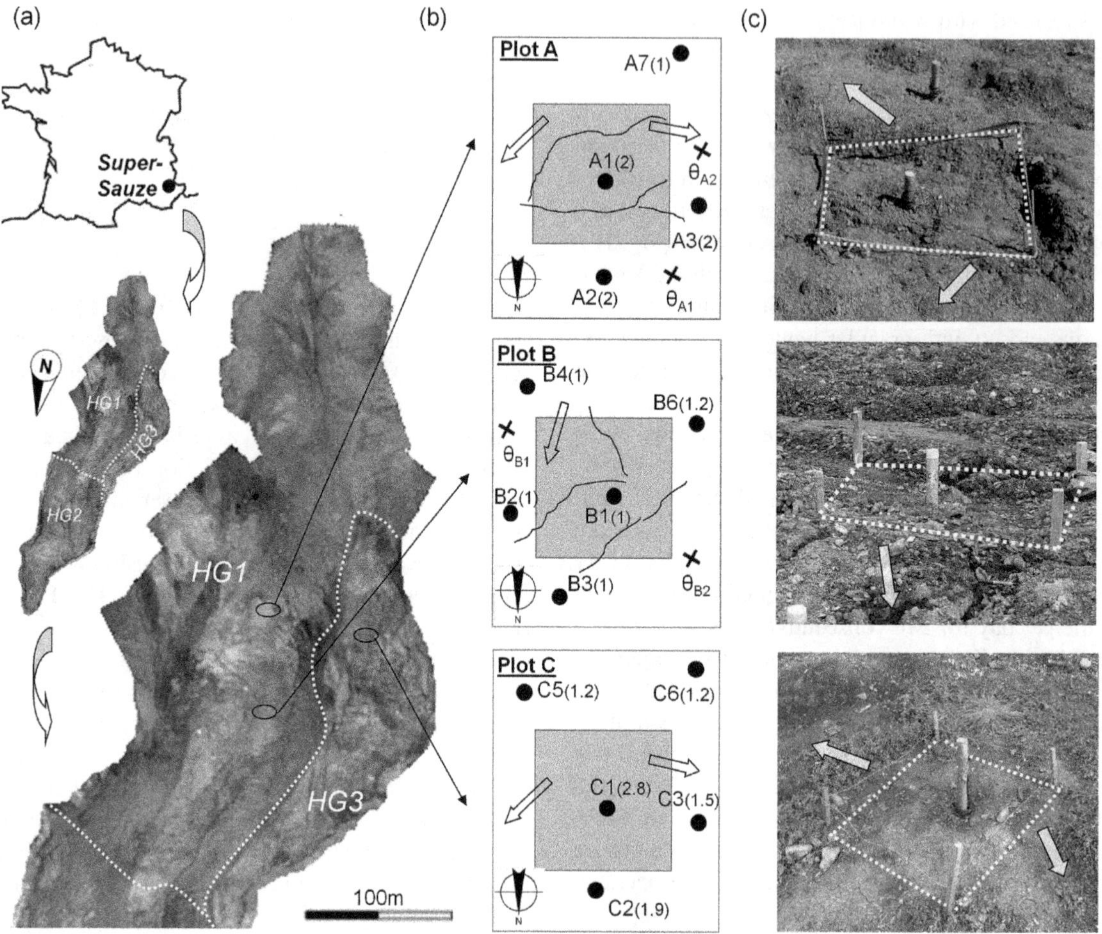

Figure 1. **(a)** The upper part of the Super-Sauze landslide with indicated location of three sprinkling tests (plot A, B and C); the white dashed lines indicate the hydro-geomorphological units (after Malet et al. 2005). **(b)** Schematic representation of the experimental setup of each area (not scaled): grey squares represent $1 \times 1\,\mathrm{m}^2$ sprinkling plots, dots represent the location of the piezometers, numbers in brackets indicate the depth of the piezometers in metres, crosses indicate the location of the theta probes, undulating lines indicate fissure distribution within the sprinkling plots, and arrows show the local slope direction in the area. **(c)** Photographs of the soil surface of each sprinkling area with arrows showing the local slope direction in the area.

tracer) flow paths and to restrain mixing processes:

$$\begin{aligned} &V(t_{\mathrm{end}}) = V_{\mathrm{PE}} + V_{\mathrm{INF}} - V_{\mathrm{SSF}} \\ &C_{\mathrm{Br^-/Cl^-}}(t_{\mathrm{end}}) \\ &\cdot V(t_{\mathrm{end}}) = C_{\mathrm{Br^-/Cl^-,PE}} \cdot V_{\mathrm{PE}} + C_{\mathrm{Br^-/Cl^-,EW}} \\ &\cdot V_{\mathrm{INF}} - C_{\mathrm{Br^-/Cl^-}}(t) \cdot V_{\mathrm{SSF}}, \end{aligned} \tag{4}$$

where $V(t_{\mathrm{end}})$, V_{PE}, V_{INF} and V_{SSF} are the estimated total water volume in the soil column at the end of sprinkling tests, the estimated volume of pre-event water, the estimated volume of infiltrated water (fraction of EW) and the estimated volume of subsurface flow (including exfiltration), respectively. The volume of pre-event (V_{PE}) was calculated based on the initial groundwater level and initial soil moisture content as follows:

$$V_{\mathrm{PE}} = A \cdot (h_{t_0} \cdot n + (z - h_{t_0}) \cdot \theta_{\mathrm{ini}}), \tag{5}$$

where A is the plot area (m^2), h_{t0} is the groundwater level (m) observed before the experiment, n is the average porosity (–), z is the depth of the analysed soil column (m) and θ_{ini} is the initial soil moisture (–).

2.3 Characteristics of experimental plots

The experimental design was tested at the highly active Super-Sauze landslide (Fig. 1a), which covers $0.17\,\mathrm{km}^2$ with an average slope of 25° and has displacement velocities varying from 0.01 to $0.40\,\mathrm{m\,day}^{-1}$, depending on the season (Malet et al., 2002). The small-scale sprinkling plots A and B are located in the upper part of the landslide, which is the most active in terms of displacement rates, abrupt changes in groundwater levels throughout the season, and changes in fissure density and openings (Fig. 1b). Plot C is located in a relatively stable part of the landslide, but still at the direct contact with the most active area, and is representative of small displacement rates, small changes in groundwater levels throughout the season and no changes in fissure characteristics (Fig. 1b). As such, the three experimental plots shall present different hydrological responses (Malet, 2003; de Montety et al., 2007). All sprinkling experiments were carried out in relatively flat areas with slopes of 5–7°. The porosity values for the experimental plots were assumed to be 0.35, 0.38 and 0.30 on average for plot A, B and C, respectively, based on gravimetric measurements (Malet, 2003). The geomorphology of each plot is detailed below:

- Plot A is located in the active area near the crown consisting of relatively fresh but heavily broken marl blocks and deposits (marly fragments of approximately 2 cm). There are wide (aperture of 0.07–0.15 m), undulating fissures observed on the surface (see Fig. 1b for the sketch), partly or totally filled with reworked marl fragments. The open depth of these fissures varies from 0.09 to 0.12 m.

- Plot B is also located in the very active area, at a secondary mudslide deposition area, that consists of gravel crust, which is characterised by coarse fragments (larger than 2 mm) overlaying a finer matrix. There are wide open (apertures around 0.10 m) fissures present within the plot area with an open depth reaching 0.50 m (see Fig. 1b for the sketch).

- Plot C is situated in the compacted, relatively stable, western part of the landslide and consists of fine-grained material with different rock fragments. No fissures are observed at the surface.

The average depth to the bedrock is around 10 m in plot A, 3 m in plot B and 5 m in plot C (Travelletti and Malet, 2012). The depth of the piezometers is different at each area. Within plot A all piezometers were installed at approximately 2 m depth. Within plot B the piezometers depths are around 1 m due to the shallow groundwater level (see also Fig. 1b–c.). Within plot C the depths of the piezometers were conditioned by the presence of rock fragments in the soil and vary from 1.2 to 3.0 m.

3 Results of sprinkling experiments – hydrological and hydrochemical responses

Within each sprinkling plot different hydrological behaviours were observed. Fig. 2a–c summarise the observed groundwater variation and tracer concentration patterns. Figure 3 shows the drawdown curves after the second day of sprinkling. For plot A and B the drawdown of the centrally located piezometers was analysed (A1 and B1), while for plot C the analysis was carried out for piezometer C2 since the groundwater level observed in piezometer C1 was strongly influenced by water sampling.

3.1 Plot A

Plot A was a dry area with no groundwater observed within the first 2 m depth (depth A1) before the experiment started. The mean initial volumetric water content of the top soil layer (up to 0.30 m depth) was 0.12 with a standard deviation of 0.03. In response to the applied sprinkling, neither overland flow nor subsurface runoff was observed. The groundwater fluctuation in A1 showed a very fast vertical movement of water (approximately ±0.25–0.30 m in 15 min). Moreover, the drawdown after each day of sprinkling lasted only 4 h (Fig. 2a).

In A3, located in the direction of the surface fissures, the groundwater level started to react during the first day, after the fourth sprinkling block (SB-04). In A2, located downslope of the sprinkling plot, but not in the direction of the surface fissures, the groundwater reaction started only during the second day after the fifth sprinkling block (SB-05). There was no response observed in A7.

The soil moisture variation observed in the soil profile in θ_{A2} (approximately 1 m distance from the sprinkling area) was negligible. In θ_{A1} no changes were observed in the first 0.60 m of soil profile. At 1 m depth soil moisture increased until saturation over the 2 days of the sprinkling experiment.

The tracer concentration in the piezometers gradually increased with the cumulated amount of applied sprinkled water. Similar to the hydrological responses, the most intense changes were observed in A1, reaching 84 % (first day) and 93 % (second day) of applied tracer concentration at the end of each 7–8 h sprinkling. Moreover, the tracer concentration decreased during the recession phase. At the start of the second day, Br^- concentration had nearly dropped back to the initial value. The same trend was observed for Cl^- during the second day. The tracer concentration in A2 and A3 followed the trend observed in A1, with the maximum measured tracer concentration reaching 26–38 % (first day) and 55–71 % (second day) of applied concentration for A3 and A2, respectively. It is important to note that in plot A, on the second day of experiment, Br^- was applied during the first four sprinkling blocks (SB-01 to SB-04) at very high concentration (461 mg L^{-1}). This incident determined the behaviour of Br^- concentration at the beginning of the second day of sprinkling: the maximum concentration of Br^- was observed after SB-06 in A1, after SB-08 in A2 and after SB-12 in A3.

3.2 Plot B

Plot B was located in an area with shallow groundwater level (0.35–0.55 m below the surface). The average initial volumetric water content in the first 0.30 m of soil was 0.25, with a standard deviation of 0.07. During the sprinkling experiment, an increase of groundwater level was observed only in B1 and B3 and it fluctuated ±0.07 m on average in response to a single sprinkling block. No groundwater level changes were registered in B2 and B6, and no changes in

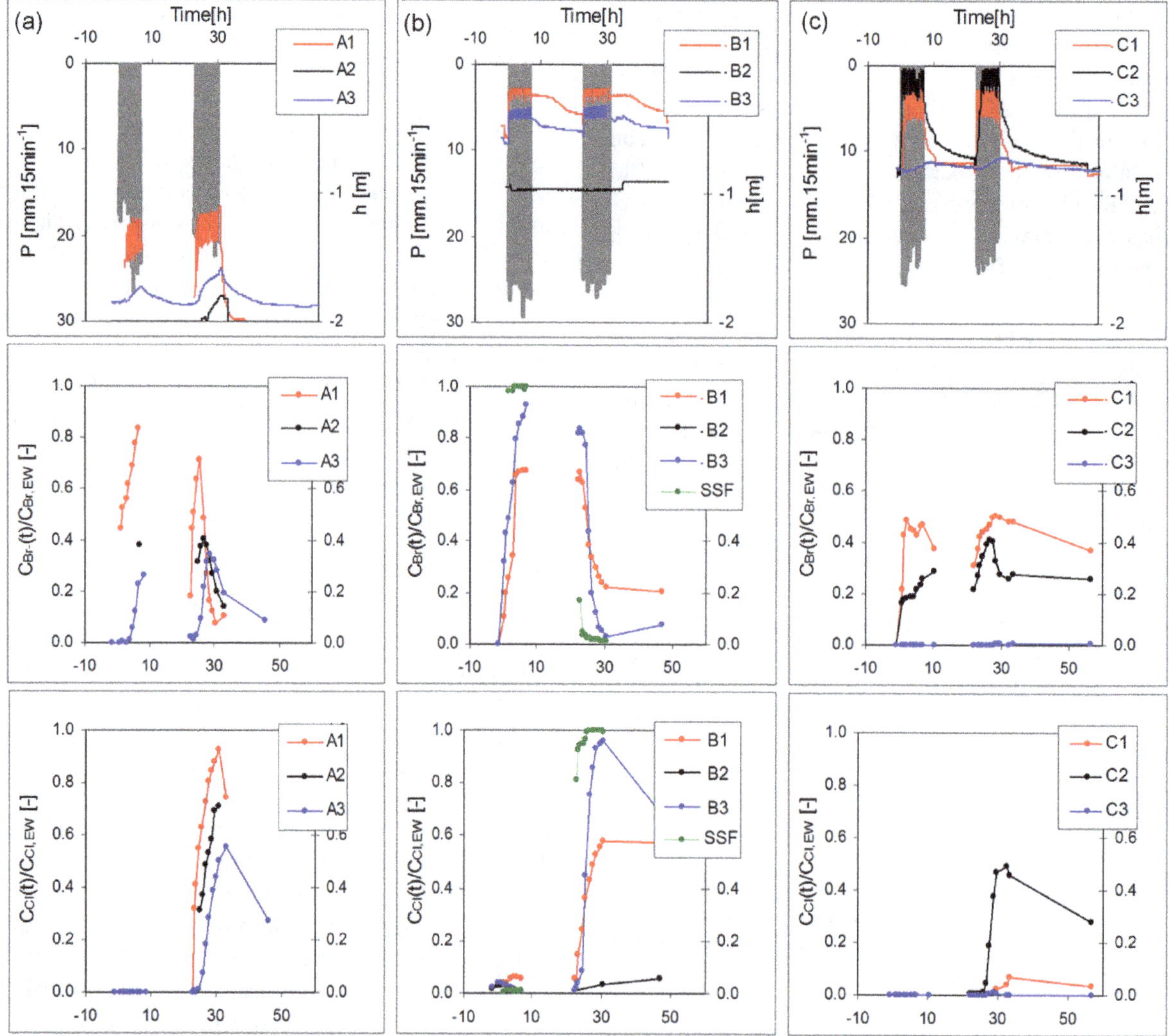

Figure 2. Monitoring results of three sprinkling experiments: **(a)** plot A, **(b)** plot B and **(c)** plot C. Upper panels show the intensity of the sprinkling (primary *y* axis) and groundwater responses in piezometers (secondary *y* axis). Middle and bottom panels show the ratio between tracer concentration measured in the piezometers or subsurface runoff (SSF) and the applied tracer concentration.

soil moisture content in θ_{B1} or θ_{B2} (located within a distance of approximately 1 m from the $1 \times 1\ m^2$ sprinkling plot) were observed.

The exfiltrating subsurface runoff was measured around 1–1.5 m downslope of the experimental plot. The volume of subsurface runoff per sprinkling block increased with time. During the first day, it started at $15.9 \times 10^{-3}\ m^3$ for SB-05 and reached $18.3 \times 10^{-3}\ m^3$ for SB-14. During the second day, it ranged from $11.4 \times 10^{-3}\ m^3$ (for SB-01) to $19.5 \times 10^{-3}\ m^3$ (for SB-14).

In B1 and B3 the relative Br^- concentration rose quickly and reached a maximum of 67 and 93 %, respectively, at the end of the first day. Moreover, it remained at a high level in between 2 days of sprinkling (Fig.2b). A similar tracer concentration behaviour was observed during the second day of the experiment, when chloride was applied. The observed Br^- concentration gradually decreased, while the Cl^- concentration increased, reaching 58 % (B1) and 99 % (B3) of applied concentration. The Cl^- concentration remained high (58 % of the applied concentration in B1 and 68 % in B3) even 20 h after the end of the experiment. It is worth noting that the concentrations of Br^- and Cl^- were most of the time higher in B3 than in B1 and that the tracer concentration in the subsurface runoff equalled (first day) or almost equalled (second day: 81–99 %) that of the sprinkling water.

3.3 Plot C

The initial groundwater level at plot C was around 0.75–1.00 m below the surface and the initial volumetric water content varied between 0.20 and 0.25 (0.23 on average) in the first 0.10 m of soil. In contrast to the dynamics observed

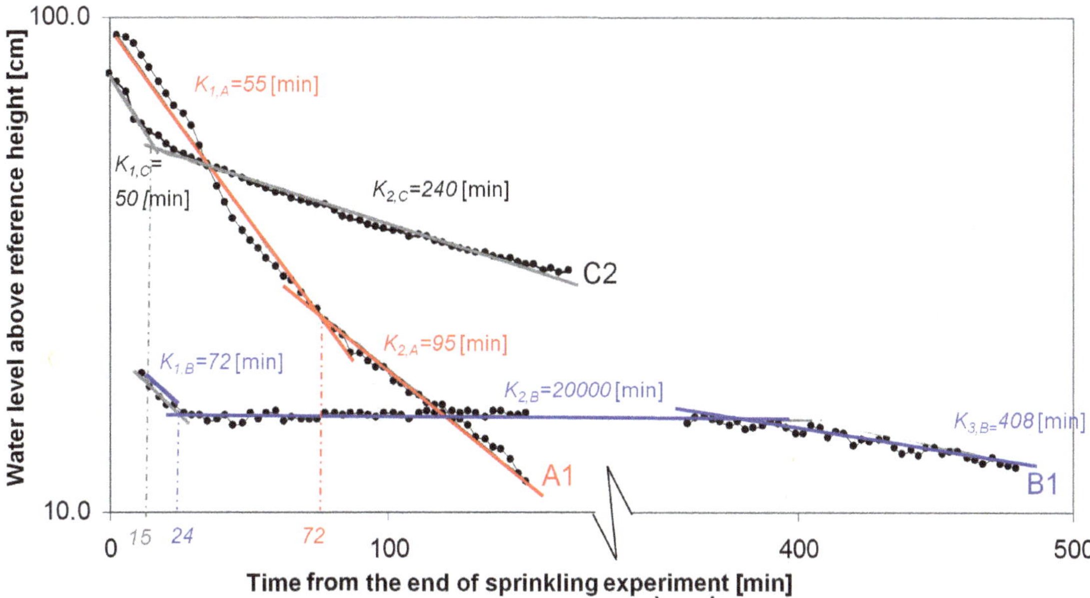

Figure 3. Drawdown curves observed in piezometers A1, B1 and C2 after the end of sprinkling experiments and corresponding depletion factors K [min].

in plot A, 75 % of the sprinkling water left the soil column as overland flow. Moreover, ponding was observed within the 1×1 m^2 plot during the entire experiment.

The groundwater level observed in C1 and C2 responded similarly: an increase of groundwater level up to 0.20 m (for C1) and 0.05 m (for C2) below the surface and fluctuations of about 0.20 m after each 15 min of sprinkling. The drawdown observed in C1 stopped after 4 h, whereas in B2 it lasted around 12 h after the sprinkling experiment. C3 showed a 0.03–0.07 m groundwater level fluctuation, and no response took place in C5 and C6. The groundwater level in the soil column went back to its initial stage within 12 hours after the sprinkling ceased.

In C1 the relative concentration of Br^- reached approximately 43–49 % of the applied tracer concentration and was relatively constant during the first day of sprinkling. At the start of the second day Br^- relative concentration was 31 %, and it rose again up to 50 % as soon as new sprinkling water (without Br^-) was applied (Fig. 2c). Similar trends were observed for C2, with the maximal tracer concentration reaching 28 and 40 % of applied concentration for the first and second day, respectively. No tracer was found in C3.

During the second day of the sprinkling experiment, the Cl^- concentration showed a very limited increase in C1 but a gradual increase up to around 50 % of the applied concentration in C2. The Cl^- concentration decreased after the second day. However, in C2 it remained relatively high even 20 h after the experiment (300 mg L^{-1}). Again, no tracer was found in C3.

4 Discussion of experimental results and model conceptualisation

4.1 Water balance and tracer mass balance analysis

The water budget was calculated for each day of the sprinkling experiment from the beginning of the first sprinkling block (SB-01) until the end of the drawdown observed in the centrally located piezometer. As a first approximation, the water balance components were estimated based on the assumption that the whole experimental area (1×1 m^2) is hydrologically active, meaning all water stored in the soil column is mobile and full mixing of pre-event water and event water occurs. The groundwater flow variations were assumed to be only due to infiltrating sprinkling water over the experimental plot (Fig. 4). This means that we assume no change in overall groundwater flow and no change in deep percolation (plot B and C) due to the sprinkling activity. In the case of plot A, where no groundwater level was observed before and shortly after the experiment, the direction of the estimated subsurface flow could not be determined. Therefore, the subsurface flow comprised both vertical deep percolation (Pe) and lateral flows (SSF). The volume of pre-event (V_{PE}) water was estimated based on Eq. (7), and the volume of infiltrated water (V_{INF}) was calculated as $V_{EW} - V_{OF}$. The volume of subsurface fluxes (V_{SSF}), which comprises all subsurface fluxes, was estimated using the measured groundwater level responses to the sprinkling blocks. The change in storage ΔS was calculated as $V_{INF} - V_{SSF}$. Evaporation (E) was assumed to be negligible, as the sprinkling plots were covered with a tent. Table 1 shows the measured (^{m}) and estimated (^{e}) water balance components.

Table 1. Measured (^{m}) and estimated (^{e}) components of water balance for each plot, with the assumption that whole experience area is hydrologically active.

	Plot A		Plot B		Plot C	
Day of experiment (duration)	1st (7 h)	2nd (8 h)	1st (7 h)	2nd (8 h)	1st (7 h)	2nd (7 h)
Assumed average porosity, n [–]	0.35	0.35	0.38	0.38	0.30	0.30
$^{(m)}$Initial average volumetric soil moisture, θ_{ini} [–]	0.12	0.12	0.25	0.27	0.23	0.25
$^{(e)}$Water in soil column, V_{PE} [m^3]	0.23	0.24	0.32	0.34	0.78	0.79
$^{(m)}$Sprinkling volume, P [m^3]	0.27	0.33	0.36	0.42	0.29	0.30
$^{(m)}$Overland flow, OF, [m^3]	Not observed		Not observed		0.22	0.23
$^{(e)}$Infiltrated water, V_{INF} [m^3]	0.27	0.33	0.36	0.42	0.09	0.08
$^{(e)}$Subsurface flow, (SSF) [m^3]	–*	0.23	0.34	0.41	0.09	0.008
– $^{(m)}$exfiltration [m^3]	Not observed		> 0.17**	0.30	Not observed	
$^{(e)}$Water in soil column, $V(t_{end})$ [m^3]	–*	0.34	0.35	0.35	0.79	0.79
$^{(e)}$Change in storage, ΔS [m^3]	–*	0.10	0.02	0.01	0.01	0.01
$^{(e)}$Final average volumetric soil moisture, $\theta(t_{end})$ [–]	–*	0.17	0.27	0.28	0.25	0.26

* Estimation not possible because of missing groundwater level observation. ** Exfiltration started after 2 h of sprinkling but was measured only from the third hour of the sprinkling experiment.

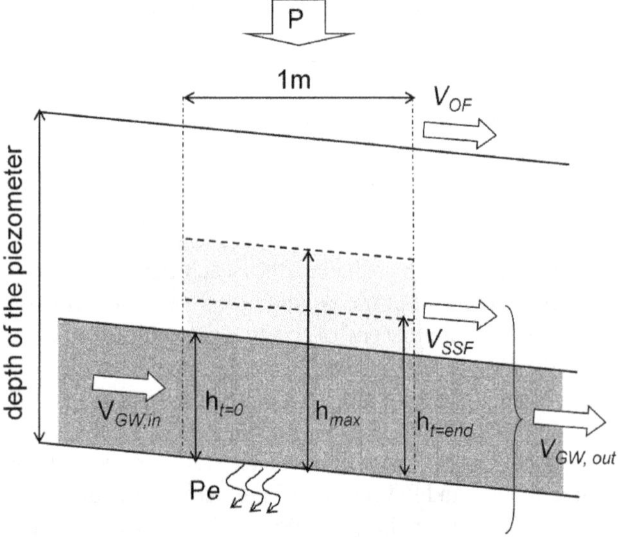

Figure 4. Schematic representation of water balance components of experimental plots. P is sprinkling volume; V_{OF} is volume of overland flow; V_{SSF} is volume of subsurface flow; Pe is volume of "deep percolation"; $V_{GW,in}$ and $V_{GW,out}$ are volumes of groundwater inflow and outflow, respectively; h_t is the groundwater level at time t ($t = 0$ – just before the sprinkling starts; t = end – at the end of sprinkling experiment); and h_{max} is maximal groundwater level observed during sprinkling experiment.

The tracer mass balance analysis was performed in order to evaluate the assumption of water mobility and full mixing within the soil column. The bromide and chloride masses were calculated from the chemical measurements and corresponding water volumes. The tracer mass remaining in the soil column was calculated in two ways: (1) based on the tracer mass balance, and (2) based on the measured tracer concentration in the groundwater at the end of the sprinkling experiments. The differences between the two allowed us to evaluate the mixing assumption. The percentage of the experimental area that is hydrologically active (x) was estimated based on the following equation:

$$\begin{aligned} & C_{Br^-/Cl^-}(t_{end}) \cdot V(t_{end}) = x \cdot C_{Br^-/Cl^-,PE} \cdot V_{PE} \\ & + C_{Br^-/Cl^-,EW} \cdot V_{INF} - x \cdot C_{Br^-/Cl^-}(t) \cdot V_{SSF} \\ & V(t_{end}) = x \cdot V_{PE} + V_{INF} - x \cdot V_{SSF}. \end{aligned} \quad (6)$$

It is important to note that V_{PE} and V_{SSF} were estimated based on groundwater level observation multiplied by the (hydrologically active) area of the experiment.

Table 2 shows tracer mass balance component and is subdivided into two parts: first, the results based on the assumption that the whole soil column is hydrologically active (i.e. full mixing), and second, the results taking into account a percentage of the soil column that is hydrologically active.

Furthermore, the influence of porosity values was evaluated. Increasing or decreasing the average porosity by 0.01 and 0.02 results in changes in the water balance components. The influence of porosity on the estimated volume of pre-event water was limited: no changes in plot A (since there was no groundwater before the experiment), ±5 % in plot B and ±3 % in plot C. The volume of subsurface flow is more sensitive for changes in soil porosity. It varies between ±35 % in plot A and ±24 % in plot C. Within plot B the change in subsurface flow volume, expressed as a percentage, is also significant (between +11 and −55 %), but it corresponds to relatively low absolute values (0.05–0.1 m^3). Consequently, the changes in volume of water stored in the soil column at the end of the experiment are the highest for plot A (between −23 and +44 %) and relatively limited for plot B (±7 %) and plot C (±14). The influence of porosity changes on the calculated percentage of hydrologically active area (Eq. 6) is limited to a maximum of ±2 %.

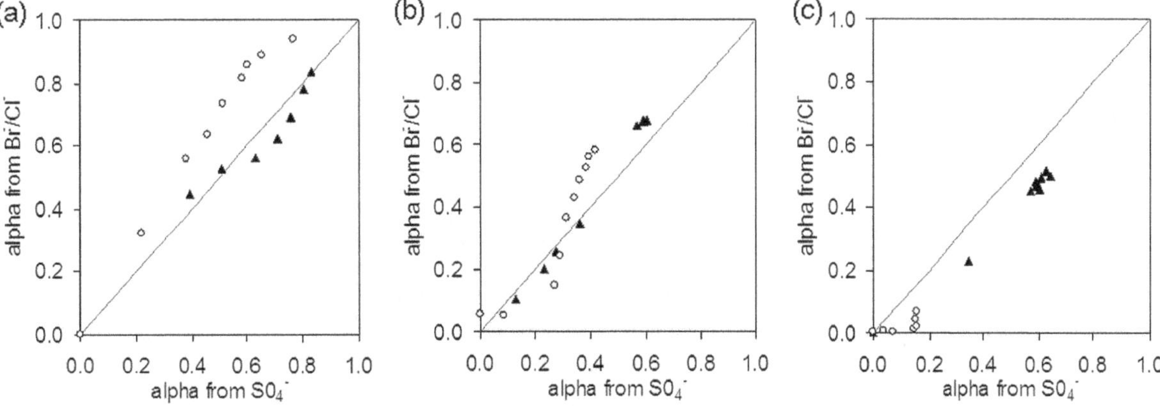

Figure 5. EMMA model results for the centrally located piezometers A1 **(a)**, B1 **(b)** and C1 **(c)**. The full triangles are estimates for the first day of the experiment and the open dots represents second day of the experiment. The grey line is the 1 : 1 line.

4.2 Hydrological and hydrochemical observation

A diverse spectrum of results clearly emerges from the experiments. However, the results also show interesting similarities. The sprinkling water infiltrates into the top soil through both matrix and preferential (fissure) flow paths. Once water enters into the soil the plots essentially show two types of drainage. First, groundwater level depletes quickly and slows down after 15–90 min. Interestingly, fast infiltration and fast drainage do not coincide. Plot A has both fast infiltration and fast drainage (both in first and second stage), whereas plot B shows high infiltration capacity and the second reservoir shows the slowest depletion of the stored subsurface water. Plot C has a low infiltration rate but seems to drain the infiltrated water relatively smoothly.

The tracer information shows similarities with the groundwater patterns. Both bromide and chloride concentrations rise, almost reaching the initial concentration of the sprinkling in the centre of plot A and around 60 % of the initial concentration of the sprinkling in the centre of plot B within the duration of the experiment. In plot C both tracers reach a maximum of 50 % relative concentration, indicating more mixing with pre-event water. The location of highest relative concentration is in the centre for plot A and downslope of plot B (in B3). Plot C results are again a bit more diverse: both piezometers C1 and C2 show mixing of sprinkling water with pre-event water for the first day, but only C2, located downslope, shows a significant increase of chloride concentration during the second day experiment.

The results of the EMMA model underline the differences in mixing processes and their dynamics observed at each plot (Fig. 5). For all plots the relation between mixing proportions estimated based on both applied tracer and sulfate concentrations do not follow the 1 : 1 line exactly. This can be an effect of soil–water interaction (dissolution of pyrite). This can be partly due to the uncertainty on the estimates of PE sulfate concentrations, which may substantially vary over short distances. Within plot A and B the mixing processes are clear: the mixing proportions change progressively during the sprinkling experiment from 0 % to more than 90 % (plot A), or around 70 % (plot B) for both tracers and sulfate. In plot C the mixing proportions increase during the first day but are limited to 64 %. Moreover, in plot A and B, the artificial and environmental tracers behave similarly over the 2 days of sprinkling experiment, showing that mixing processes can be explained with two end members only: mixing of event water with pre-event water. This is also the case for the first day in plot C. During the second day of the experiment, a sharp dilution of sulfate was observed in C1, while the Cl^- concentration remained low and the Br^- concentration increased. This indicates that in plot C, both event waters (EW1 during first day and EW2 during second day of sprinkling) contributed independently in mixing with pre-event water. However, it is important to stress that EMMA results are based only on tracer concentrations and give a relative mixing proportion and do not give the absolute mass of the mixed tracers.

The three plots show different spatial responses. In the case of plot A, three piezometers show a response in water level and in tracer concentration. In plots B and C only two piezometers react to the sprinkling in groundwater level and water quality. This suggests that plots B and C have structured flow paths whereas plot A is more permeable in all directions. Subsurface flow often follows the slope gradient; however, the presence of fissures and macroporosity (plot A and B) strongly influence flow direction.

Based on the interpretation of the sprinkling experiment, three types of hydrological behaviour are identified:

- Fast input, fast output (plot A) and very fast infiltration as well as fast drainage.

- Fast input, slow output (plot B) and fast infiltration but very slow drainage.

Table 2. Measured (^{m}) and estimated (^{e}) tracer mass balance components and the evaluation of the hydrologically active area assumptions.

	Plot A		Plot B		Plot C	
Day of experiment (applied tracer)	1st (Br^-)	2nd (Cl^-)	1st (Br^-)	2nd (Cl^-)	1st (Br^-)	2nd (Cl^-)
Tracer in applied water						
– $^{(m)}$concentration, $[\,]_{EW}$ [g m^{-3}]	118	1035	122	128	123	1047
– $^{(e)}$mass [g]	32.3	343.7	44.1	53.5	35.6	322.5
$^{(e)}$Infiltrated tracer [g]	32.3	343.7	44.1	53.5	10.1	82.5
assuming that whole area is hydrologically active						
$^{(e)}$ Tracer out of soil column via						
– overland flow, [g]	Not observed		Not observed		22.3	237.3
– subsurface flow, [g]	–*	162.8	> 33**	36.2	4.2	2
$^{(e)}$Mass of tracer remaining in soil column based on mass budget [g] ($mass_{in}$–$mass_{out}$)	–*	180.9	< 11.1	17.3	5.9	82.6
$^{(m)}$Tracer concentration in the groundwater, [ ] (t_{end}) [g m^{-3}]	91.4	768.5	81.4	73.6	45.8	50.34
$^{(e)}$Mass of tracer remaining in soil column based on measured concentration, [g] ($V(t_{end}) \times [\,]((t_{end}))$)	–*	261.3	27.7	25.8	35.7	39.7
assuming that x percentage of the area is hydrologically active						
$^{(e)}$Percent of the plot area that is hydrologically active, x[%]	–*	53	24	60	17	210
$^{(e)}$Tracer out of the soil column via subsurface, [g]	–*	86.3	> 25.7**	34.2	0.7**	4.3
$^{(e)}$Mass of tracer remaining in soil column based on mass budget [g]	–*	257.4	< 18.4**	19.3	9.4**	82.7
$^{(e)}$Mass of tracer remaining in soil column based on measured concentration, [g]	–*	257.7	18.4	18.9	9.5	80.8

* Estimation/measurements not possible because of the missing groundwater level observation. ** Exfiltration started after 2 h of sprinkling but was measured only from the third hour of the sprinkling experiment.

- Fast but limited input, moderate output (plot C), limited infiltration and relatively slow drainage (when compared with plot A).

4.2.1 Flow regime 1: fast input–fast output

Plot A represents a fast input–output type of hydrological response: a very fast response to the onset of sprinkling, as well as a sudden groundwater level drop after the sprinkling has finished. The sharp groundwater level decrease in A1 (see Fig. 3) after the end of the sprinkling test is an indication of drainage from a highly permeable fraction of the subsurface, e.g. the fissure fraction. Moreover, the second part of the depletion curve is quite rapid as well, indicating that the matrix fraction is also highly permeable. The very high permeability is confirmed by the fact that groundwater responses are observed not only in the centre of the experimental plot (A1) but also in two directions downslope: relatively quick response in A3 (direction of fissures observed on the surface) and delayed in A2.

There is 0.7 m^3 of pre-event water (approximately 33–36 % of maximal storage capacity) storage in the soil column. Of this pre-event water, 50–54 % could mix and readily move with the infiltrating sprinkling water (Tables 1 and 2). The incident with the accidental application of high concentration Br^- at the beginning of the second day demonstrates the dominance of fast preferential flow through the plot and short residence times of water within the soil column.

4.2.2 Flow regime 2: fast input–slow output

The hydrological responses in plot B can be described as fast input–delayed output. The presence of a largely open (up to 14 cm) fissure system influences the distribution of infiltrating water. However, more than 70 % of the infiltrated water is flowing out of the soil column and exfiltrating downslope. The observed hydrological response is a combination of fast vertical infiltration, fast subsurface flow and much slower matrix flow. The shape of the drawdown curve (Fig. 3) also indicates the combination of mainly preferential flow with some matrix flow.

The behaviour of the tracer concentration indicates complex mixing processes in plot B. The changes in the Br^- and Cl^- concentration also show that infiltrating water of the first day replaced the pre-event water and is temporarily stored until the new source of water (sprinkling of second day) appears. The relatively low concentration of tracer in B1 shows that a significant amount of pre-event water (approximately 80–84 % of maximum storage) is stored in the matrix and 24–61 % of this water is involved in the mixing process (Tables 1 and Table 2). The spatial distribution of tracer concentration (lower concentration in B1; higher in B3 and in subsurface flow) indicates a well-structured subsurface (including fissure system) which can provide direct drainage for infiltrated water. The fast flow domain is isolated from the matrix (no or poor connection). When the groundwater level is high, a well-connected preferential flow system becomes active and the applied water drains directly ($K_{1,B}$; Fig. 3). However, once the water level has dropped by several centimetres, the drainage stops (e.g. dead-end fissure) and the system maintains high groundwater levels for several hours ($K_{2,B}$; Fig. 3). The last drainage phase ($K_{3,B}$; Fig. 3) can be interpreted as matrix flow occurring after saturation, connecting the wet fissure areas.

4.2.3 Flow regime 3: fast-but-limited input–moderate output

The general observation of the water balance component (Table 1) and drawdown curves (Fig. 3) indicates that plot C represents a matrix-like infiltration behaviour; however, the influence of preferential flow cannot be neglected. A significant drainage at the end of each 15 min sprinkling block and a depletion almost 3 times higher in the steep part of the depletion curve (K_{1C}) with respect to the gentle slope part (K_{2C}) indicate the presence of preferential flow paths (fissures, macropores), which influence the hydrological behaviour at studied scale.

There is a significant amount of pre-event water (approximately 92–94 % of maximum storage) is stored in the soil column. The tracer mass balance for the first day of the experiment (Table 2) indicates that only around 16–18 % of the pre-event water stored in the subsurface is actively mixed with the infiltrated sprinkling water. The opposite conclusion can be drawn when analysing the mass balance for the second day of the sprinkling experiment. Under the assumption that all infiltrated sprinkling water is stored in the $1 \times 1\ m^2$ plot, a double amount of pre-event water should be involved in the mixing processes in order to match the measured Cl^- concentration in C1. However, the concentrations of Cl^- in C2 (located outside the sprinkling plot) indicate that there is a significant amount of tracer stored outside the experimental plot, due to surface ponding and subsurface water flow. Moreover, assuming that the hydrologically active area during the second day of experiment is the same as during the first day of experiment (around 20 %), only 1–2 % of infiltrated tracer mass is enough to reach the measured tracer concentration in the groundwater at the end of experiment. This indicates the presence of preferential drainage. Nevertheless, the presence of Br^- in C1 (middle of the sprinkling plot) during the second day, when only Cl^- was applied, confirms that matrix flow dominates in the area and that piston flow processes occurred. The rise of the Br^- concentration, both in C1 and C2, observed at the beginning of the second day of sprinkling might be explained by the tracer settling over soil surface during water ponding on the first day of sprinkling and being mobilised by "new" sprinkled water.

4.3 Discussion of conceptual models for the Super-Sauze landslide

The improvement of hydrological modelling of the Super-Sauze landslide is not a direct aim of this paper: the small number of sprinkling experiments and their small scale in relation to the landslide area of 0.17 ha are not sufficient to cover the whole landslide. Therefore, the results cannot yet be up-scaled to a complete distributed hydrological model of the landslide.

However, the components of the hydrological system, identified based on small-scale sprinkling tests, are in line with the conceptual model of the Super-Sauze landslide proposed by Malet et al. (2005) and de Montety et al. (2007). The former proposed a distributed conceptual model of the hydrological behaviour of the Super-Sauze landslide dividing the landslide into three hydro-geomorphological units (Fig. 1a). The upper unit (HG1), where plot A and B are located, is very active and characterised by very rapid responses and large groundwater level fluctuation (up to 0.5 m) at the event scale. The western part of upper unit (HG3), where plot C is located, is the most stable part of the landslide, with very limited groundwater level fluctuation (centimetres). Our results confirm this hydrological concept, but they also stress clear differences in the hydrological response in the upper unit (Fig. 6), which was not presented as clearly by Malet et al. (2005). However, it is important to note that the hydrochemical behaviour observed in plot C is strongly influenced by the presence of small fissures and cannot be compared with general the hydrological concept of Malet et al. (2005).

The hydrological interpretation of the Super-Sauze landslide presented by de Montety et al. (2007), based on the long-term observation of spatial distribution of major cations and anions, defines dominant hydrological processes along the landslide profile. The upper part of the landslide (directly below the main scarp) is the "transition" zone, while the middle part of the landslide is dominated by preferential flow. This is in agreement with our observed fast input–fast output behaviour in plot A and fast input–slow output behaviour in plot B. The stable part of the landslide, where plot C was located, was not considered in the work of de Montety et al. (2007). De Montety et al. (2007) stressed the limitation of their investigation having only a qualitative assessment of the water fluxes and underlined the need for more detailed investigations. Our experiments show the potential for more

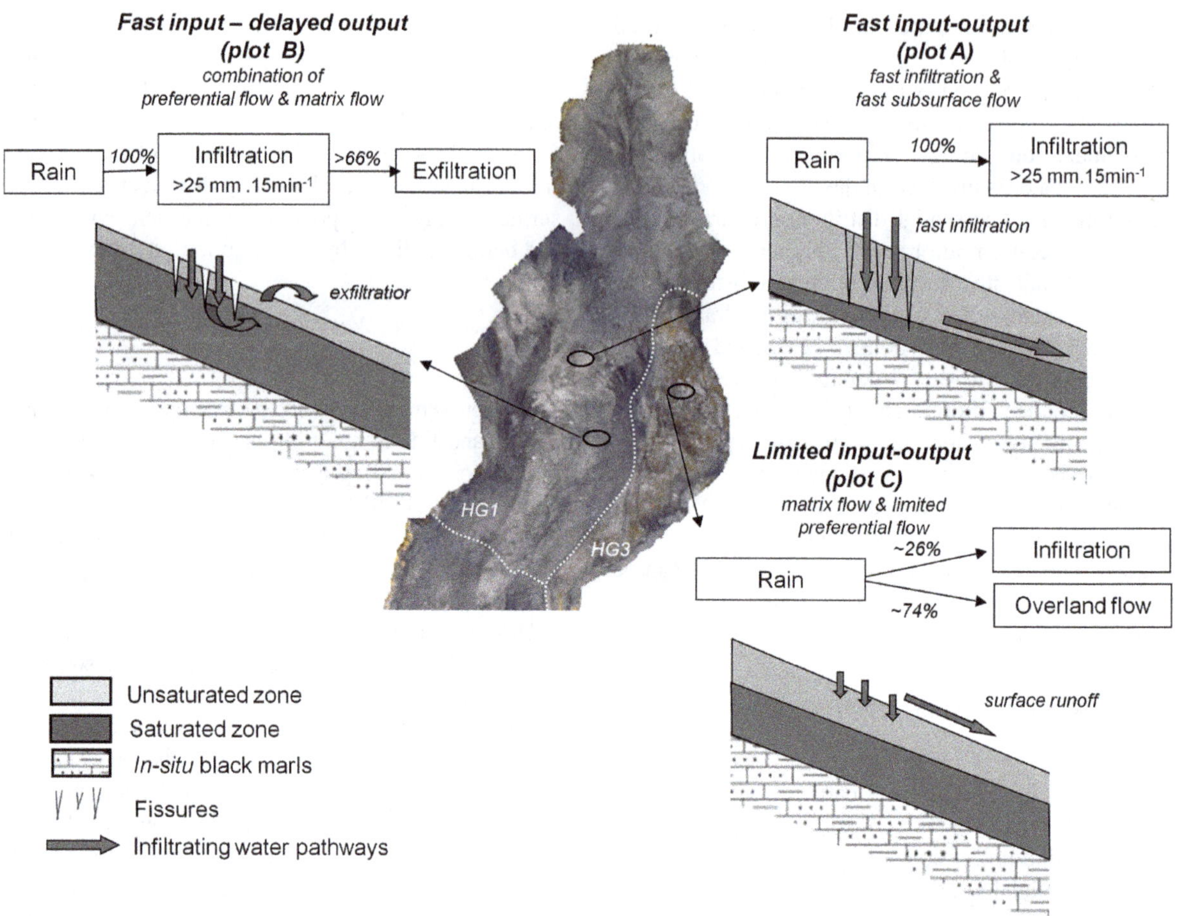

Figure 6. Flow regimes derived from hydrological and hydrochemical analysis of small-scale sprinkling experiment and their distribution across the upper part of the Super-Sauze landslide. The white dashed lines indicate the hydro-geomorphological (HG) units defined by Malet et al. (2005).

quantitative analyses of the components of the hydrological processes acting on the landslide and possible extension of the conceptual model with the identification of surface hydrological processes such as exfiltration and runoff.

Finally, our results are comparable with the large-scale sprinkling experiment performed previously in the area where plot B was located at the Super-Sauze landslide (Debieche et al., 2012). They confirm that hydrodynamic and hydrochemical responses cannot be fully inferred from surface area characteristics only. The sprinkling water infiltrates into the soil both through the matrix and preferential flow paths. The groundwater flow follows the overall slope direction, but the presence of fissures and subsurface structures strongly influences the exact direction of the subsurface water flow. Furthermore, unweathered marly blocks, characterised by relatively low permeability, decrease the percolation rate and create areas of limited hydrological activity.

4.4 Broader implications for landslide studies

Our findings outline the spatial heterogeneity existing in slow-moving slopes due to the interaction between displacements and hydraulic properties of the soil and thus hydrological behaviour. They reflect clear different topographic features in the landslide mass (i.e. the hydro-geomorphic units including fissure types). Linking the knowledge of geomorphological landslide development (e.g. Travelletti and Malet, 2012) with a hydrological characterisation derived by the small-scale hydrological and hydrochemical sprinkling experiments seems a logical next step. Secondary mass movements, like a local mudflow or landslide on the overall slow-moving landslide, will change the local hydrological regime. However, how long this processes would last and how they would (re-)develop over time is hard to say. Hereto the sprinkling experiments should be spatially distributed, but should as well be redone after major displacement events. Our limited set of experiments and observations does not allow for far-reaching conclusions on the relationship between hydrological regime and landslide development. However, it is tempting to speculate on it: as displacements in a landslide occur, the moving mass changes/adapts its hydraulic properties and, consequently, the pore pressure field. This influences the geomechanical behaviour of the moving mass: it could stabilise the mass due to drainage, but could also result in acceleration due to excess pore pressure during transient undrained condition. These feedbacks and the interplay of processes seem very relevant for the behaviour of a slope or landslide. This constitutes an example of how to study the co-evolution of landscape development and hydrology, as advocated in recent studies on catchment hydrology (Ehret et al., 2014).

5 Conclusions

This paper shows the potential of combined hydrological and hydrochemical analyses by means of small-scale $1 \times 1\ m^2$ sprinkling experiments for studying the spatial differences in hydrological response to precipitation input. The approach was applied at the specific environment of the highly heterogeneous Super-Sauze landslide (French Alps).

Dual or multiple permeability systems can be found in many hillslopes and they steer the hydrological dynamics of the hillslope. In such cases, laboratory tests for characterising hydraulic soil parameters are insufficient and in situ measurements or experiments are necessary. Small-scale sprinkling experiments performed with the use of artificial tracers and in situ observations of hydrological and hydrochemical response showed to be very effective in unravelling complex hydrological systems. A 2-day sprinkling experiment also had the clear advantage of allowing for more in-depth analyses of mixing processes to be performed (pre-event–infiltrated water). These analyses confirmed that the presence of fissures increases the vertical infiltration rate and controls the direction of subsurface water flow (e.g. McDonnell, 1990; Uchida et al., 2001). Furthermore, our results support the concept that antecedent water storage influences the initiation of preferential flow, as found, for example, in Trojan and Linden (1992), Zehe and Fluhler (2001) and Weiler and Naef (2003).

Presented experiments are relatively inexpensive, can be deployed throughout the landslide area and do not need long-term monitoring programmes. This paves the road for more widespread applications in order to better understand the spatial differences and similarities of hydrological processes across a landslide area. In order to extend the application of small-scale sprinkling experiments and to overcome current shortcomings, the following should be considered:

- detailed measurements of soil characteristics, their heterogeneity in the analysed soil profile, and their high temporal resolution monitoring during the sprinkling experiment;
- applying non-destructive measures to provide more detailed characteristics of subsurface fissure system, especially in vertical directions. Grandjean et al. (2012) and Travelletti et al. (2012) presented promising results based on seismic azimuth tomography or ERT measurements. However, both methodologies need further improvement to provide the unique characteristics of subsurface flow paths.

Although we performed only a limited amount of experiments, we showed that small-scale sprinkling experiments were capable of capturing and monitoring the hydrological processes across the landslide. Additionally, they show a potential for quantifying subsurface flow processes.

Acknowledgements. This work was supported by the European Commission within the Marie Curie Research and Training Network "Mountain Risks: from prediction to management and governance" (2007–2010, contract MCRTN-035798) and by the French National Research Agency (ANR) within the project "Ecou-Pref-Ecoulements préférentiels dans les versants marneux" (contract ANR05-ECCO-007-04).

Edited by: V. Vanacker

References

Allaire-Leung, S. E., Gupta, S. C., and Moncreif, J. F. : Water and solute movement in soil as influenced by macropore characteristics: I. Macropore continuity, J. Contamin. Hydrol., 41, 283–301, 2000.

Allaire S. E., Roulier S., and Cessna A. J. : Quantifying preferential flow in soils: a review of different techniques, J. Hydrol., 378, 179–204, 2009.

Anderson, L.: Fracture Mechanics: Fundamentals and Applications, 3rd Edition, Taylor & Francis, 2005.

Angulo-Jaramillo, R., Gaudet, J.-P., Thony, J.-L., and Vauclin, M.: Measurement of hydraulic properties and mobile water content of a field soil, Soil Sci. Soc. Am. J., 60, 710–715, 1996.

Beven, K. and Germann, P.: Macropores and water flow in soils, Water Resour. Res., 18, 1311–1325, 1982.

Bogaard, T. A., Buma, J. T., and Klawer, C. J. M.: Testing the potential of geochemical techniques for identifying hydrological systems within landslides in partly weathered marls, Geomorphology, 58, 323–338, 2004.

Boll, J., Steenhuis, T. S., and Selker, J. S.: Fiberglass wicks for sampling of water and solutes in the vadose zone, Soil Sci. Soc. Am. J., 56, 701–707, 1992.

Christophersen, N. and Hooper, R. P.: Multivariate analysis of stream water chemical data – the use of principal component analysis for the end-member mixing problem, Water Resour. Res., 28, 99–107, 1992.

Coe, J. A., Michael, J. A., Crovelli, R. A., Savage, W. Z., Laprade, W. T., and Nashem, W. D.: Probabilistic assessment of precipitation-triggered landslides using historical records of landslide occurrence, Seattle, Washington, Environ. Engin. Geosci., 10, 103–122, 2004.

Collins, R., Jenkins, A., and Harrow, M.: The contribution of old and new water to a storm hydrograph determined by tracer addition to a whole catchment, Hydrol. Proc., 14, 701–711, 2000.

Cras, A., Marc, V., and Travi, Y.: Hydrological behaviour of sub-Mediterranean alpine headwater streams in a badlands environment, J. Hydrol., 339, 130–144, 2007.

Debieche, T.-H., Bogaard, T. A., Marc, V., Emblanch, C., Krzeminska, D. M., and Malet, J.-P.: Hydrological and hydrochemical processes observed during a large-scale infiltration experiment at the Super-Sauze mudslide (France), Hydrol. Proc., 26, 2157–2170, 20112.

De Montety, V., Marc, V., Emblanch, C., Malet, J.-P., Bertrand, C., Maquaire, O., and Bogaard, T. A. : Identifying the origin of groundwater and flow processes in complex landslides affecting black marls: insights from a hydrochemical survey, Earth Surface Proc. Land., 32, 32–48, 2007.

Ehret, U., Gupta, H. V., Sivapalan, M., Weijs, S. V., Schymanski, S. J., Blöschl, G., Gelfan, A. N., Harman, C., Kleidon, A., Bogaard, T. A., Wang, D., Wagener, T., Scherer, U., Zehe, E., Bierkens, M. F. P., Di Baldassarre, G., Parajka, J., Van Beek, L. P. H., Van Griensven, A., Westhoff, M. C., and Winsemius, H. C.: Advancing catchment hydrology to deal with predictions under change, Hydrol. Earth Syst. Sci., 18, 649–671, 2014, http://www.hydrol-earth-syst-sci.net/18/649/2014/.

Flury, M., Fluhler, H., Jury, W. A., and Leuenberger, J.: Susceptibility of soils to preferential flow of water: a field study, Water Resources Research, 30, 1945–1954, 1994.

Garel, E., Marc, V., Ruy, S., Cognard-Plancq, A.-L., Klotz, S., Emblanch, C., and Simler, R.: Large scale rainfall simulation to investigate infiltration processes in a small landslide under dry initial conditions: the Draix hillslope experiment, Hydrol. Proc., 26, 14 pp., doi:10.1002/hyp.9273, 2012.

Grandjean, G., Bitri, A., and Krzeminska, D.M.: Characterisation of a landslide fissure pattern by integrating seismic azimuth tomography and geotechnical testing, Hydrol. Proc., 26, 2120–2127, 2012.

Greco, R.: Preferential flow in macroporous swelling soil with internal catchment, model development and applications, J. Hydrol., 269, 150–168, 2002.

Gerke, H. H.: Preferential flow descriptions for structured soils, J. Plant Nutrit. Soil Sci., 169, 382–400, 2006.

Gwo, J. P., Jardine, P. M., Wilson, G. V., and Yeh, G. T.: A multiple-pore-region concept to modeling mass transfer in subsurface media, J. Hydrol., 164, 217–237, 1995.

Haneberg, W. C.: Observation and analysis of short-term pore pressure fluctuations in a thin colluvium landslide complex near Cincinnati, Ohio, Engin. Geol., 31, 159–184, 1991.

Hornberger, G. M., German, P. F., and Beven, K. J.: Throughflow and solute transport in an isolated sloping soil block in a forested catchment, J. Hydrol., 124, 81–99, 1991.

IA PAS: Manual for Field Operated Meter (FOM), Institute of Agrophysics, Polish Academy of Science Lublin, 1–34, 2006.

James, A. L. and Roulet, N. T.: Investigating the applicability of end-member mixing analysis (EMMA) across scale: A study of eight small, nested catchments in a temperate forested watershed, Water Resour. Res., 42, 8, doi:10.1029/2005WR004419, 2006.

Jarvis, N. J.: A review of non-equilibrium water flow and solute transport in soil macropores: principles, controlling factors and consequences for water quality, European Journal of Soil Science, 58, 523–546, 2007.

Kabeya, N., Katsuyama, M., Kawasaki, M., Ohte, N., and Sugimoto, A.: Estimation of mean residence times of subsurface waters using seasonal variation in deuterium excess in a small headwater catchment in Japan, Hydrol. Proc., 21, 308–322, 2007.

Kienzler, P. and Naef, F.: Temporal variability of subsurface stormflow formation, Hydrol. Earth Syst. Sci., 12, 257–265, 2008, http://www.hydrol-earth-syst-sci.net/12/257/2008/.

Kirchner, J. W.: A double paradox in catchment hydrology and geochemistry, Hydrol. Proc., 17, 871–874, 2003.

Larsbo, M. and Jarvis, N.: Information content of measurements from tracer microlysimeter experiments designed for parameter identification in dualpermeability models, J. Hydrol., 325, 273–287, 2006.

Linsley, R. K., Kohler, M. A. and Paulhus, J. L. H.: Hydrology for Engineers, Third Edition, New York, McGraw-Hill Book Company, 1982.

Malet, J.-P., Maquaire, O., and Calais, E.: The use of Global Positioning System techniques for the continuous monitoring of landslides: application to the Super-Sauze earthflow (Alpes-de-Haute-Provence, France), Geomorphology, 43, 33–54 2002.

Malet, J.-P.: Les glissements de type écoulement dans les marnes noires des Alpes du Sud. Morphologie, fonctionnement et modélisation hydro-mécanique, Ph.D. thesis, Université Louis Pasteur, Strasbourg, 2003.

Malet, J.-P., Auzet, A. V., Maquaire, O., Ambroise, B., Descroix, L., Esteves, M., Vandervaere, J. P., and Truchet, E.: Soil surface characteristics influence on infiltration in black marls: application to the Super-Sauze earthflow (Southern Alps France), Earth Surf. Proc. Landf., 28, 547–564, 2003.

Malet, J.-P., van Asch, T. W. J., van Beek, L. P. H., and Maquaire, O.: Forecasting the behaviour of complex landslides with a spatially distributed hydrological model, Natural Haz. Earth Syst. Sci., 5, 71–85, 2005.

Mali, N., Urbanc, J., and Leis, A.: Tracing of water movement through the unsaturated zone of a coarse gravel aquifer by means of dye and deuterated water, Environ. Geol., 51, 1401–1412, 2007.

McDonnell, J. J.: The influence of macropores on debris flow initiation, Quarterly J. Engin. Geol. Hydrogeol., 23, 325–331, 1990.

Mikovari, A., Peter, C., and Leibundgut, Ch.: Investigation of preferential flow using tracer techniques, in: Tracer Technologies for Hydrological Systems, Proceedings of a Boulder Symposium, July 1995, 1995.

Mulholland, P. J. and Hill, W. R.: Seasonal patterns in streamwater nutrient and dissolved organic carbon concentrations: Separating catchment flow path and in-stream effects, Water Resour. Res., 33, 1297–1306, 1997.

Niethammer, U., James, M. R., Rothmund, S., Travelletti, J., and Joswig, M.: UAV-based remote sensing of the Super-Sauze landslide: Evaluation and results, Engin. Geol., 128, 2–11, 2012.

Pierson, T. C.: Soil pipes and slope stability, Quart. J. Engin. Geol., 16, 1–15, 1983.

Savage, W. Z., Godt, J. W. and Baum, R. L.: A model for spatially and temporally distributed shallow landslide initiation by rainfall infiltration, in: Proceedings of the third international conference on debris flow hazards mitigation: mechanics, prediction, and assessment, edited by: Rickenmann, D. and Chen, C.-L., Davos, Millpress, Rotterdam, 179–187, 2003.

Schulson, E. M. and Duval, P.: Creep and Fracture of Ice, Cambridge University Press, New York, 2009.

Šimůnek, J., Jarvis, N. J., van Genuchten, M. T., and Gardenas, A.: Review and comparison of models for describing non-equilibrium and preferential flow and transport in the vadose zone, J. Hydrol., 272, 14–35, 2003.

Sivapalan, M., Jothityangkoon, C., and Menabde, M.: Linearity and nonlinearity of basin response as a function of scale: discussion of the alternative definitions, Water Resour. Res., 38, 2, doi:10.1029/2001WR000482, 2002.

Soulsby, C., Rodgers, P., Smart, R., Dawson, J., and Dunn, S.: A tracer-based assessment of hydrological pathways at different spatial scales in a mesoscale Scottish catchment, Hydrol. Proc., 7, 759–777, doi:10.1002/hyp.1163, 2003.

Stumpf, A., Malet, J.-P., Kerle, N., Niethammer, U., and Rothmund, S.: Image-based mapping of surface fissures for the investigation of landslide dynamics, Geomorphology, 15, 12–27, 2012.

Travelletti, J. and Malet, J.-P.: Characterization of the 3D geometry of flow-like landslides: A methodology based on the integration of heterogeneous multi-source data, Engeen. Geol., 128, 30–48, 2012.

Travelletti, J., Sailhac, P., Malet, J.-P., Grandjean, G., and Ponton, J.: Hydrological response of weathered clay-shale slopes: water infiltration monitoring with time-lapse electrical resistivity tomography, Hydrol. Proc., 26, 2106–2119, 2012.

Trojan, M. and Linden, D.: Micro relief and rainfall effects on water and solute movement in earthworm burrows, Soil Sci. Soc. Am. J., 56, 727–733, 1992.

Tromp-van Meerveld, H. J. and McDonnell, J. J.: Threshold relations in subsurface stormflow: 2. the fill and spill hypothesis, Water Resour. Res., 42, 3, doi:10.1029/2004WR003778, 2006.

Uchida, T., Kosugi, K., and Mizuyama, T.: Effects of pipeflow on hydrological process and its relation to landslide: a review of pipeflow studies in forested headwater catchments, Hydrol. Proc., 15, 2151–2174, 2001.

Walter, M., Arnhardt, C., and Joswig, M.: Seismic monitoring of rockfalls, slide quakes, and fissure development at the Super-Sauze mudslide, French Alps, Engin. Geol., 128, 12–22, 2012.

Weiler, M. and Naef, F.: An experimental tracer study of the role of macropores in infiltration in grassland soils, Hydrol. Proc., 17, 477–493, 2003.

Weiler, M. and McDonnell, J. J.: Conceptualizing lateral preferential flow and flow networks and simulating the effects on gauged and ungauged hillslopes, Water Resour. Res., 43, 3, doi:10.1029/2006WR004867, 2007.

Wienhöfer, J., Lindenmaier, F., and Zehe, E.: Challenges in Understanding the Hydrologic Controls on the Mobility of Slow-Moving Landslides, Vadose Zone Journal, 10, 496–511, 2011.

Van Bavel, M. and Nichols, C.: Theta and Profiler Soil Moisture Probes – Accurate Impedance Measurement Devices – New Applications, Technical Report, 2002.

Van Beek, L. P. H. and van Asch T. W. J.: A combined conceptual model for the effects of fissure-induced infiltration on slope stability, in: Process Modelling and Landform Evolution, Hergarten S, Neugebauer HJ (eds), Lecture Notes in Earth Sciences, 78, 147–167, 1999.

Van Schaik, L.: The role of macropore flow from PLOT to catchment scale, Ph.D. thesis, in: Netherlands Geographical Studies, University of Utrecht, Netherlands, 2010.

Zehe, E. and Fluhler, H.: Preferential transport of isoproturon at a plot scale and a field scale tile-drained site, J. Hydrol., 247, 100–115, 2001.

10

Ice flow models and glacial erosion over multiple glacial–interglacial cycles

R. M. Headley[1] and T. A. Ehlers[2]

[1]Geosciences Department, University of Wisconsin-Parkside, Kenosha, WI, USA
[2]Department of Geosciences, University of Tübingen, Tübingen, Germany

Correspondence to: R. M. Headley (headley@uwp.edu)

Abstract. Mountain topography is constructed through a variety of interacting processes. Over glaciological timescales, even simple representations of glacial-flow physics can reproduce many of the distinctive features formed through glacial erosion. However, detailed comparisons at orogen time and length scales hold potential for quantifying the influence of glacial physics in landscape evolution models. We present a comparison using two different numerical models for glacial flow over single and multiple glaciations, within a modified version of the ICE-Cascade landscape evolution model. This model calculates not only glaciological processes but also hillslope and fluvial erosion and sediment transport, isostasy, and temporally and spatially variable orographic precipitation. We compare the predicted erosion patterns using a modified SIA as well as a nested, 3-D Stokes flow model calculated using COMSOL Multiphysics.

Both glacial-flow models predict different patterns in time-averaged erosion rates. However, these results are sensitive to the climate and the ice temperature. For warmer climates with more sliding, the higher-order model yields erosion rates that vary spatially and by almost an order of magnitude from those of the SIA model. As the erosion influences the basal topography and the ice deformation affects the ice thickness and extent, the higher-order glacial model can lead to variations in total ice-covered area that are greater than 30 % those of the SIA model, again with larger differences for temperate ice. Over multiple glaciations and long timescales, these results suggest that higher-order glacial physics should be considered, particularly in temperate, mountainous settings.

1 Introduction

Over geological time, mountainous topography is formed through a combination of erosional and tectonic processes. In many regions, mountain topography has felt the effects of glacier erosion, in addition to other geomorphic processes. Quantifying the effects of glaciation on topography requires consideration of the physics and rheological properties governing glacial erosion. This study builds upon previous work and evaluates how different assumptions and levels of complexity used in glacial-flow models impact the magnitude of erosion over multiple glacial–interglacial cycles and geologic timescales. This type of study is important for evaluating what level of model complexity (and computational sophistication) is required to quantify glacial erosion processes and sediment production in mountainous regions.

Numerical models have often been used to study the influence of glacial erosion on landscape development. These models have ranged from simple 2-D glacial profile models (Anderson et al., 2006; MacGregor et al., 2009) to more complex 3-D models that incorporate a variety of processes and mechanisms (Kessler et al., 2008; Egholm et al., 2012a; Pedersen and Egholm, 2013). Other studies have incorporated glacial erosion into landscape dynamic models that also include fluvial and hillslope erosional processes (Herman and Braun, 2008; Egholm et al., 2011; Yanites and Ehlers, 2012). The evolution and continued development of glacial-flow and erosion models has resulted in the simulation of increasingly complex processes such as the influence of subglacial hydrology (Egholm et al., 2011; Herman et al., 2011; Iverson, 2012). Despite these advances, other mechanisms are still represented by simplified assumptions and approximations, particularly the underlying physics of ice flow. Within the field of glaciology, as computing power has increased, higher-order glacial-flow models (Pattyn et al., 2008) have been made more accessible. These higher-order processes are often simulated over the timescales useful for glacial and climatic studies (10^3 to 10^4 years), yet still shorter than the timescale of orogen topographic development and Quaternary glaciations (10^5 to 10^7 years). The incorporation of these higher-order models into orogenic timescale models can be useful to better represent the glacial flow in alpine settings, where the effects of longitudinal and lateral stresses on glacial flow and erosion should be important (Egholm et al., 2012a, b).

In many orogenic-scale models, glacial flow and erosion have been represented using simplifying assumptions, such as the shallow ice approximation (SIA) for glacial flow (Kessler et al., 2008; Iverson, 2012). This approximation simplifies the ice flow equation (Glen flow law) by only considering the first-order simple shear stresses (Cuffey and Paterson, 2010). While this approximation is appropriate for some specific glacial settings, particularly ice sheets where surface slopes are shallow, ice thickness is small compared to ice extent, and sliding velocities are small compared to deformational velocities; for alpine glaciers this assumption misses effects that result from glacial flow through narrow and steep topography (Egholm et al., 2011). While even the simplest approach has its merits, defining the conditions under which a higher-order model should be used warrants more detailed consideration. Recent work has shown that, over the length and timescales of glacial valley formation, higher-order (HO) glacial-flow models have important feedbacks (Egholm et al., 2009, 2012b; Pedersen and Egholm, 2013). Specifically, Egholm et al. (2012a, b) showed that, on the glacier valley scale and over a single glacial cycle, the incorporation of lateral and longitudinal stresses can provide an important mechanism for suppressing potential runaway problems that can come from using the SIA. In addition, larger, regional models have been investigated, and other investigators have stressed that the physics and form of the glacier are certainly important, and that models of alpine glaciers and their erosional patterns are influenced by the choice of physics, particularly as the landforms and valley profiles evolve (Pedersen and Egholm, 2013).

Previous work has addressed differences between SIA and higher-order models, which includes work that focuses on glaciers in steep terrain (Egholm et al., 2011) and glaciers in landscape evolution models (Egholm et al., 2012b). For glaciers in alpine terrain and in landscape evolution scales, the SIA approach was found to be less accurate in predicting sliding velocities and patterns of basal shear stress, since this approximation fails to incorporate lateral and longitudinal stresses in glacier valleys. The SIA led to positive feedbacks where enhanced erosion caused deeper and steeper topography that in turn led to higher sliding and erosion rates. hese feedbacks tend to be dampened when used within full Stokes and higher-order models which incorporate at least approximations of the lateral and longitudinal stresses. Other studies have also addressed comparisons between SIA and HO ice flow models, including the benchmarks developed and tested in Pattyn et al. (2008). However, these benchmarks pertain mostly to 3-D continental ice sheets. The benchmarks themselves only include 2-D flow-line models for an alpine glacier and focus only on non-evolving ice flow and basal shear stress.

This study complements previous work by evaluating the effect of the glacial-flow physics model on predicted variations in glacial erosion over both single and multiple glacial cycles. This study can be differentiated from other work due to the following reasons. (1) The timescale investigated in this study is significantly longer, and extends to geologic timescales of 400 000 years, including three full glaciations, with a range of different climate scenarios. Thus, we report here the effects of differences between the two approaches over temporally varying, and multiple, glacial–interglacial cycles. This effect is potentially important because catchment hypsometry evolves over long timescales, as does the thermal state of the glacier. Furthermore, (2) previous work by Egholm et al. (2011, 2012b) used a 3-D second-order shallow ice approach (see also Pedersen and Egholm, 2013) for comparison to a traditional SIA model. A full-stress Stokes flow model was used only in 2-D for comparison to the SIA and could not evaluate 3-D topographic effects on ice mechanics. We evaluate the strengths and weaknesses of SIA and Stokes flow within a 3-D landscape evolution model and address how a HO glacial model might affect topography and its evolution over multiple glacial–interglacial periods. Our end goal is to add to the understanding of when and under what conditions more simplified models, such as the SIA are sufficient, as applicable to larger-scale problems such as sediment production, ice-sheet stability, and tectonic–erosional interactions under alpine glaciers and continental ice sheets. While many simplifications are used in this comparison, such as the choice of erosion and sliding laws, which are discussed more thoroughly in Sect. 4.4, this comparison ultimately yields some quantitative evidence of how HO glacial-flow physics can influence the evolution of topography. This comparison also provides more evidence that the choice of glacial-flow physics in landscape evolution models should be made at least partially based upon the climate of the model and the timescales of interest, along with the topography and glaciation style (i.e., ice sheets versus alpine glaciers).

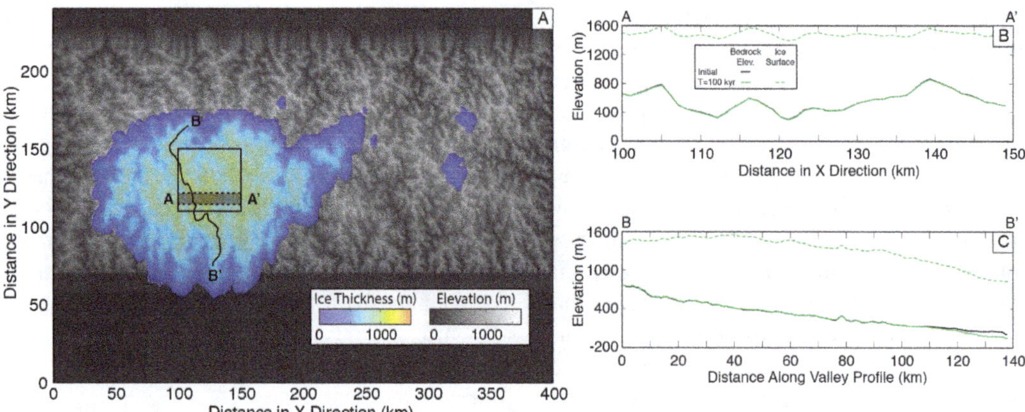

Figure 1. **(a)** Model domain with ice coverage from glacial maximum for Simulation 2 with hillshade topography at $T = 100$ kyr. Glacial maxima for the other simulations generally follow the same form. The black box outlines region of nested, higher-order physics domain, used in Figs. 5 and 6. The shaded region (A–A') shows the location of the orogen-parallel swath used for comparisons in Figs. 7 and 9. Line B–B' gives the valley profile used for comparisons in Figs. 8 and 10. **(b)** Bedrock topography and ice surface at glacial maxima of $T = 100$ kyr and initial bedrock topography along A–A'. **(c)** Bedrock topography and ice surface at the same glacial maxima and initial bedrock topography along B–B'.

2 Methods

Here we build upon previous work and investigate the influence of glacial-flow physics on a developing orogen over geologic timescales using a modified version of the ICE-Cascade landscape evolution model. In order to compare the importance of the choice of ice physics, simulations are run using both the SIA model and the nested HO model. We repeat these comparisons over different climate scenarios in order to highlight temperature-dependent effects. The different climate simulations are used because glacial erosion is dependent upon the existence of liquid water at the base of the glacier, a temperature-dependent property. For brevity, a summary is given of ICE-Cascade, its major components, the physics governing the ice flow, and the modeling framework behind the HO glacial-flow model. All relevant model parameters used for ICE-Cascade are presented in Table 1; readers are referred to the associated references for additional details.

2.1 Simulations

Three separate simulations based on climatic and ice temperature conditions are performed using both the SIA and the HO physics models described in Sect. 2.3. These simulations are summarized in Table 2, and the sea-level temperature patterns over time are shown in Fig. 3.

- Simulation 1 uses a sinusoidal temperature pattern with amplitude of 6 °C and sea-level minimum temperature of 2 °C, with a wavelength of 100 kyr.
- Simulation 2 has a similar pattern but with the sea-level minimum shifted to 0 °C, so there are more instances of cold ice where the base is frozen.
- Simulation 3 uses the same temperature pattern as Simulation 1; however, in this simulation sliding occurs everywhere the ice thickness is greater than 10 m, and the ice temperature at the base of the glacier is not factored into this calculation.

Each simulation was run for over 400 kyr, with glaciations major occurring every 100 kyr. Over three full glaciations are captured during this time interval. Figure 1 shows Simulation 2 (SIA) with hillshade topography and the ice coverage and thickness at $T = 100$ kyr in the simulation. Compared to any individual Pleistocene glacial pulse, the simulated glaciations are extensive and long-lived, particularly with the sinusoidal climate forcing. This is to emphasize long-term evolution effects versus influence from a single, quick glaciation.

2.2 ICE-Cascade orogen development model and climate parameters

ICE-Cascade allows us to model topographic evolution over geologic timescales, with the influences of both constructive (tectonics and sediment deposition) and destructive (erosion) processes (Herman and Braun, 2008; Yanites and Ehlers, 2012). At each time step, the topography is uplifted according to an input rate (Table 1) along with a component based on flexural isostasy. The isostatic response of the landscape is affected by loading and unloading due to erosion, sediment deposition, and the ice thickness. Following the uplift, the landscape is eroded according to the rates from the fluvial, hillslope, and glacial modules. Where glacier ice is nonexistent or thinner than 10 m, fluvial and hillslope processes erode and redistribute sediment (Yanites and Ehlers, 2012). Sediment transport by the glaciers occurs in regions of ice

Table 1. Landscape evolution and orographic precipitation model parameters.

Description and parameter	Value [units]	Reference
	Glacial parameters	
Flow rate factor for temperate ice A	6.8×10^{24} [$\mathrm{Pa^{-n}\,yr^{-1}}$]	Cuffey and Paterson (2010)
Glen's flow law exponent n	3	Cuffey and Paterson (2010)
Sliding constant A_{sl}	1×10^{-15} [$\mathrm{Pa^{-m}\,s^{-1}}$]	Braun et al. (1999)
Sliding exponent m	3	Braun et al. (1999)
Erosion rate constant K	0.001 [$(\mathrm{yr\,m^{-1}})^{l-1}$]	Humphrey and Raymond (1994)
Erosion rate exponent l	1	
Constriction constant	1000	Braun et al. (1999), Egholm et al. (2011)
Density of ice ρ_i	910 [$\mathrm{kg\,m^{-3}}$]	
Ice thermal conductivity	2.4 [$\mathrm{W\,m^{-1}\,K^{-1}}$]	Herman and Braun (2008)
Snow stability angle for avalanching	35 [°]	Kessler et al. (2008)
	Tectonic parameters	
Vertical rock uplift rate	0.25 [$\mathrm{mm\,yr^{-1}}$]	
Geothermal heat flux	0.05 [$\mathrm{W\,m^{-2}}$]	Herman and Braun (2008)
Flexural plate length	1000 [km]	Braun and Sambridge (1997)
Elastic plate thickness	15 [km]	Braun and Sambridge (1997)
Young's modulus	1×10^{11} [Pa]	Braun and Sambridge (1997)
Poisson's ratio	0.25	Braun and Sambridge (1997)
Density of crust	2750 [$\mathrm{kg\,m^{-3}}$]	Braun and Sambridge (1997)
Density of asthenosphere	3300 [$\mathrm{kg\,m^{-3}}$]	Braun and Sambridge (1997)
	Climate parameters	
Sinusoidal temperature period	100 [kyr]	
Atmospheric lapse rate	6.5 [$\mathrm{°C\,km^{-1}}$]	Yanites and Ehlers (2012)
Positive-degree-day melting coefficient	8.0×10^{-3} [$\mathrm{K\,m\,yr^{-1}}$]	Braithwaite (1995)
Annual temperature variation	15 [°C]	Yanites and Ehlers (2012)
a0	0.3 [$\mathrm{m\,yr^{-1}}$]	Roe et al. (2003)
a1	110 [$\mathrm{m\,yr^{-1}}$ per $\mathrm{m\,s^{-1}}$]	Roe et al. (2003)
Alf	100 [$\mathrm{s\,m^{-1}}$]	Roe et al. (2003)
Average wind speed	0.6 [$\mathrm{m\,s^{-1}}$]	Roe et al. (2003)
Wind direction (angle from x axis)	90 [degrees]	Roe et al. (2003)
	Fluvial and hillslope parameters	
Hillslope diffusivity	2×10^{-6} [$\mathrm{km^2\,yr^{-1}}$]	Braun and Sambridge (1997)
Threshold hillslope landsliding	35 [degrees]	Burbank et al. (1996), Stolar et al. (2007)
Fluvial erosion coefficient	3.5×10^{-4}	Braun and Sambridge (1997)
Fluvial erosion length scale	1000 [m]	Braun and Sambridge (1997)
Channel width scaling coefficient	0.1 [$(\mathrm{yr\,m^{-1}})^{0.5}$]	Yanites and Ehlers (2012)
Discharge threshold	4 [$\mathrm{m\,km^2\,yr^{-1}}$]	Braun and Sambridge (1997)
Alluvium length scale	100 [m]	Braun and Sambridge (1997)

Table 2. Landscape evolution and orographic precipitation model parameters.

Simulation	Pattern	Amplitude	Sea-level minimum	Other factors
1	Sinusoidal	6 °C	2 °C	
2	Sinusoidal	6 °C	0 °C	
3	Sinusoidal	6 °C	2 °C	Temperature-independent sliding

thicker than 10 m; bed material is eroded and immediately transferred to the fluvial system that emanates from the toe of the glacier. River discharge is calculated based upon the upstream precipitation and water-equivalent ice melt from upstream regions. Fluvial erosion processes are calculated from this discharge and the sediment supply, local topographic slope, and the channel width, all of which also are input into a linear sediment cover model (Braun and Sambridge, 1997). Hillslope processes are simulated from diffusion and a threshold landsliding algorithm when hillslopes steepen beyond a certain threshold (Burbank et al., 1996; Stolar et al., 2007).

Within ICE-Cascade, the climate simulation is a combination of the inputs that govern the pattern of sea-level temperature and an orographic precipitation model (Yanites and Ehlers, 2012; Roe et al., 2003). The temperature and moisture content variations over all elevations is calculated using an input lapse rate and the Clausius–Clapeyron relation, and these values, along with inputs governing the wind speed and direction, are then used to calculate annual precipitation (Roe et al., 2003). When the temperature is below freezing, the precipitation takes the form of snow. A positive-degree-day algorithm is used to determine the number of days above freezing in any given year (Braithwaite, 1995), and this, in turn, is used in calculate the amount of melt of snow and ice. For any point in the landscape, the annual mass balance is then simply the difference between the amount of snowfall and the amount of melt. Climate parameters governing these processes are given in Table 1.

For these simulations, the initial topography is an equilibrium landscape generated using only fluvial and hillslope processes. This topography was built from an earlier simulation (simulation m01 from Yanites and Ehlers, 2012) that started with random topography and was allowed to develop over 16 Myr with the same hillslope, fluvial, uplift, and climate parameters as found in Table 1.

2.3 Glacial models

At the beginning of the simulations, glaciations evolve where a persistent positive mass balance exists. Glaciers flow from the orogen and its valleys onto the continental slope, where they form piedmont lobes. The shelf can be seen in Fig. 1, extending to the edges of the y domain from 0 to 70 km and from 225 to 245 km. The shelves, with a slope of 0.001, were added to ensure that the ice velocities at the orogen-parallel boundaries are numerically stable (Yanites and Ehlers, 2012). The shelf edges ($y = 0$ and 245 km) have Dirichlet boundary conditions, where their elevation is fixed to sea level and does not change throughout the simulations. During these simulations, the ice never reached these boundaries.

The depth-averaged and sliding velocities are calculated from the glacial-flow physics (see Sects. 2.3.1 and 2.3.2 for more details). The sliding velocity is only calculated where the temperature at the base of the glacier is at the pressure-melting point associated with the local ice thickness (except in Simulation 3, where there is sliding everywhere). The basal temperature is determined from a conductive heat model for ice, where the upper boundary is the surface temperature, and the lower boundary is the geothermal heat flux (Table 1). Beyond glacial flow, other ice feedbacks are incorporated in the model. On steep slopes avalanching occurs when the snow surface slope exceeds a critical stability angle (Table 1), in which case all the snow at the point is redistributed to the next downstream node. Where the glacier terminates below sea level on the shelf, iceberg calving occurs and is a function of the depth of ice below sea level. In addition to calving, a buoyancy feedback is also incorporated for when the glacier erodes an overdeepening below sea level (Yanites and Ehlers, 2012). In an overdeepened, below-sea-level reach, sea level is assumed to be the equipotential ground water surface, such that the glacial erosion rate decreases as the ice approaches buoyancy. For these model comparisons, no below-sea-level overdeepenings develop. While this is a simplification of the complex feedbacks among the glacial flow, subglacial and englacial hydrology, and the subglacial sediment (Hooke, 1991; Alley et al., 2003), glaciated valleys are produced that fit the expectations that overdeepened glaciers tend to reach critical angles. For the model comparisons, all processes except for glacier sliding and deformation are calculated within the ICE-Cascade model not within the HO nested model. Certain processes, such as avalanching and hillslope processes can, occur in different locations between the HO and SIA model if the ice extent is significantly different.

Glacial erosion is performed by two major mechanisms: abrasion and quarrying (Hallet, 1979, 1996; Iverson, 1991). These processes operate on spatial and temporal scales that are orders of magnitude shorter than those of the glacial valley, the climate cycle, and the orogen. The erosional processes are often simply upscaled in landscape evolution models (Tomkin, 2003; Herman and Braun, 2008). Following the methods of many existing studies of glacial erosion on orogenic timescales, we use a simple relationship between erosion and sliding. For both models, glacial erosion, $\frac{\partial z_b}{\partial t}$, is proportional to the sliding velocity, which comes from modeling and empirical studies of glacial erosion (Harbor et al., 1988; Humphrey and Raymond, 1994), such that

$$\frac{\partial z_b}{\partial t} = K|u_{sl}|^l, \quad (1)$$

where K is a constant that characterizes the erodibility of the subglacial material (Laitakari et al., 1985; MacGregor et al., 2009; Duhnforth et al., 2010) and l is another constant generally equal to 1 (Table 1). While a few recent studies have used more sophisticated rules for erosion (MacGregor et al., 2009; Iverson, 2012), this is the same rule as used in previous ICE-Cascade and other glacially influenced, landscape evolution models (e.g., Braun et al., 1999; Herman and Braun,

2008; Kessler et al., 2008; Egholm et al., 2012b; Yanites and Ehlers, 2012).

2.3.1 Shallow ice approximation

The shallow ice approximation (SIA) simplifies full-stress glacial flow by assuming that the ice is significantly wider than it is thick, and the surface slopes are not large, i.e., it is shallow (Fowler and Larson, 1978; Hutter, 1983). In this assumption, all stresses but the simple shear stresses in the direction of ice flow are assumed to be negligible; longitudinal and lateral stresses, including drag, are assumed insignificant. This considerably simplifies how ice flow can be modeled, which is particularly useful when used on orogenic length and timescales. However, the assumptions based on glacial geometry and surface slope, while originally derived for use on large, shallow ice sheets, are not necessarily true when used for alpine glaciers with considerable sliding and where narrow valleys channelize flow and large surface and basal slopes are present (Hutter, 1983; Egholm et al., 2011).

Ice thickness, H, is computed from mass-balance equation

$$\frac{\partial H}{\partial t} = \nabla \cdot F + M, \tag{2}$$

where M is the mass balance and F is the vertically integrated ice flux.

The ice velocity, u, is simplified to just two dimensions where vertical velocities are deemed negligible, so that $u = u\hat{i} + v\hat{j}$. Each directional component is the sum of two components of glacial motion, which are designated using the SIA in this definition: deformation $u_{\mathrm{d}}^{\mathrm{sia}}$ and sliding $u_{\mathrm{sl}}^{\mathrm{sia}}$ for the velocity in the x direction (v is defined in a parallel fashion for the velocity in the y direction).

$$u = u_{\mathrm{d}}^{\mathrm{sia}} + u_{\mathrm{sl}}^{\mathrm{sia}} \tag{3}$$

The deformation component is defined

$$u_{\mathrm{d}}^{\mathrm{sia}} = -\frac{2A\beta}{n+2}(\rho g)^n H^{n+1} |\nabla (H + z_{\mathrm{b}})|^{n-1} \nabla (H + z_{\mathrm{b}}), \tag{4}$$

where A, which is pressure- and temperature-dependent but treated as a constant for most of this analysis and is furthered discussed in Sect. 4.4, and n come from the Glen flow equation (Cuffey and Paterson, 2010), which relates the strain rate, $\dot{\varepsilon}$, to the stress, σ, via

$$\dot{\varepsilon} = A\sigma^n. \tag{5}$$

β is a constriction factor that is dependent upon the gradient of the subglacial topography. In a standard SIA model, β is unity. For this version of ICE-Cascade, the SIA model is modified to incorporate this factor to partially account for the stresses in steep alpine glaciated valleys where the SIA is too simple (Herman and Braun, 2008):

$$\beta = \left(1 + k_{\mathrm{c}} \frac{\partial^2 z_{\mathrm{b}}}{\partial x_{\mathrm{f}}^2}\right)^{-1}. \tag{6}$$

The constant, k_{c}, has been evaluated over a 1 km wide glacial valley by Egholm et al. (2011). A value of 1000 was found to have reasonable agreement with a higher-order approximation, with comparisons of sliding and deformation velocities. Using the constriction factor, overpredictions of erosion rate from the standard SIA were diminished to underpredictions in comparison to the erosion rates from the higher-order approximation (Egholm et al., 2011, 2012b).

The sliding velocity $u_{\mathrm{sl}}^{\mathrm{sia}}$ is defined in a similar form to the deformation, dependent upon the shear stress at the bed of the glacier,

$$u_{\mathrm{sl}}^{\mathrm{sia}} = -\frac{2A_{\mathrm{sl}}}{N-P}(\rho g H)^n |\nabla (H + z_{\mathrm{b}})|^{n-1} \nabla (H + z_{\mathrm{b}}), \tag{7}$$

incorporating the sliding flow factor A_{sl}, the constriction factor discussed above β (Table 1), and simple subglacial hydrology as the effective pressure, the difference between the ice overburden pressure and the water pressure, $N - P$. Water pressure and its change over time are influences on the sliding velocity and has been used in a variety of other glacial erosion and landscape evolution models (e.g., MacGregor et al., 2000; Tomkin, 2003; Herman and Braun, 2008; Kessler et al., 2008; Egholm et al., 2011). However, in order to focus on just the physics of the ice flow, we do not consider variable subglacial hydrology in this model and treat $N - P$ as 80 % of the ice overburden pressure; Sect. 4.4 contains more discussion of this choice.

2.3.2 Higher-order physics

While a variety of various higher-order simplifications have been used to represent flow in other models (Pattyn, 1996; Egholm et al., 2011), we opt for a full-stress solution nested into the larger ICE-Cascade framework. Nesting of this modeling within the SIA model is required due to computational considerations, and our analysis is mostly focused on differences between the two model setups over a limited region. Figure 1 shows an example of topography during a glacial maximum with the nested region highlighted (from 100 to 150 km in the x direction and 100 to 160 km in the y direction). In this region, we use the full-stress equations and treat ice as a nonlinear viscous flow. The sliding velocity is calculated for this region and then passed back to ICE-Cascade. To ensure that the pattern of sliding velocity is smooth, a linear interpolation is used between the sliding velocity at the edges of the nested domain and those in the nested domain. This small region surrounding the nested domain has a width of less than 1 km.

Conservation of mass is given by

$$\nabla \cdot \boldsymbol{u} = 0, \tag{8}$$

where $\boldsymbol{u}$ is the 3-D velocity vector $\boldsymbol{u} = u\hat{i} + v\hat{j} + w\hat{k}$, and is also subdivided into deformational and sliding portions: $\boldsymbol{u} = u_{\mathrm{d}}^{\mathrm{ho}} + u_{\mathrm{sl}}^{\mathrm{ho}}$.

Ice flows as an incompressible laminar material. The Glen flow equation, when written in viscosity form for a single stress component, is

$$\sigma'_{ij} = A\eta\dot{\varepsilon}_{ij}, \tag{9}$$

and the viscosity η is given as

$$\eta = \frac{1}{2}A^{-1/n}\dot{\varepsilon}^{(1-n)/n}, \tag{10}$$

where $\dot{\varepsilon}$ is the effective strain rate, the second invariant of the strain rate. A is a flow constant dependent upon the temperature of the ice but treated as a constant in this analysis.

The full strain rate tensor is defined as

$$\begin{pmatrix} \dot{\varepsilon}_{xx} & \dot{\varepsilon}_{xy} & \dot{\varepsilon}_{xz} \\ \dot{\varepsilon}_{yx} & \dot{\varepsilon}_{yy} & \dot{\varepsilon}_{xy} \\ \dot{\varepsilon}_{zx} & \dot{\varepsilon}_{zy} & \dot{\varepsilon}_{zz} \end{pmatrix} \tag{11}$$

$$= \begin{pmatrix} \frac{\partial u}{\partial x} & \frac{1}{2}\left(\frac{\partial u}{\partial y} + \frac{\partial v}{\partial x}\right) & \frac{1}{2}\left(\frac{\partial u \partial z + \partial w}{\partial x}\right) \\ \frac{1}{2}\left(\frac{\partial u \partial y + \partial v}{\partial x}\right) & \frac{\partial v}{\partial y} & \frac{1}{2}\left(\frac{\partial v}{\partial z} + \frac{\partial w}{\partial y}\right) \\ \frac{1}{2}\left(\frac{\partial u}{\partial z} + \frac{\partial w}{\partial x}\right) & \frac{1}{2}\left(\frac{\partial v}{\partial z} + \frac{\partial w}{\partial y}\right) & \frac{\partial w}{\partial z} \end{pmatrix}, \tag{12}$$

and the second invariant that is used to define the effective strain rate in the viscosity term, Eq. (10), is

$$\dot{\varepsilon}^2 = \sum_{ij} \frac{1}{2}\dot{\varepsilon}_{ij}\dot{\varepsilon}_{ij}. \tag{13}$$

The shear stress (τ_{ij}) can be found from Eq. (9). With Eqs. (9), (10), and (12) combined, the full shear stress acting in the x direction, τ_{xz}, is then defined

$$\tau_{xz} = \frac{1}{2}A^{-1/n}\dot{\varepsilon}^{(1-n)/n}\dot{\varepsilon}_{xz}, \tag{14}$$

and this is the term (along with the shear stress in the y direction τ_{yz}), when defined at the bed of the glacier τ_{b}, upon which the basal sliding is dependent. The sliding velocity is defined in both the x and y directions, similar to Eq. (7). For the x direction, the sliding velocity is

$$u_{\mathrm{sl}}^{\mathrm{ho}} = \frac{A_{\mathrm{sl}}\tau_{xz}^{m}}{N-P}, \tag{15}$$

where A_{sl} is the same constant as in Eq. (7), and m is 3 in this case (Table 1). As in the SIA model, the effective pressure $(N-P)$ is 80 % the ice overburden pressure.

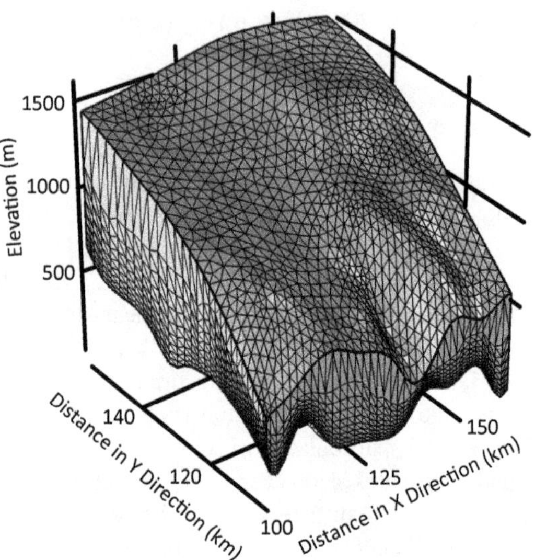

Figure 2. Graphical representation of the mesh used within the HO nested model. The width and depth correspond to those of the black box in Fig. 1. The height corresponds to the ice thickness, and the bottom edge is at the elevation of the topography. The mesh is explained further in Sect. 2.5.

2.4 COMSOL Multiphysics

COMSOL Multiphysics has been used for modeling ice flow in 2-D flow-band and flow-line profiles (Johnson and Staiger, 2007; Campbell, 2009; Headley et al., 2012). Here, we modify the steady-state viscous flow module using the Glen flow law. The flow is represented in equation form by the 3-D incompressible laminar flow, and we set the viscosity to be dependent upon the strain rates (Eq. 9). The geometry is defined from the bed topography and the ice thickness within the nested domain. The boundary conditions on the edges of the domain are open boundaries. These boundaries are only used within COMSOL simulations, as the velocity distribution over depth is not output back into ICE-Cascade. The top surface is a free surface, and the shear stress at the bottom boundary, along with the basal temperature, is used to calculate the sliding velocity per Eq. (14). Within COMSOL, proper meshing is important for solution convergence. In this case, we set the mesh size along the top and bottom ice surface as COMSOL setting "Coarse". From the nonlinear ice flow law, Eq. (5), velocity gradients are larger closer to the bed. For more efficient computations, we use nine boundary layers perpendicular to the bed and surface, with decreasing vertical dimension approaching the bed (z direction), using order 10^4 tetrahedral elements; this corresponds to a resolution at the bed spaced between 500 and 800 m in the x and y directions. Figure 2 shows an example of the mesh used in COMSOL for the nested model. When the sliding and surface velocities and basal shear stress are input back into the ICE-Cascade model, the velocity profile of both the HO region and the surrounding SIA region is linearly interpolated over a small region (< 1 km wide) bordering the nested region to ensure that the final results are numerically smooth. The velocities and basal shear stress are interpolated to the node spacing used in the ICE-Cascade model (Herman and Braun, 2008), which is variable to down to 100 m but initially calculated with a spacing of about 1 km. To investigate whether any significant numerical differences had arisen when the models were combined, a preliminary comparison was performed where no erosion took place but the sliding and ice deformation only influenced the glaciated area; no statistical significance in landscape features was found in this comparison, though glaciated area was still varied on a similar scale between the two models.

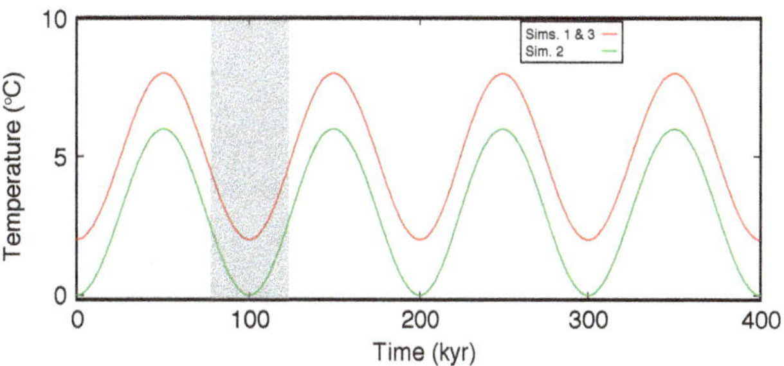

Figure 3. Climate variations over time for the different simulations. Over three complete glaciations are simulated. Simulation 3 has the same climate as Simulation 1 and is not specifically shown. The grey box shows the single glaciation used in the analysis presented in Figs. 5, 7, and 8.

3 Results

The ICE-Cascade model outputs used in this study include the glacial erosion rate, subglacial topography, glacier thickness, and sediment thickness; these primary outputs are analyzed for the rest of the study. Figure 1 shows example output of the topography and ice coverage at a glacial maximum. The size of the nested model region (black box in Fig. 1) was set due to computational limitations in the Stokes flow simulations. The two profiles highlighted in Fig. 1 are used for comparison of the models: one profile is orogen parallel (A–A'), and the other follows a valley profile (B–B'). Three separate simulations are performed and compared. The three climate and thermal settings (Simulations 1–3) are further discussed in Sect. 2.2, summarized in Table 2, and shown in Fig. 3. For these three simulations, each was performed using both the ICE-Cascade with SIA glacial-flow physics and with the HO nested subdomain. Erosion rate is generally used instead of sliding velocity or basal shear stress because erosion rate not only reflects the patterns of both of these and freezing at the bed of the glacier (Eqs. 1, 7, and 14) but can also be compared to other destructive and constructive processes incorporated into the orogenic model.

3.1 Influence of glacial-flow physics on glacial area and volume

Large-scale properties of glaciations can be used to compare the different climate scenarios and the different glacial-flow models. In this case, we use the glacier-covered area and the maximum ice thickness for each time step. As expected, colder climates lead to significantly larger glacial-covered areas and persistent thick glaciers (Fig. 4), regardless of glacial-flow model. The glaciations are progressively smaller for each subsequent glaciation even though the temperature amplitude is not changing.

The overall pattern of ice cap growth is similar between the SIA and HO models, and the differences between the higher-order glacial-flow physics and the shallow ice approximation models are generally difficult to see when comparing the full magnitudes of ice-covered area and maximum ice thickness (Fig. 4). However, comparing the percent difference between the two ($\frac{A_{\mathrm{SIA}} - A_{\mathrm{HO}}}{A_{\mathrm{SIA}}} \times 100$ and $\frac{H^*_{\mathrm{SIA}} - H^*_{\mathrm{HO}}}{H^*_{\mathrm{SIA}}} \times 100$, where A is the ice-covered area and H^* is the maximum thickness, and the denominators are those values averaged over the glaciation center around the 100 kyr) shows significant variation (Fig. 4), reaching a difference of over 30 % for some of these simulations. The percent differences between the two glacial-flow models show that even a higher-order nested model that does not cover the full ice-covered domain can have an effect on the fully glaciated area.

At the start of the simulations, the percent differences in the area are small for all the simulations (Fig. 4), and the maximum ice thickness follows a similar pattern (Fig. 5). Ice thickness tends to grow rapidly, reaches a maximum when the maximum area and minimum temperature are reached, and then tapers slightly before decreasing rapidly. As time progresses, changes are exacerbated, and the difference between SIA and HO is more readily apparent. In general, the simulations that contain warmer temperatures and/or have more sliding (i.e., Simulation 3) have a much larger differences in ice thickness and glaciated area due to the HO glacial-flow physics, while Simulation 2, the coldest, has minimal differences.

A null simulation was also performed where the SIA and HO models were used but glacial erosion was turned off. In this case, for both the maximum ice thickness and the glaciated area, differences between the two models were on the order of 5 % or smaller during every glaciation. The differences did not increase as time progressed. This result is presented to emphasize that the differences between the two models for the three simulations come mainly from the models' influence on the glacial erosion rate and are not inherent from the model nesting, model choice, or other erosional processes.

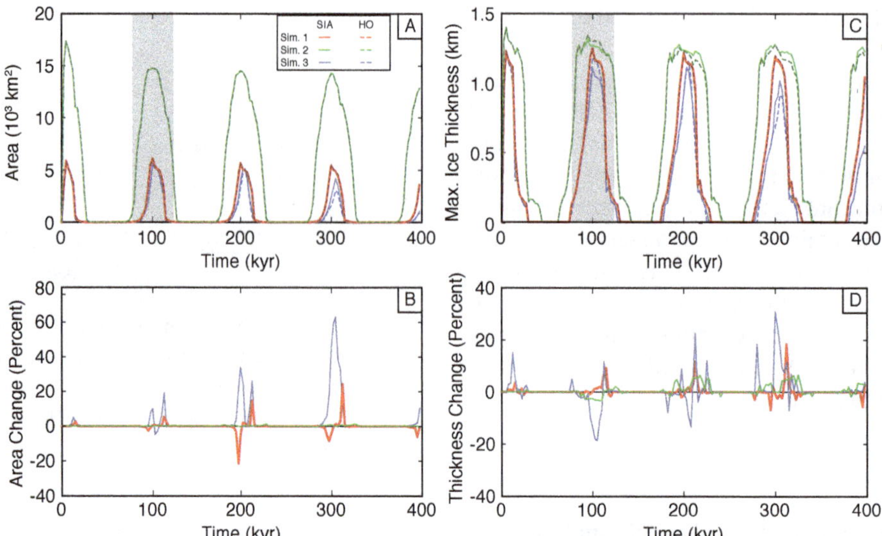

Figure 4. Variations in glaciated area and maximum thickness over 400 kyr. The colors correspond to the simulations in Table 2 and described in Sect. 2.2. For **(a)** and **(c)**, solid lines indicate the SIA model, and dashed lines the HO. **(a)** The area is defined at each time step by the area covered by ice greater than 10 m thick. Solid lines indicate the SIA model, and dashed lines the HO. **(b)** The percent difference between the SIA and HO. **(c)** The maximum ice thickness is defined at each time step when ice greater than 10 m thick exists. **(d)** The percent difference between the SIA and HO.

3.2 Simulation 1: the influence of glacial-flow physics on glacial erosion pattern

Results and comparisons of the pattern of erosion rate from Simulation 1 are shown in detail in Figs. 5 and 6. Because Simulation 1 uses climate parameters (Fig. 3 and Table 2) that are generally in the middle of the range covered by the simulations, the results from this show the effects of the different physics models without extreme behavior, such as sliding everywhere in Simulation 3 or the extensive frozen bed in Simulation 2. However, since the erosion rate is dependent upon the basal temperature such that, in very cold excursions and at high altitudes, the glacier can be frozen to the bed and those regions are protected from erosion. We look in detail at the pattern of erosion over the area of the nested model (Fig. 1) and compare these patterns averaged over a single glaciation and over the full simulation (400 kyr).

Figure 5 shows the erosion rate averaged over the first full glaciation centered around 100 kyr (grey-shaded region in Figs. 3 and 4) for the nested subdomain, and the corresponding averaged erosion rates over the full simulation are shown in Fig. 6. The patterns of erosion are similar over the two timescales for each of the SIA (Figs. 5a and 6a) and HO (Figs. 5b and 6b) simulations. Within a given model, the erosion rates are more than half as low on the long term, due to the averaging including the interglacial periods when no glaciers are present. On the long term (Fig. 6), the area actively eroded by the glacier is more extensive than over the single glaciation (Fig. 5), as the regions under frozen ice over a single glaciation are not necessarily always under frozen ice over the entire 400 kyr.

Figures 5c and 6c show the differences between the SIA modeled erosion rates and those from the HO model. When comparing the SIA model with the HO model, we focus on the full 400 kyr average (Fig. 6c), as results for the single glaciation average (Fig. 5) follow a similar pattern. In Fig. 6b, the erosion rate in the HO model peaks at over 4 times that of the SIA, yet the bed is frozen over more of the HO model domain. Differences are noticeably larger in regions where the basal temperature significantly varies between the two simulations, i.e., where one model or the other has no sliding (thus no erosion) occurring. While for both glacial physics models there is generally a region of highest erosion rate in the SE corner of the model, the SIA modeled pattern has only one broad region of high erosion rates around $x = 150$ km and $y = 112$. The HO modeled erosion rate pattern shows three specific smaller areas of higher erosion rate in the region between $x = 140$ and 150 km and between $y = 110$ and 130 km.

3.3 Influence of glacial-flow physics on glacial erosion rate in different climates

In order to compare the erosion rates over both the 400 kyr and the single glaciation timescales for all climate simulations, we average the erosion rates over the profiles within the nested domain (Fig. 1). Figure 7 shows the erosion rates for the orogen-parallel (A–A') swath, and Fig. 8 along the valley profile (B–B'). Similar to the 2-D contour plots for Simulation 1, the SIA modeled erosion rates (Figs. 7a and 8a) are generally larger and more variable than those from the HO model (Figs. 7b and 8b). The differences between the two models (Figs. 7c and 8c) can be up to almost an order of

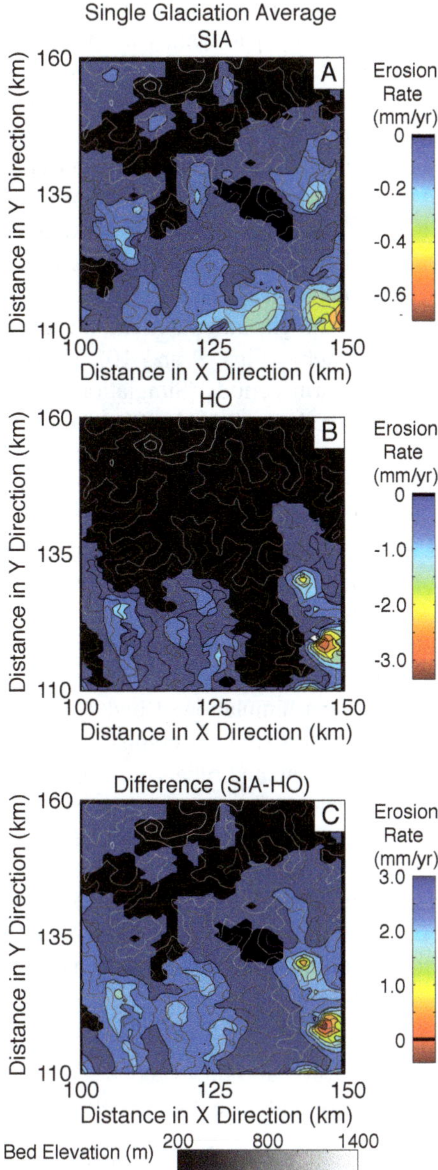

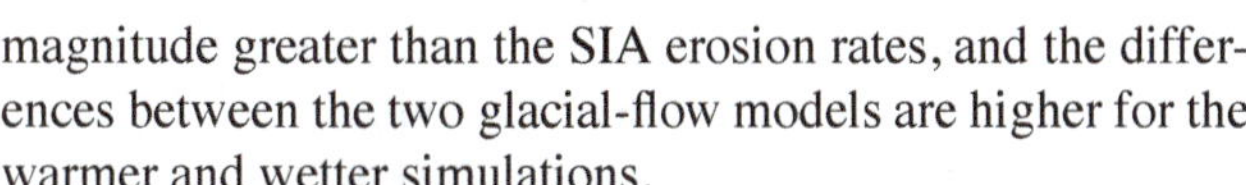

Figure 5. Pattern of erosion rates in nested region (Figs. 1 and 2) averaged over a single glaciation (grey-shaded areas in Figs. 3 and 4). The grey-scale topographic lines show the elevation of the underlying topography. **(a)** SIA erosion rates, where regions of no erosion are shown in white. **(b)** HO erosion rates, where regions of no erosion are shown in white. **(c)** Difference between the time-averaged erosion rates (SIA–HO).

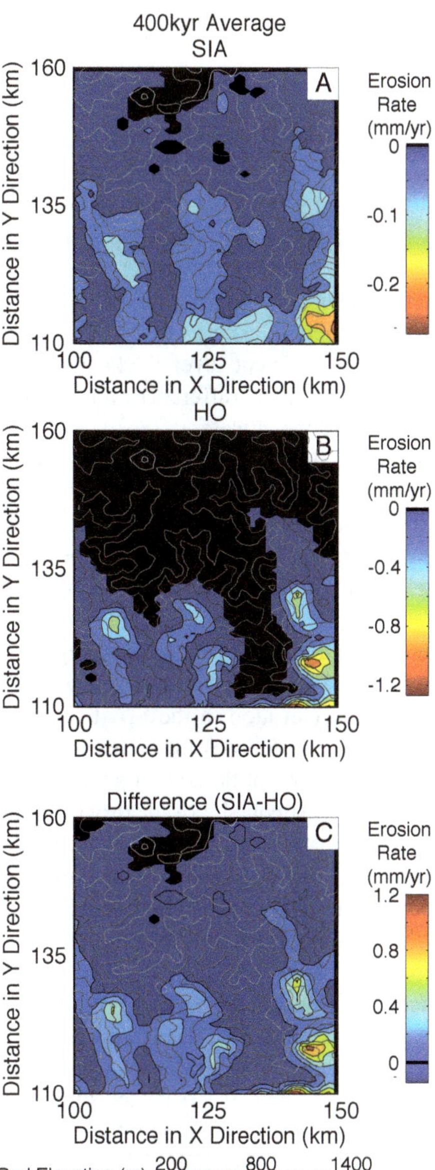

Figure 6. Pattern of erosion rates in nested region (Figs. 1 and 2) averaged over the full simulation. The grey-scale topographic lines show the elevation of the underlying topography. **(a)** SIA erosion rates, where regions of no erosion are shown in white. **(b)** HO erosion rates, where regions of no erosion are shown in white. **(c)** Difference between the time-averaged erosion rates (SIA–HO).

magnitude greater than the SIA erosion rates, and the differences between the two glacial-flow models are higher for the warmer and wetter simulations.

When comparing the differences, those purely related to the climate are also striking. The patterns of erosion within a given glacial-flow model are generally similar in the simulations, and only the magnitudes differ. The differences are largest in the valleys, where ice is thickest and moving fastest, around $x = 105$, 120–127, and 150 km (Figs. 7c and 10a–b). For the valley profile B–B', there is little erosion at high elevations due to the frozen bed (Figs. 8 and 10). Simulation 3, however, particularly accentuates the role of temperature dependence in the sliding law in how the glacial erosion is partitioned over the landscape. Where sliding occurs even under thin, cold ice, Simulation 3 shows significant variation in the pattern of these differences, with the largest erosion rate differences occurring around $x = 100$ and 132 km (Fig. 7c).

3.4 Influence of glacial-flow physics on subglacial topography and sedimentation

Subglacial topography (Figs. 9a–b and 10a–b) is composed of not only bedrock but also sediment deposited by proglacial fluvial processes. The effects of the choice of glacial-flow physics can be seen in comparing the topography (bedrock elevation and sediment thickness combined), the bedrock topography, and the sediment layer thickness. These are all related, but there are many differences among the different physical models and the climate.

The topography shown in Figs. 9a–b and 10a–b is influenced by glacial, fluvial, and hillslope erosion along with sediment that is accumulated when the fluvial system lacks the carrying capacity to transport it. Hillslopes and steep areas are regions of net erosion, particularly seen in the orogen parallel swath (Fig. 9, around $x = 117$ km and $x = 130$–140 km), even when glacial erosion rates might be small or nonexistent (Figs. 7 and 8) due to a frozen bed. When comparing the bedrock elevation in the SIA to the HO models (Figs. 9d and 10d), the bedrock differences (SIA–HO) have a similar pattern to those of the total topography, though the magnitude is slightly subdued. For the orogen parallel swath (Fig. 9, around $x = 120$–125 km), significant valley fill only occurs down glacier of the swath, particularly for Simulation 3, where material is eroded everywhere on the glacier, including the regions frozen to the bed in the other simulations.

The differences between the SIA and HO models for total topography, bedrock elevation, and sediment layer thickness (Figs. 9c–e and 10c–e) show similar patterns to the erosion rate (Figs. 7 and 8). In Figs. 9 and 10, the simulations generally show very similar patterns in both the elevation and the differences between the physical models. Warmer and wetter runs are associated with larger differences between the physical models. Simulation 3 shows the most extreme changes to the topography as well as the most extreme sediment accumulation (Figs. 9a–b and 10a–b) and differences between the physical models (Figs. 9c–e and 10c–e), whereas Simulation 2 shows the least change except for a large amount of deposition in the SIA model around $x = 110$ km (Fig. 9e).

4 Discussion

If a more complex model and a simpler model can be in agreement when the assumptions and approximations in the simpler model are valid, then the simpler model with fewer free parameters is preferred. Model choice also depends upon how this similarity is defined and what the area of interest is, as well as what features or processes are being modeled and what time and length scales are of interest. When looking at a full orogen, it seems that the modified SIA can reproduce features similar to those found in actual orogens. However, in mountainous topography, particularly at sub-polar latitudes, glacier dynamics are influenced by the physical constraints of valleys and fjords and also are a strong control on the erosion and sliding rates.

The different glacial-flow models have an effect on the glacier and the topographic evolution of the orogen, although the magnitude of this effect is variable. Additionally, these effects are dependent upon the climate. When the glacier is mostly frozen (Simulation 2), the physics chosen have only a small influence on the glacial extent and thickness (Fig. 4) and on the topography (Figs. 9 and 10). However, if the glacier has large warm periods (Simulation 1) or is forced to be wet-based (Simulation 3), even if cold temperatures are reached over large periods, then the model choice is considerably more important.

Sliding velocity, which is dependent upon basal shear stress and the basal temperature, is ultimately one of the most important functions in the simulations. As such, erosion rate is used in many of the comparisons and can be considered a stand-in for sliding velocity since our chosen law for erosion rate scales directly with sliding velocity (Eq. 1). Comparisons between Simulations 1 and 3, which have the same climate parameters (Fig. 3), emphasize how important the choice of sliding law and reliance on basal temperature are, regardless of glacial-flow model. For example, Figs. 7 and 8 show maximum erosion rates in Simulation 3 of more than double those of Simulation 1. Simulation 1 generally has more than double the erosion rates compared to those of Simulation 2 (Figs. 7 and 8), which stresses how important the temperature can be even without extreme erosion laws like in Simulation 3.

4.1 Glacial properties and uncertainty in differing climates and over different timescales

Regions like the Gamburtsev Subglacial Mountains underneath the East Antarctic Ice Sheet (Young et al., 2011) illustrate that, no matter how large and thick ice coverage might be, as in Simulation 2 (Fig. 4), if the basal temperature is regularly below freezing, there will be little modification of subglacial topography because there is no sliding. It follows that if the ice is generally below freezing, the glacial-flow model does not tremendously matter if interest is on the evolution of the landscape; in that case, an SIA model could be a reasonable and computationally efficient choice. However, if interest is in ice extent or coverage, which would be important for climate or sea-level-rise studies, then the glacial-flow model can be important even for completely cold-based ice when timescales are long enough (Fig. 4, particularly Simulation 2). While it might be predicted that, over multiple glacial–interglacial cycles on geologic timescales, differences between the two glacial-flow physics might be averaged out due to the other erosional processes reshaping the landscape during interglacial periods, that does not appear to

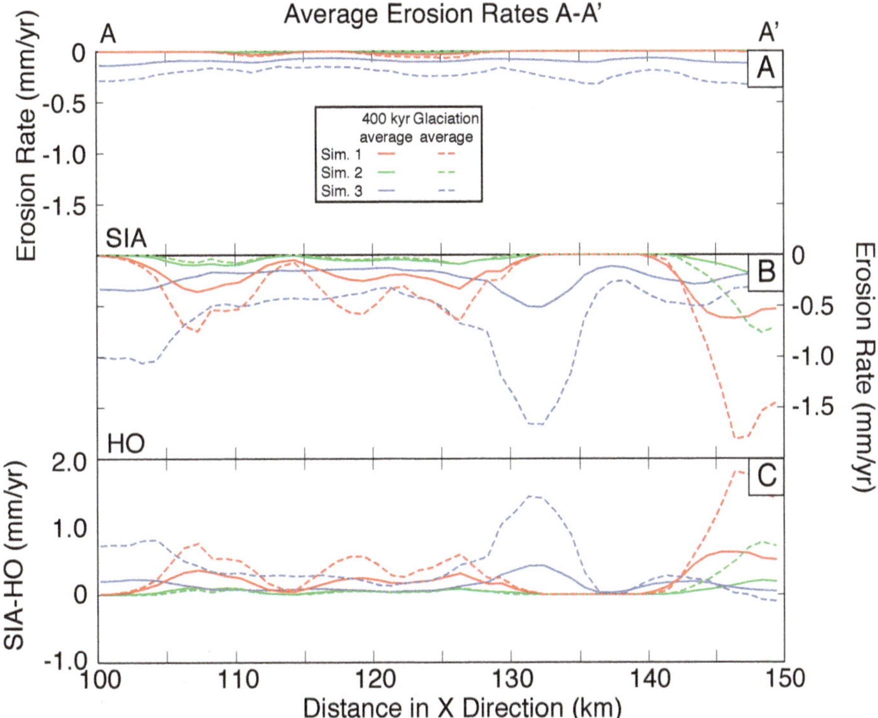

Figure 7. Erosion rates over the A–A' orogen-parallel swath profile (Fig. 1). **(a)** SIA erosion rates averaged over the 100 kyr glaciation (dashed line; grey-shaded area in Figs. 3 and 4) and over the full simulation (solid line). **(b)** HO erosion rates averaged over the 100 kyr glaciation (dashed line) and over the full simulation (solid line). **(c)** Difference between the time-averaged erosion rates (SIA–HO).

be the case. Cumulative effects of sediment deposition and transport due to the non-glacial processes in our simulations do not noticeably moderate the final topography when comparing between the glacial-flow physics.

Comparing the simulations among the different climate scenarios and not just between the two physical models allows us to consider the influence of glacial-flow physics versus climate. The climate simulations (Fig. 3) are considerably different, and their effects on the erosional pattern and topographic evolution are substantial: the ice-covered area varies by over a factor of 3 (Fig. 4a), and the erosion rates can vary by over a factor of 2 (Figs. 7a–b and 8a–b). Particularly for the orogen-wide properties (ice-covered area and maximum thickness), the variations from the climate are substantially larger than those from the choice of glacial-flow physics model (Fig. 4). However, when the erosion rates and topographic evolution are compared over the swath profiles (Figs. 7–10), the differences from the choice of glacial-flow physics model are generally of the same magnitude or larger than those from different climates. These results emphasize that the choice of glacial-flow physics model is less important than the climate if interest is only on larger, orogen-wide properties. For properties like bed topography and erosion rate on the valley scale, the choice of glacial-flow physics can make a more significant difference than even very different climate models.

4.2 Evolution of subglacial topography and sediment thickness

The subglacial topography and deposited sediment are important metrics to consider, as in real landscapes these are the relics from previous glaciations, formed through erosion and other geomorphological processes. We evaluate not just how the glacial-flow physics model influences the erosion of topography but also how this topography and the eroded sediment work within the other geomorphological systems. Fluvial erosion is an important mechanism, and the fluvial network is also responsible for the transport and redistribution of glacially eroded sediment once it has left the toe of the glacier (Hallet et al., 1996; Alley et al., 2003). In some cases, the existing rivers do not have the carrying capacity to support the evacuation of all the available eroded material. This material impacts the landscape not only by protecting the bedrock from further erosion but also by impacting the glacial flow, subglacial hydrology, and erosion patterns (Humphrey and Raymond, 1994; Hallet et al., 1996).

The sediment thicknesses produced vary slightly between the SIA and the HO models. Along the swath profiles, the pattern of deposition varies between the physical models. Generally, the HO model leads to no change or smaller sediment thicknesses by 5–20 m (Figs. 9c and 10c). However, Simulation 3 differs considerably, with HO model producing significantly more deposition (over 25 m) around $x = 147$ km

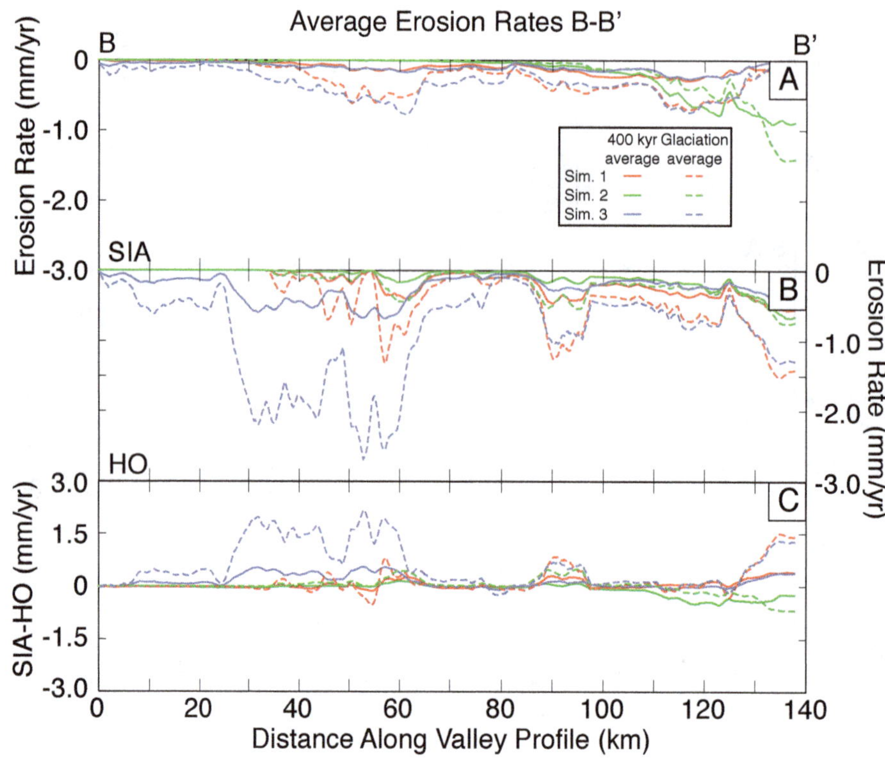

Figure 8. Erosion rates following the B–B' valley profile (Fig. 1). **(a)** SIA erosion rates averaged over the 100 kyr glaciation (dashed line; grey-shaded areas in Figs. 3 and 4) and over the full simulation (solid line). **(b)** HO erosion rates averaged over the 100 kyr glaciation (dashed line) and over the full simulation (solid line). **(c)** Difference between the time-averaged erosion rates (SIA–HO).

in the orogen parallel swath and significantly less (almost 20 m) deposition around $x = 120$–125 km in the same swath. Along the valley profile B–B' (Fig. 10) there is also considerably less deposition for Simulation 3 in the HO model lower in the profile (Fig. 10c), despite significantly more erosion occurring upstream. These variations in amount of sedimentary fill have implications for the structure of future drainage areas, glacial flow, and isostasy.

4.3 Comparison to other models

This study compliments a variety of other research on the effects of glaciers on developing topography. In general, this study compares well with models that also test different ice physics models. However, the magnitude of the effects of climate and ice physics can also be compared to the role of other processes or properties, glacial or otherwise, in landscape evolution models, such as choice of subglacial hydrology, erosion law, and initial conditions.

As mentioned briefly in the model setup section, the sliding rule used is a simple one that does not incorporate subglacial fluvial water pressures or pressure changes. For both sliding and erosion, subglacial water has increasingly been shown to be important (Clarke, 2005; Cohen et al., 2006; Bartholomaus et al., 2008). While much current research, modeling or field-based, is focused on understanding the dynamics of the subglacial fluvial system on the short-term scale, how this system can be meaningfully scaled up to glaciological or geological timescales is still an open question. In previous work, hydrology has been shown to make a marked difference in the development of topography. Systems with a coupled, dynamic hydrologic system in general led to more sliding and therefore more erosion (Egholm et al., 2012a). Feedbacks developed among the water, topography, and glacier, and more erosion occurred over a shorter period of time than in a control study. In addition, hydrology was found to be linked to the development of topography associated with previously glaciated regions, including hanging valleys and overdeepenings (Egholm et al., 2012a). The effects of the hydrologic system were in general on the order of or larger than most effects discussed within this current study. To isolate just the influence of ice physic, the hydrology has been kept simple.

The choice of erosion law, Eq. (1), is based upon another assumption, and various other erosion laws tie the erosion rate to other powers of the sliding velocity (Hallet, 1979; Iverson, 1995; MacGregor et al., 2009), the ice flux (Kessler et al., 2008), or the basal shear stress (Pollard and Deconto, 2007), and a different choice would influence the patterns of erosion and the locations of maximum erosion rate. Comparing the magnitude of effects of erosion laws on developing topography can be difficult for a number of reasons: scaling

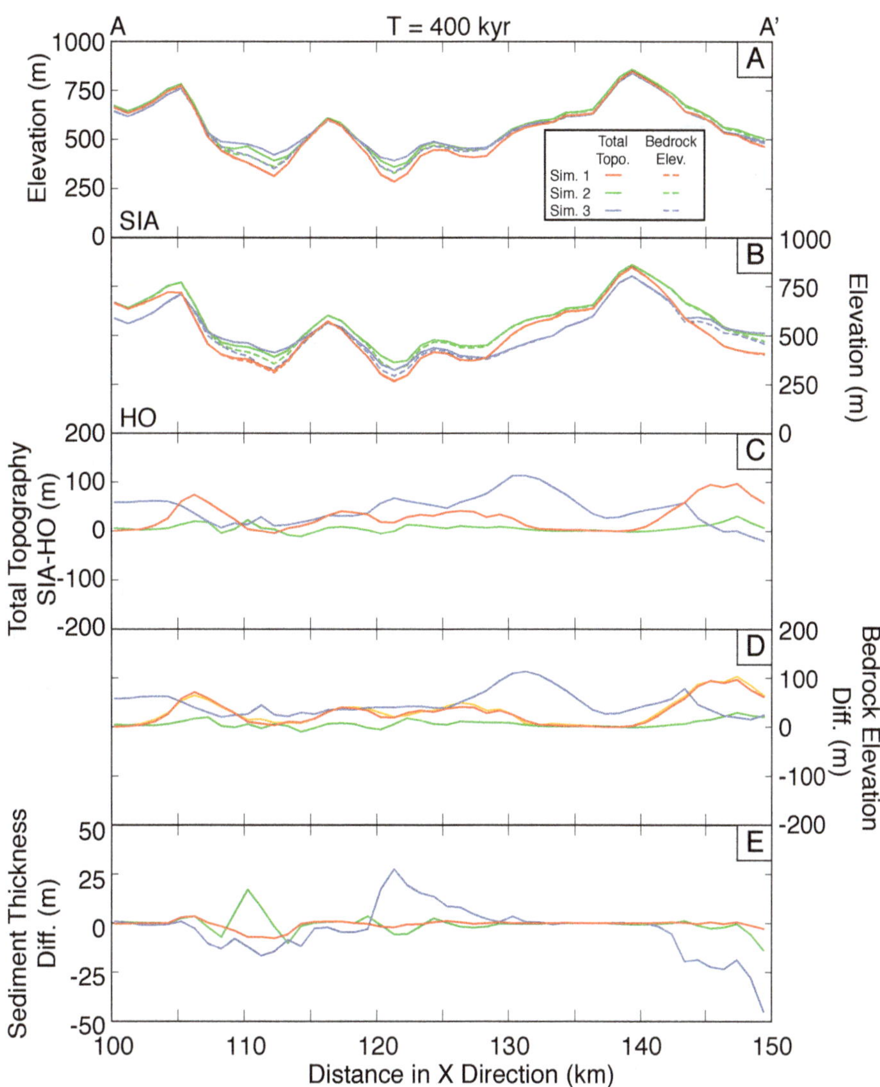

Figure 9. **(a–c)** Topographic swath profiles over the A–A' orogen-parallel swath (Fig. 1). Solid lines indicate the ice-bed topography contact, and the dashed lines indicate the bedrock elevation (sediment is allowed to accumulate in ICE-Cascade). **(a)** SIA topographic profiles at the end of the full simulation. **(b)** HO topographic profiles at the end of the full simulation. **(c)** Differences between the topographic profiles erosion rates (SIA–HO). **(d)** Differences between the bedrock topography (SIA–HO). **(e)** Differences between the sediment thicknesses (SIA–HO).

factors and constants might vary; the ice physics, which can add further complications, determine the basal shear stress; and subglacial hydrology is still quite complicated. MacGregor et al. (2009) determined that the use of a composite erosion law, incorporating both quarrying and abrasion, in a varying climate led to the creation of more rugged terrain with higher-frequency roughness and a deeper cirque (by over 40 %) due to the different erosion laws focusing erosion in different regions along the glacier valley length compared to using the empirical law (Eq. 1). In terms of magnitude of change, the use of different erosion laws and different dependencies on hydrology within a non-evolving flow-line model led to differences in erosion magnitude of over 20 % along with shifting of erosion peaks by tens of kilometers (Headley et al., 2012). Over long timescales, this could significantly influence the development of the valley. This variation in erosion rate due to the incorporation of different or composite erosion laws could certainty lead to variations in developing topography larger than those just from changing the ice physics within a fixed glacial erosion law.

Recent work has also emphasized the importance of initial conditions on landscape evolution in glaciated regions (Pedersen et al., 2013). By using topography with more or less relief or that which has been preconditioned by rivers versus glaciers, the final landscape can significantly vary. The initial topography within this study is an equilibrium land-

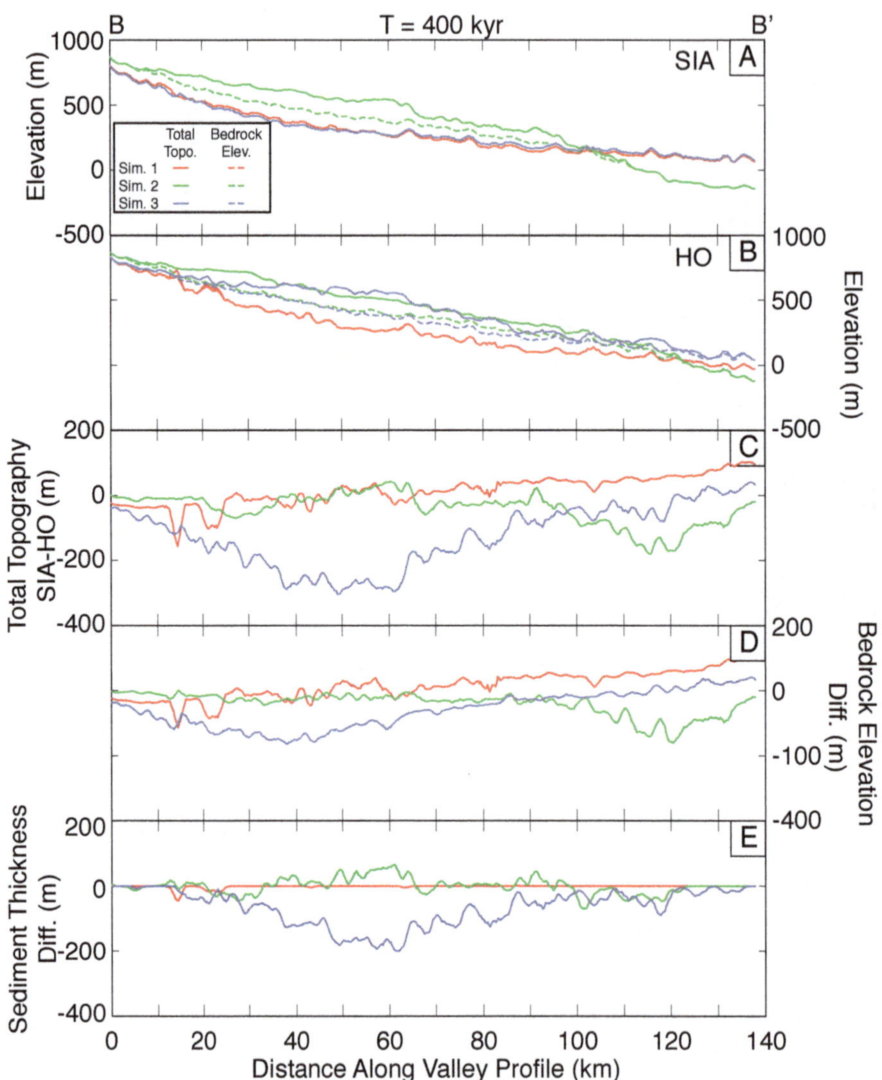

Figure 10. **(a–c)** Topographic profiles following the B–B' valley profile (Fig. 1). Solid lines indicate the ice-bed topography contact, and the dashed lines indicate the bedrock elevation (sediment is allowed to accumulate in ICE-Cascade). **(a)** SIA topographic profiles at the end of the full simulation. **(b)** HO topographic profiles at the end of the full simulation. **(c)** Differences between the topographic profiles erosion rates (SIA–HO). **(d)** Differences between the bedrock topography (SIA–HO). **(e)** Differences between the sediment thicknesses (SIA–HO).

scape generated using only fluvial and hillslope processes (Yanites and Ehlers, 2012). Using glacial steady-state landscapes or non-equilibrium landscapes might lead to different landforms emerging due to the changing focus of glacier erosion. Glaciers occupying fluvial landscapes tend to erode quickly at high altitudes, creating basins within which future glaciers form but which also lower over time so as to create a negative feedback among the glacial erosion and extent (Pedersen et al., 2013). This negative feedback is likely the same as seen in Fig. 4; as the landscape developments, the glaciations become smaller and smaller. This effect might have been further exacerbated had glacial topography been the initial condition of the model comparisons. However, this negative feedback exists within this current study regardless of model choice, so the influences of ice physics would likely still stand out on their own.

4.4 Model caveats and limitations

The SIA and HO models used in this study have several caveats and limitations worth mention. First, this study provides only minimal estimates for the effects of the higher-order flow model on the topographic and glacial evolution, particularly over the full domain. Since the higher-order model is nested within the standard model, not all regions of the glacier and the bed feel the effects of the flow, though the nested region covers the ice divide and spans multiple valleys (Fig. 1b and c). However, we suspect that incorporating all of the glacial ice would lead to larger differences in how topography develops. The full orogen comparisons (Fig. 4) might then be viewed as conservative estimates as to how much the HO model can influence the glacier extent and thickness. Many of the simplifications that are incorporated herein act to further emphasize the singular role of choice of glacial physics without incorporating other complicated feedbacks.

The topography falls within the limits considered generally outside of the scope of the SIA in other studies (Egholm et al., 2011). For topography used herein, there is on average about 300 m of valley-ridge relief in the mountain range, with a maximum elevation around 1600 m. The smaller valleys where glaciers flow are generally between 5 and 7 km wide. Ice thickness reaches a maximum of over 1 km and during glaciations is generally hundreds of meters thick. This range of thickness is on the same scale as, or only an order of magnitude smaller than, the valley width. One way of determining whether the SIA is applicable can be through comparison of aspect ratios, such as ice thickness to width or ice thickness to characteristic length (Hutter, 1983). The ratios, particularly of ice thickness to valley width, in this topography generally fall outside the range of usage of the SIA, where the aspect ratio should be 0.1 or lower. In addition, the topography is sufficiently steep beyond the reaches of the SIA. The topography used in this analysis can overall be described as a triangular, east–west-trending mountain range with a large shelf to the south and a smaller shelf to the north, with an average slope around 4° on the southern slope and a steeper 5° for the northern slope. The slope at the front of the range, representing the continental shelf, is set at around 0.05° (a gradient of 0.001). On the valley scale and smaller, the topography has a maximum slope around 20°, with over 50 % of the topography steeper than 1°. As the model evolves, the orogen-scale relief and slopes do not vary significantly on the timescales under consideration.

One simplification in this comparison is that of the choice of non-thermomechanically coupled ice flow. The constant A from the Glen flow Law, Eq. (5), is treated as constant for all simulations, though this term is actually temperature and pressure dependent. Preliminary comparisons between thermomechanically coupled and uniform temperature simulations found long-term differences between the simulations to be minimal, an order of magnitude less than the differences between the simulations with different glacial physics.

Another erosional feedback that would influence the evolution of the glacier and the topography would come out of the choice of erodibility, K in Eq. (1). In this model, all bedrock material is treated with the same erodibility, with no accounting for fracture mechanics or different rock types. In addition, sediment that is eroded then deposited is reincorporated into the model with the same erodibility as the original bedrock. A variety of research related to modeling and field data of different erosional systems has shown that the rock erodibility, whether this varies due to fracturing, rheology, or composition, can have a significant effect on the ultimate form of the landscape (Duhnforth et al., 2010; Ward et al., 2012).

On geologic timescales, glaciations can significantly change not just the extent of a future glaciation but also how future glaciers grow and retreat, which further influences the future patterns of erosion (Pedersen and Egholm, 2013). Similarly, in this study, the differences between the physical models become more evident the longer the topography develops, as multiple glaciations erode the landscape and the fluvial and hillslope processes do not completely overprint these glacial signals. As models move to higher-order physical representations, the question of how well a given model represents reality needs to be further addressed. While this current study focuses on model comparisons, examples of regional hypsometry, patterns of glacial erosion and ice thickness, and patterns of deposition might be used to better determine whether more sophisticated models better reproduce more realistic topography.

5 Conclusions

This research shows a comparison of a nested HO glacial-flow model with a SIA glacial-flow model within the ICE-Cascade landscape evolution model. The simulations incorporate constant tectonic uplift rate, orographic precipitation, hillslope erosion, fluvial erosion, and sedimentation in addition to the glacial erosion processes. Using multiple climate simulations, the effect of glacial-flow physics is evaluated over 400 kyr of topographic evolution. In general, the glacial-flow model choice makes a difference in the development of a glaciated landscape over the long-term, which corresponds to what other studies have shown in simulations without fluctuating temperatures (Egholm et al., 2012b). We have evaluated a variety of properties of landscape evolution in a glaciated orogen, from the ice-covered area and ice thickness to the bed topography and sediment thickness.

Two major conclusions can be reached. First, as expected, the climate, and particularly whether the glacier is mostly cold-based or mostly sliding, has a large influence on glacial development; a few degrees' difference in the minimum sea level can more than triple the glaciated area over an orogen. Changing the climate parameters can lead to large variations in average erosion rates and topography after multiple glaciations have occurred, generally by factors of 3 to 4 between warm and wet glaciers and cold-based glaciers (Figs. 7–10). Second, though these climate influences are large, comparing a HO glacial-flow model to a SIA model show that the choice of model can have as large, if not larger, an influence on the developing orogeny as climate alone for glaciers with more warm beds and less influence for colder glaciers. As topography develops, even with other processes (fluvial, hillslope) dominating over large time periods, the deviation between SIA and HO grows over time to make a difference of over 30 % for completely wet-based glaciers (Fig. 4). The subglacial topography is similarly affected by the incorporation of the HO model, as long-term erosion rates can vary by over an order of magnitude between the HO and the SIA models (Figs. 5–8), and the subglacial topography can vary by over 100 m between the physics models over 400 kyr (Figs. 9–10). Differences between the models for the colder climate are significantly subdued. Therefore, for modeling studies of landscape dynamics in cold (generally below freezing) climates, glacier physics based upon the SIA approximation would be the most efficient choice; any sacrifice of more realistic glacier dynamics is not significant. However, over orogenic timescales and relatively warm climates, or if interests are on glacial extent and ice thickness regardless of climate, the choice of higher-order glacial-flow models should be considered, as, compared to the SIA model, such models can lead to large differences in the glacial coverage, developing topography, sediment deposition, and averaged erosion rates.

Acknowledgements. The authors would like to thank B. Yanites for his help in model setup, coupling, and general discussion. W. Kappler is thanked for his assistance in program modification and trouble shooting. This work was funded by the German Research Foundation grant to T. Ehlers (DFG-EH 329/1-1). We would also like to thank D. Egholm and multiple anonymous reviewers for useful comments on versions of this manuscript.

Edited by: J. Braun

References

Alley, R. B., Lawson, D. E., Larson, G. J., Evenson, E. B., and Baker, G. S.: Stabilizing feedbacks in glacier-bed erosion, Nature, 424, 758–760, doi:10.1038/nature01839, 2003.

Anderson, R. S., Molnar, P., and Kessler, M. A.: Features of glacial valley profiles simply explained, J. Geophys. Res., 111, F01004, doi:10.1029/2005JF000344, 2006.

Bartholomaus, T. C., Anderson, R. S., and Anderson, S. P.: Response of glacier basal motion to transient water storage, Nat. Geosci., 1, 33–37, doi:10.1038/ngeo.2007.52, 2008.

Braithwaite, R.: Positive degree-day factors for ablation on the Greenland ice sheet studied by energy-balance modelling, J. Glaciol., 41, 153–159, 1995.

Braun, J. and Sambridge, M.: Modelling landscape evolution on geological time scales: a new method based on irregular spatial discretization, Basin Res., 9, 27–52, doi:10.1046/j.1365-2117.1997.00030.x, 1997.

Braun, J., Zwartz, D., and Tomkin, J. H.: A new surface-processes model combining glacial and fluvial erosion, Ann. Glaciol., 28, 282–290, doi:10.3189/172756499781821797, 1999.

Burbank, D. W., Leland, J., Fielding, E., Anderson, R. S., Brozovic, N., Reid, M. R., and Duncan, C.: Bedrock incision, rock uplift and threshold hillslopes in the northwestern Himalayas, Nature, 379, 505–510, 1996.

Campbell, A. J.: Numerical Model investigation of Crane Glacier response to collapse of the Larsen B ice shelf, Antarctic Peninsula, Portland, OR, 83 pp., 2009.

Clarke, G. K. C.: Subglacial Processes, Annu. Rev. Earth Planet. Sci., 33, 247–276, doi:10.1146/annurev.earth.33.092203.122621, 2005.

Cohen, D., Hooyer, T. S., Iverson, N. R., Thomason, J. F., and Jackson, M.: Role of transient water pressure in quarrying: A subglacial experiment using acoustic emissions, J. Geophys. Res., 111, F03006, doi:10.1029/2005JF000439, 2006.

Cuffey, K. M. and Paterson, W. S. B.: Physics of Glaciers, 4th Edn., Academic Press, New York, USA, 2010.

Duhnforth, M., Anderson, R. S., Ward, D., Stock, G. M., and Dühnforth, M.: Bedrock fracture control of glacial erosion processes and rates, Geology, 38, 423–426, doi:10.1130/G30576.1, 2010.

Egholm, D. L., Nielsen, S. B., Pedersen, V. K., and Lesemann, J.-E.: Glacial effects limiting mountain height, Nature, 460, 884–887, doi:10.1038/nature08263, 2009.

Egholm, D. L., Knudsen, M. F., Clark, C. D., and Lesemann, J. E.: Modeling the flow of glaciers in steep terrains: The integrated second-order shallow ice approximation (iSOSIA), J. Geophys. Res., 116, F02012, doi:10.1029/2010JF001900, 2011.

Egholm, D. L., Pedersen, V. K., Knudsen, M. F., and Larsen, N. K.: Coupling the flow of ice, water, and sediment in a glacial landscape evolution model, Geomorphology, 141–142, 47–66, doi:10.1016/j.geomorph.2011.12.019, 2012a.

Egholm, D. L., Pedersen, V. K., Knudsen, M. F., and Larsen, N. K.: On the importance of higher order ice dynamics for glacial landscape evolution, Geomorphology, 141–142, 67–80, doi:10.1016/j.geomorph.2011.12.020, 2012b.

Fowler, A. C. and Larson, D. A.: On the Flow of Polythermal Glaciers, I. Model and Preliminary Analysis, P. Roy. Soc. Lond. A, 363, 217–242, 1978.

Hallet, B.: A theoretical model of glacial abrasion, J. Glaciol., 23, 39–50, 1979.

Hallet, B.: Glacial quarrying: a simple theoretical model, Ann. Glaciol., 22, 1–8, 1996.

Hallet, B., Hunter, L., and Bogen, J.: Rates of erosion and sediment evacuation by glaciers: A review of field data and their implications, Global Planet. Change, 12, 213–235, doi:10.1016/0921-8181(95)00021-6, 1996.

Harbor, J. M., Hallet, B., and Raymond, C. F.: A numerical model of landform development by glacial erosion, Nature, 333, 347–349, doi:10.1038/333347a0, 1988.

Headley, R., Hallet, B., Roe, G., Waddington, E. D., and Rignot, E.: Spatial distribution of glacial erosion rates in the St. Elias range, Alaska, inferred from a realistic model of glacier dynamics, J. Geophys. Res., 117, 1–16, doi:10.1029/2011JF002291, 2012.

Herman, F. and Braun, J.: Evolution of the glacial landscape of the Southern Alps of New Zealand: Insights from a glacial erosion model, J. Geophys. Res., 113, F02009, doi:10.1029/2007JF000807, 2008.

Herman, F., Beaud, F., Champagnac, J.-D. J., Lemieux, J.-M., and Sternai, P.: Glacial hydrology and erosion patterns: A mechanism for carving glacial valleys, Earth Planet. Sc. Lett., 310, 498–508, doi:10.1016/j.epsl.2011.08.022, 2011.

Hooke, R. L. E. B.: Positive feedbacks associated with erosion of glacial cirques and overdeepenings, Geol.

Soc. Am. Bull., 103, 1104–1108, doi:10.1130/0016-7606(1991)103<1104:PFAWEO>2.3.CO;2, 1991.

Humphrey, N. F. and Raymond, C. F.: Hydrology, Erosion and Sediment Production in a Surging Glacier - Variegated Glacier, Alaska, 1982–83, J. Glaciol., 40, 539–552, 1994.

Hutter, K.: Theoretical Glaciology, Reidel, Dordrecht, the Netherlands, 1983.

Iverson, N. R.: Potential effects of subglacial water-pressure fluctuations on quarrying, J. Glaciol., 37, 27–36, 1991.

Iverson, N. R.: Processes of erosion, vol. 1, edited by: Menzies, J., Butterworth-Heinemann, Oxford, 241–259, 1995.

Iverson, N. R.: A theory of glacial quarrying for landscape evolution models, Geology, 40, 679–682, doi:10.1130/G33079.1, 2012.

Johnson, J. V. and Staiger, J. W.: Modeling long-term stability of the Ferrar Glacier, East Antarctica: Implications for interpreting cosmogenic nuclide inheritance, J. Geophys. Res., 112, F03S30, doi:10.1029/2006JF000599, 2007.

Kessler, M. A., Anderson, R. S., and Briner, J. P.: Fjord insertion into continental margins driven by topographic steering of ice, Nat. Geosci., 1, 365–369, doi:10.1038/ngeo201, 2008.

Laitakari, I., Aro, K., and Fogelberg, P.: The effect of jointing on glacial erosion of bedrock hills in southern Finland, Fennia, 163, 369–371, 1985.

MacGregor, K. R., Anderson, R. S., and Waddington, E. D.: Numerical modeling of glacial erosion and headwall processes in alpine valleys, Geomorphology, 103, 189–204, doi:10.1016/j.geomorph.2008.04.022, 2009.

MacGregor, K. R., Anderson, R. S., Anderson, S. P., and Waddington, E. D.: Numerical simulations of glacial-valley longitudinal profile evolution, Geology, 28, 1031–1034, doi:10.1130/0091-7613(2000)28<1031:NSOGLP>2.0.CO;2, 2000.

Pattyn, F.: Numerical modelling of a fast-flowing outlet glacier: experiments with different basal conditions, Ann. Glaciol., 23, 237–246, 1996.

Pattyn, F., Perichon, L., Aschwanden, A., Breuer, B., de Smedt, B., Gagliardini, O., Gudmundsson, G. H., Hindmarsh, R. C. A., Hubbard, A., Johnson, J. V., Kleiner, T., Konovalov, Y., Martin, C., Payne, A. J., Pollard, D., Price, S., Rückamp, M., Saito, F., Souček, O., Sugiyama, S., and Zwinger, T.: Benchmark experiments for higher-order and full-Stokes ice sheet models (ISMIP-HOM), The Cryosphere, 2, 95–108, doi:10.5194/tc-2-95-2008, 2008.

Pedersen, V. K. and Egholm, D. L.: Glaciations in response to climate variations preconditioned by evolving topography, Nature, 493, 206–210, doi:10.1038/nature11786, 2013.

Pollard, D. and Deconto, R. M.: A Coupled Ice-Sheet/Ice-Shelf/Sediment Model Applied to a Marine-Margin Flowline: Forced and Unforced Variations, edited by: Hambrey, M. J., Christoffersen, P., Glasser, N. F., and Hubbard, B., Glacial Sediment. Process. Prod., Blackwell Publishing Ltd., Oxford, UK, 37–52, doi:10.1002/9781444304435.ch4, 2007.

Roe, G. H., Montgomery, D. R., and Hallet, B.: Orographic precipitation and the relief of mountain ranges, J. Geophys. Res., 108, 2315, doi:10.1029/2001JB001521, 2003.

Stolar, D., Roe, G., and Willett, S.: Controls on the patterns of topography and erosion rate in a critical orogen, J. Geophys. Res., 112, F04002, doi:10.1029/2006JF000713, 2007.

Tomkin, J. H.: Erosional feedbacks and the oscillation of ice masses, J. Geophys. Res., 108, 2488, doi:10.1029/2002JB002087, 2003.

Ward, D. J., Anderson, R. S., and Haeussler, P. J.: Scaling the Teflon Peaks: Rock type and the generation of extreme relief in the glaciated western Alaska Range, J. Geophys. Res., 117, 1–20, doi:10.1029/2011JF002068, 2012.

Yanites, B. J. and Ehlers, T. A.: Global climate and tectonic controls on the denudation of glaciated mountains, Earth Planet. Sc. Lett., 325–326, 63–75, doi:10.1016/j.epsl.2012.01.030, 2012.

Young, D. A., Wright, A. P., Roberts, J. L., Warner, R. C., Young, N. W., Greenbaum, J. S., Schroeder, D. M., Holt, J. W., Sugden, D. E., Blankenship, D. D., van Ommen, T. D., and Siegert, M. J.: A dynamic early East Antarctic Ice Sheet suggested by ice-covered fjord landscapes, Nature, 474, 72–75, doi:10.1038/nature10114, 2011.

Multiple knickpoints in an alluvial river generated by a single instantaneous drop in base level: experimental investigation

A. Cantelli[1] and T. Muto[2]

[1]Shell International Exploration and Production, Houston, Texas, USA
[2]Graduate School of Fisheries Science and Environmental Studies, Nagasaki University, 1–14 Bunkyomachi, Nagasaki 852-8521, Japan

Correspondence to: A. Cantelli (alessandro.cantelli@shell.com)

Abstract. Knickpoints often form in bedrock rivers in response to base-level lowering. These knickpoints can migrate upstream without dissipating. In the case of alluvial rivers, an impulsive lowering of base level due to, for example, a fault associated with an earthquake or dam removal commonly produces smooth, upstream-progressing degradation; the knickpoint associated with suddenly lowered base level quickly dissipates. Here, however, we use experiments to demonstrate that under conditions of Froude-supercritical flow over an alluvial bed, an instantaneous drop in base level can lead to the formation of upstream-migrating knickpoints that do not dissipate. The base-level fall can generate a single knickpoint, or multiple knickpoints. Multiple knickpoints take the form of cyclic steps, that is, trains of upstream-migrating bedforms, each bounded by a hydraulic jump upstream and downstream. In our experiments, trains of knickpoints were transient, eventually migrating out of the alluvial reach as the bed evolved to a new equilibrium state regulated with lowered base level. Thus the allogenic perturbation of base-level fall can trigger the autogenic generation of multiple knickpoints which are sustained until the alluvial reach recovers a graded state.

1 Introduction

Knickpoints are zones of locally steepened bed slope in the long profiles of rivers. They are most commonly observed in rivers incising into bedrock (e.g. Crosby and Whipple, 2006) or a cohesive substrate (e.g. Papanicolaou et al., 2008). When sufficiently sharp, they take the form of waterfalls or head cuts (e.g. Hayakawa and Matsukura, 2003; Bennett and Alonso, 2006). They may also be manifested, however, as less steep knickzones (Hayakawa and Oguchi, 2006). Although the position of a knickpoint can stabilize in some cases (e.g. Bennet et al., 2000), a migrating knickpoint invariably recedes upstream (e.g. the analysis of 236 knickpoints in Crosby and Whipple, 2006). Shen et al. (2012) reports a rapid response in the lower Mississippi River to sea level forcing. Terrace formation after a sea-level fall in the Mississippi River basin extended to 600 km upstream within 10 ky, which happened over a very short time compared to the basin response time. The study, even though the flow is supercritical in the experiment, demonstrates how the system responds to a base-level fall, generates multiple knickpoints, and reaches a new equilibrium, and also show time fraction of individual knickpoint migration relative to the basin equilibrium timescale. Knickpoint formation and migration has been modeled successfully in laboratory experiments using a model bedrock or cohesive material. Knickpoints are usually initiated by emplacing an initial knickzone along a channel, making a step elevation drop at the downstream end or impulsively lowering base level at some point during the experiment. Gardner (1983) found that knickpoints in homogeneous model bedrock generated by successive base level drops tended to decay as they migrated upstream. Numerous studies, however, have shown that in cases with heterogeneous model bedrock (e.g. alternating cohesive and noncohesive layers), knickpoints can evolve to an approximately self-preserving state as they migrate upstream (e.g. Holland and Pickup, 1976; Stein and LaTray, 2002). Hasbargen and Paola (2000) observed the autogenic formation and regression of successive knickpoints in a homogeneous model bedrock as base level was lowered at a constant rate. These knickpoints seem likely due to mechanism of incisional cyclic steps studied theoretically by Parker and Izumi (2000), observed in the field by Fildani (2006) experimentally in a homogeneous model bedrock by Brooks (2001) and formed by turbidity currents by Toniolo and Cantelli (2007).

Experiments that aim to produce knickpoints or knickzones have also been performed in noncohesive sediment. Brush and Wolman (1960) found that an oversteepened reach dissipated under such conditions. Lee and Hwang (1994), supporting this notion, further found that an initially vertical face created by base-level lowering tended to decay rapidly in time as the bed degraded, so that a knickpoint was not formed. Cantelli et al. (2004, 2008) verified this result in the course of experimental and numerical studies of dam removal.

Under the right conditions of Froude-supercritical flow, multiple knickpoints can form under quasi-equilibrium conditions (Winterwerp et al, 1992; Taki et al., 2005; Yokokawa et al., 2011). These knickpoints take the form of cyclic steps, that is, trains of upstream-migrating bedforms, each bounded by hydraulic jumps and can form under conditions of constant stationary base level, with new steps forming at the downstream end of the reach as previously formed steps migrate out of the reach. Muto et al. (2012) have observed similar quasi-equilibrium trains of cyclic steps on the alluvial top set of prograding deltas.

Here we document experiments demonstrating the autogenic formation of transient trains of upstream-migrating knickpoints in noncohesive alluvium, triggered by an impulsive lowering of base level. That is, multiple knickpoints, in the form of cyclic steps, can be autogenously generated in response to the perturbation of a sudden base-level drop. Such autogenic knickpoints eventually disappear as the alluvial reach regrades to lowered base level.

2 Experimental setup and procedures

Physical experiments to explore the autogenic response of alluvial rivers to a discrete drop in base level were performed with experimental facilities at Nagasaki University. An alluvial river was built within a narrow, transparent acrylic flume with a length of 4.3 m and a width of 2.0 cm. The bed of the flume had a slope of 0.218 (7.1°); it defined the basement on top of which an alluvial reach with a lower slope formed. The flume was suspended inside a larger glass-walled tank with a length of 4.5 m, a depth of 1.3 m and a width of 1.0 m (Fig. 1a). The flow in the flume was subaerial, with water

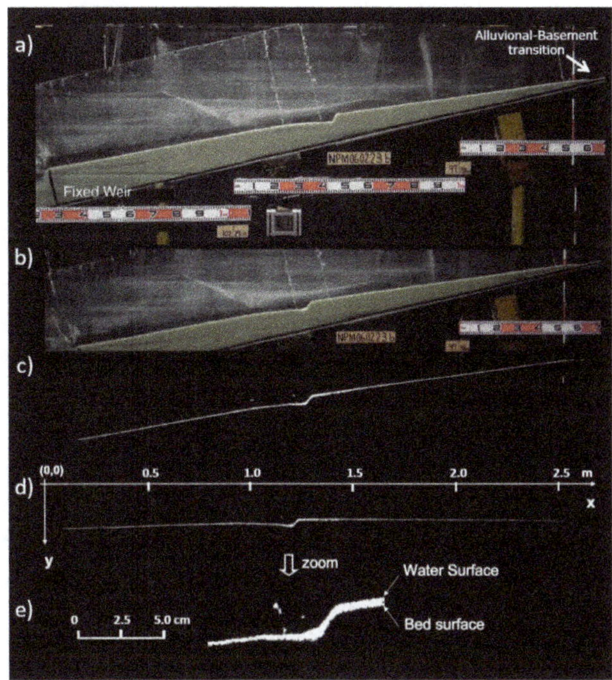

Figure 1. **(a)** Experimental setup showing the flume, the deposit and the weir; **(b)** cropped image used for data imaging elaboration; **(c)** digital image obtained by applying the color threshold to the original image, showing the contrast between white and black that allowed extraction of water depth; **(d)** the same image, rotated to be parallel to the top set before weir removal; **(e)** magnified image of the knickpoint area; some extraneous spots are present due to reflections from the glass of the flume wall.

draining into ponded water in the tank over a free overfall created by a weir.

This weir was located 2.8 m downstream of the upstream end of the flume. It was composed of two essential parts: the 15 cm high lower portion that was fixed to the flume floor, and the removable upper portion that was 2.5 cm, 5.0 cm, 7.5 cm, 10.0 cm or 15.0 cm in height. Removal of this upper portion was a simulation of an instantaneous base-level fall that varied runs.

Prior to the onset of each run, the alluvial river upstream of the weir aggraded with sediment and water supplied at constant rates from the upstream end of the flume. This was done until the alluvial river became graded, i.e. a perfect sediment-bypass system associated with no net erosion and no net deposition. This experimental alluvial configuration was similar, but not identical, to that used by Muto and Swenson (2006) who used the same tank facilities to build a graded river in a deltaic system. In the experiments reported here, base level was maintained by weir as a free overfall, whereas in the experiments of Muto and Swenson (2006), water surface functioned as base level and was continuously lowered at a constant rate. In the latter experiments, a graded state was attained in a moving-boundary system, in which sediment

Table 1. List of the experimental conditions.

	Sediment feed rate Q_s (g s^{-1})	Upstream water discharge Q_w (g s^{-1})	Q_s/Q_w	Mean grain size D_{50} (mm)	Drop of base level ΔH (cm)	Graded Slope %
Run 1	0.90	3.74	0.24	0.2	5.0	7.5
Run 2	0.90	6.34	0.14	0.2	5.0	7.2
Run 3	0.90	7.42	0.12	0.2	5.0	6.1
Run 4	0.90	15.34	0.06	0.2	5.0	4.0
Run 5	0.90	7.25	0.12	0.2	2.5	6.2
Run 6	0.90	7.19	0.13	0.2	7.5	5.7
Run 7	0.90	7.30	0.12	0.2	10.0	5.3
Run 8	0.90	7.14	0.13	0.2	15.0	5.3

fed to the alluvial top set of the model river allowed continued delta progradation.

The alluvial deposit always had a slope that was significantly lower than the basement. Because of this, the upstream end of the alluvial reach was characterized by a relatively sharp alluvial–basement transition (Fig. 1a; Muto, 2001). This transition migrated either upstream or downstream during the approach to grade, but stabilized when grade was attained. This inherent behavior of the alluvial–basement transition was recognized in the experiments of Muto and Swenson (2005a, 2006) where alluvial grade was sustained in a deltaic, moving-boundary system.

In each run of the present experiments, after grade was attained, the upper portion of the weir was removed so as to cause sudden base-level drop, and force the generation of a knickpoint at the downstream end of the alluvial reach. Water and sediment feed rates were kept constant at particular (Table 1) values during the entire run (i.e. both before and after weir removal). Each run was continued until the alluvial river recovered a graded configuration that was regulated with the top of the new, lower weir. The recovery of grade was judged by the standstill of alluvial–basement transition and the straightness of the river profile (Muto and Swenson, 2005b).

The behavior of the knickpoint as generated in each experiment was recorded in digital images that were taken every five seconds. These images were later analyzed using digital processing tools available in Matlab. More specifically, artificial light was allocated to each pixel of the image, so as to enhance the color contrast between the flowing water and the bed. The numeric matrix of each pixel color was filtered with a threshold chosen to transform the layer of flowing water to white and the remainder of the image to black (Fig. 1b and c).

The image was then rotated by an angle equal to the initial delta top set slope in order to highlight the elevation drop across a knickpoint (Fig. 1d). A close-up image of the knickpoint highlights the spatial change in water depth (Fig. 1e). Due to the small scale of the experiment no direct flow measurements were performed. However, flow depths detected by digital image analysis were found to be within the range of 1–4 mm.

All runs were performed using sediments with a density of 2.65 g cm^{-3} and a median size D_{50} of 0.2 mm. Sediment discharge Q_s was kept constant at 0.90 g s^{-1}. Eight runs were conducted using different drops in weir height and different water discharges Q_w. In four of these (Runs 1 to 4), the drop in weir height ΔH was held constant at 5.0 cm, but Q_w was varied. In the other four (Runs 5–8), Q_w was held constant but the drop in weir height was varied. Table 1 summarizes the combinations of experimental conditions. Due to the narrowness of the flume, the alluvial bed was always completely inundated by water.

General reproducibility of the experiments was verified with another set of runs that were conducted repeatedly under the same initial conditions as Runs 1–8, though the number of multiple knickpoints generated in supplementary runs were not precisely the same as here reported. The latter could arise from minor instability of the experimental conditions adopted. As for Runs 1 and 8 and their supplementary runs, there occurred only a single knickpoint during the entire run time.

3 Results of the experiments

In all eight experiments, a single perturbation caused by the sudden removal of the top portion of the weir generated multiple knickpoints. A common pattern of knickpoint formation and subsequent upstream migration is well represented by Run 3 (Fig. 2). At time $t = 30$ s, significant erosion without a definable knickpoint occurs in the proximity of the weir right after removal of the top weir (Fig. 2a). This event represents the initiation of knickpoint formation. At $t = 220$ s (Fig. 2b) erosion and lowering of the alluvial bed propagates upstream, but the alluvial slopes are still gentle overall, with no sharp knickpoint. By $t = 370$ s (Fig. 2c), a knickpoint with a sharp face develops, and by $t = 450$ s (Fig. 2d), the knickpoint is well defined. It continues to propagate upstream until it reaches the alluvial–basement transition, as documented in Fig. 2e–g ($t = 520$ s, 670 s, 760 s). The arrival of this first

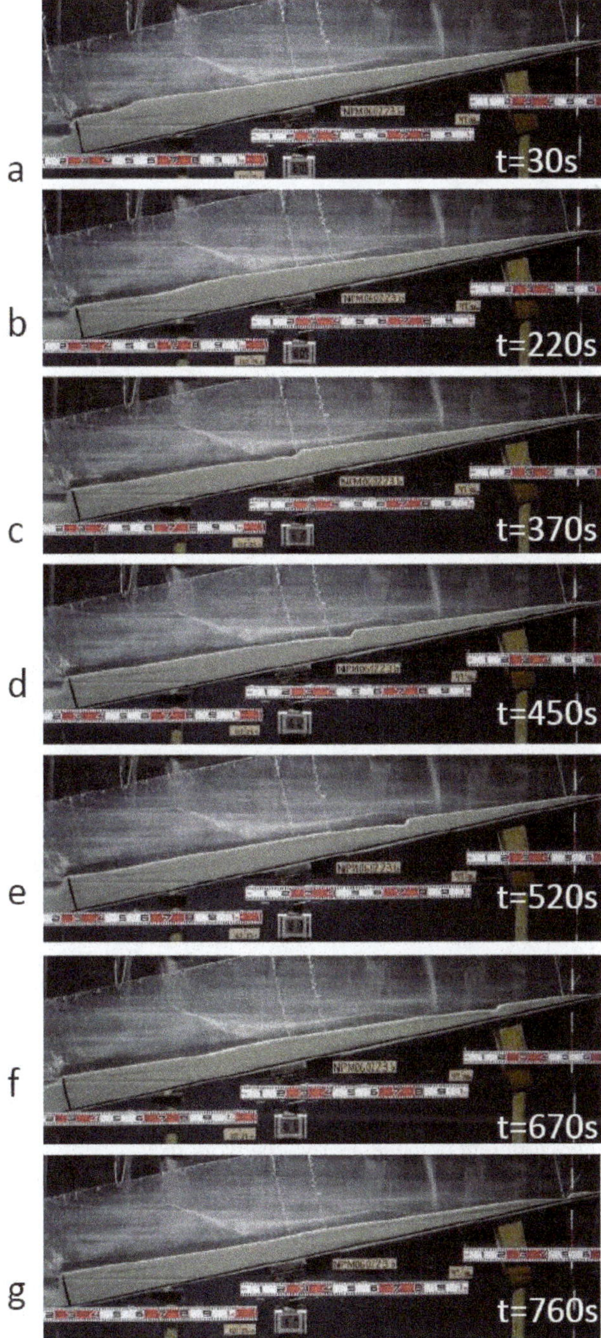

Figure 2. Sequential images of river profile in Run 3, illustrating the formation and upstream migration to the alluvial–basement transition of a knickpoint.

knickpoint at the alluvial–basement transition, however, does not correspond to the attainment of a new graded state of the river. Instead, it is followed by intermittent arrival of multiple knickpoints, with the river eventually recovering grade only after the last knickpoint has migrated out of the reach.

Substantially the same process of multiple knickpoint formation and migration was observed in Runs 2, 4, 5, 6 and 7. In Runs 1 and 8, where only a single knickpoint was created by the removal of the upper weir, the bed slope first became steeper, and then became gentler again to re-approach grade. While the reason for this different behavior is not clear, it is worth noting that (a) Run 1 was conducted with a very low value of Q_w, and (b) Run 8 was conducted with the highest value of weir height lowering ΔH, i.e. 15 cm, causing the loss of a very large volume of sediment as soon as the weir was removed.

Figure 3 shows the geomorphic evolution of the alluvial reach observed in Run 2. In the figure, change in elevation relative to the initial alluvial plane is expressed in terms of index color. The coordinate system has been rotated to be parallel to the initial alluvial graded profile in order to highlight the elevation difference at each knickpoint. For the purposes of plotting, the datum of the coordinate system has been arbitrarily placed 0.98 m below and parallel to the initial bed profile. Thus an "elevation" of,for example, 0.94 m corresponds to 4 cm downward normal from the initial bed profile. The bottom axis shows time, the left vertical axis shows distance downstream from the feed point, and the color bar next to the right axis documents "elevation" above the arbitrary datum. The plot includes a time interval of 120 s before $t = 0$, when the weir was lowered by 5 cm to an "elevation" of 0.93 m. This has been done to show a graded profile prevailing before the lowering of the weir. A series of five prominent knickpoints, along with two weaker knickpoints, can be seen to be migrating upstream.

Figure 3 also shows that the first knickpoint (knickpoint 1) took 370 s to migrate upstream over the entire alluvial reach (2.3 m long) to the alluvial–basement transition. The average rate of migration was thus 6 mm s^{-1}. Immediately after the knickpoint reached the alluvial–basement transition, the alluvial reach was modestly reduced in length to 2.20 m.

Four more prominent knickpoints formed: knickpoint 2 at $t = 250$ s, knickpoint 3 at $t = 510$ s, knickpoint 4 at $t = 1125$ s and knickpoint 5 at $t = 1810$ s. The migration rate of these knickpoints varied between 5 and 7.5 mm s^{-1}. Two weak knickpoints, which were generated between knickpoints 3 and 4 (Fig. 3), are labeled as knickpoints 3a and 4a. These were accompanied by a relatively insignificant change in elevation. They formed in the middle of the alluvial reach rather than at the downstream weir, and had a higher migration velocity of about 10 mm s^{-1}.

The elevation and length of the alluvial reach decreased during this process, and then gradually returned to the values characteristic of the graded state. The time interval and distance between knickpoint occurrences increased in time, as illustrated in Fig. 3 with black arrows.

Table 2 summarizes our observations of the knickpoints. The headings have the following meanings. In most cases, "start time" is the time when the knickpoint was initiated at the weir. Some knickpoints, however, formed upstream of the weir. "End time" is the time when the knickpoint vanished; this usually but not always corresponds to the time

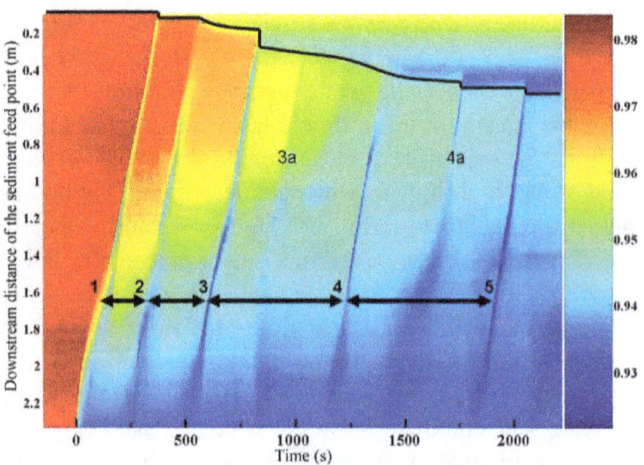

Figure 3. Geomorphic evolution of the alluvial reach in Run 2. Plot illustrating the upstream migration of five prominent knickpoints, and two weak knickpoints, in response to the lowering of the weir by 5 cm at $t = 0$. The run in question is Fig. 2. The bottom axis is time elapsed after the removal of the upper weir. For 120 s proceeding to $t = 0$, the entire alluvial river was graded (i.e. antecedent graded profile). Color index bar to the right documents relative bed elevation, as explained in the text.

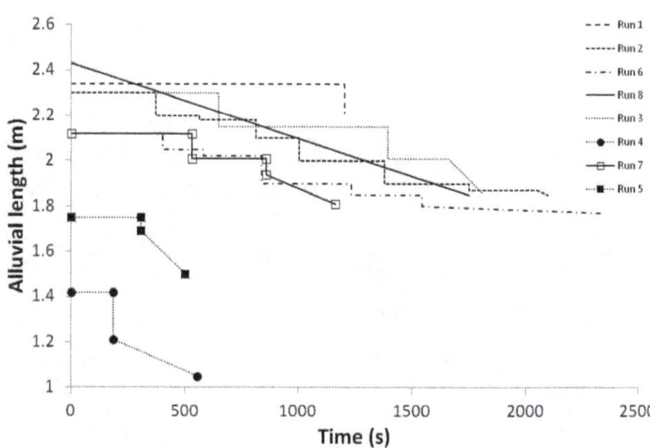

Figure 4. Plots of the time variation of alluvial length for each experiment.

when the knickpoint reached the alluvial–basement transition. "Runout" denotes the distance the knickpoint traveled. "Volume" denotes the total volume of sediment removed from the initial graded state to final the graded state.

The numbers of knickpoints that formed are as follows. Run 1 – 1; Run 2 – 5 strong and 2 weak; Run 3 – 2; Run 4 – 2; Run 5 – 1; Run 6 – 6; Run 7 – 2; and Run 8 – 1. Thus multiple knickpoints formed in most of the runs. The column "Alluvial length" in Table 2 documents how the alluvial reach shortened as knickpoints formed and migrated upstream. The alluvial length at the final grade is shorter than that at antecedent grade because for the same bed slope, a decrease in weir height results in a shorter distance from weir to the intersection with the basement. Otherwise, there was no significant difference between the final and antecedent graded states of the river.

Runs 1–4 were conducted with the same weir height lowering (5 cm), but with different water discharges Q_w (ranging from 3.74 to 15.34 mL s^{-1} in order of run). As noted above, the alluvial reach must be shorter at final equilibrium than antecedent equilibrium, because a lower weir height at the same alluvial bed slope causes the alluvial reach to intersect the steeper basement farther downstream. In the absence of knickpoints, it might be thought that the alluvial length would decrease continuously in time. Figure 4 shows that this was indeed the case for Run 8, for which only one weak knickpoint formed, and later dissipated before reaching the alluvial–bedrock transition (see Table 2). In all the other runs, however, the decrease of alluvial length was sharply discontinuous in time, with a sudden drop corresponding to the arrival time of a step at the alluvial–basement transition.

4 Discussion and conclusions

This work demonstrates that a single base-level drop can generate multiple knickpoints that lead to a new equilibrium profile. Although detailed measurements of flow were not conducted during the experiments, the Froude-supercritical flow was manifest, both at the graded and transient states. Froude-supercritical flow can generate a variety of bed states in alluvium, including plane beds, upstream- or downstream-migrating antidunes and cyclic steps. For a phase diagram of these bedforms see Muto et al. (2012). Under the right conditions, upstream-migrating cyclic steps of permanent (rather than transient) form can manifest themselves in terms of trains of sharp knickpoints (Taki and Parker, 2005; Yokokawa et al., 2011). In the case of sea level drop, alluvial rivers are largely connected with subcritical flow. Applicability of this work is therefore to be search in different cases, such as fault exposures or at small drainage basin scale where different branches of river network incise a valley, and create a local drop in level with the confluent creeks merging with the river. The latter situation, in particular, is well documented by Hasbargen and Paola (2000). In a more engineering application, this work can be linked with dam removal where an important question is related to the morphology consequences to dam removal and how and if the final new equilibrium profile can be reached and in which way. Last but not least in the deep-water environment, the seabed presents many examples of knickpoints (e.g., off-shore California) (Fildani et al., 2006), and often subsidence due to, for example, salt tectonic in the gulf of Mexico can generate the right condition to trigger similar mechanisms with supercritical turbidity currents.

In our experiments, the graded state was in the plane-bed regime. The transient trains of knickpoints caused by sudden base-level drop, however, also manifested themselves as sharp cyclic steps, each bounded by hydraulic jumps that were readily evident with the naked eye. We thus

Table 2. Characteristics of the knickpoints.

Knickpoint #	Start time [s]	End time [s]	Runout [m]	Alluvial length [m]	Velocity of migration [m s^{-1}]	Weir height [m]	Volume [m^3]	Note
Run 1		0		2.34		5.00×10^{-2}	0.11375	A single knickpoint that reached the alluvial–basement transition
1	0	1205		2.21				
Run 2		0		2.3		5.00×10^{-2}	0.10375	
1	0	370	2.18	2.2	5.90×10^{-3}			
2	250	562.5	2.19	2.18	7.00×10^{-3}			
3	520	814	2.10	2.1	7.13×10^{-3}			
3a	810	1004	1.55	2	7.99×10^{-3}			
4	1135	1380	1.91	1.9	7.78×10^{-3}			
4a	1575	1750	1.30	1.87	7.43×10^{-3}			
5	1812	2051	1.84	1.85	7.70×10^{-3}			
Run 3		0		2.3		5.00×10^{-2}	0.104	
1	0	650	2.15	2.15	3.31×10^{-3}			
2	605	1395	3.41	2.01	4.32×10^{-3}			
	1395	1810		1.86				
Run 4		0		1.42		5.00×10^{-2}	0.06175	
1	0	185	1.20	1.21	6.49×10^{-3}			
2	230	555	3.75	1.05	1.15×10^{-2}			
Run 5		0		1.75		2.50×10^{-2}	0.04063	
1	0	305	2.30	1.69	7.54×10^{-3}			
2	305	500	2.20	1.5	1.13×10^{-2}			
Run 6		0		2.11		7.50×10^{-2}	0.1455	
1	0	400	2.20	2.05	5.50×10^{-3}			
2	100	580	2.20	2.02	4.58×10^{-3}			
3	500	840	2.17	1.9	6.38×10^{-3}			
4	800	1235	2.06	1.85	4.75×10^{-3}			
5	960	1545	2.00	1.8	3.41×10^{-3}			
6	1480	2355	2.75	1.77	3.15×10^{-3}			
Run 7		0		2.12		1.00×10^{-1}	0.1965	
1	0	530	4.06	2.01	7.67×10^{-3}			
2	515	860	2.81	1.94	8.15×10^{-3}			
	860	1165		1.81				
Run 8		0		2.43		1.50×10^{-1}	0.321	A single knickpoint that vanished before reaching the alluvial–basement transition
1	0	1750		1.85				

demonstrate for the first time that a single, sudden lowering of base level can generate a transient series of knickpoints. Although our experiments were in the Froude-supercritical range, they were evidently outside the range for the formation of permanent trains of knickpoints. The allogenic perturbation of impulsive base-level lowering, however, triggered an autogenic, transient response in terms of multiple knickpoints, taking the form of cyclic steps.

In the experiments reported here, an alluvial reach of a model river was bounded upstream by a transition to a steeper basement platform, and downstream by a weir. We first formed a graded channel with constant water and sediment feed rates, and then impulsively reduced the height of the weir. The model river eventually evolved to exactly the same graded state as the antecedent one, the only difference between the initial and final states of the river being a shortening of the alluvial reach. This shortening was mediated by lower base level; the same slope profile originating from a lower base level intersects the steeper basement platform farther downstream.

In two out of eight experiments, the evolution to the new equilibrium was mediated by the upstream migration of a single knickpoint. In one of these two experiments, the knickpoint was weak, and dissipated before reaching the alluvial–basement transition; in the second of these, the knickpoint reached the alluvial–basement transition. In the rest (i.e. six runs), transient evolution of the alluvial reach was mediated by two to six consecutively forming knickpoints, all of which eventually reached the alluvial–basement transition.

When multiple steps formed, the process by which equilibrium was restored at lower base level did not consist of a continuous shortening of the alluvial reach, but instead was characterized by a series of discrete reductions, each associated with the arrival of a step at the alluvial–basement transition. Thus there can be at least three modes of response of an alluvial river to impulsively lowered base level: bed degradation without knickpoints (Gardner, 1983), degradation mediated by a single transient knickpoint (Gardner, 1983, and also the present experiment illustrated with Runs 1 and 8), and bed degradation mediated by a train of transient knickpoints (as illustrated with Runs 2–7).

Despite an attempt to link the bed evolution with the characteristics of the flow, the data did not acquire the results sufficiently. In particular, estimation of water depth from the image analysis has an error range of 20–30 % due to pixel resolution. Moreover, high sediment concentration (with a max of over 50 % of water discharge) and apparent cohesiveness are not negligible in the present phenomenon, which makes it difficult to model in terms of application to natural systems. A key control of the system is the sediment-transport capacity of the flow that generates knickpoints. Where a single knickpoint is insufficient to remove all the material to reach the new equilibrium state, there may well be room that multiple knickpoints are generated. The more distantly the river is from the equilibrium with a single passage of knickpoint, the more likely there occur another knickpoints on the river. In case, however, the initial drop of base level exceeds a particular magnitude, subsequent occurrence of knickpoints is unlikely and the river obtains a concave-downward profile that progressively approaches a new equilibrium. Further experiments are required to investigate this hydrodynamics of the physical process.

In summary, the set of experiments presented here highlights the possibility that multiple knickpoints are generated in response to a single discrete drop in base level, though an explanation from the fluid mechanics point of view is not sufficiently available yet.

Acknowledgements. This paper is a contribution to the National Center for Earth-surface Dynamics, a Science and Technology Center funded by the US National Science Foundation (EAR-0120914). This work was also financially supported in part by a 2003–2006 Japanese Grant-in-Aid for Scientific Research B (15340171) to Tetsuji Muto, which provided travel allowance to Alessandro Cantelli. The authors highly appreciate the helpful advice and encouragement provided by Gary Parker and the great contribution provided by the reviewers Kim Wonsuck, Tom Coulthard and Brandon McElroy.

Edited by: F. Metivier

References

Bennett, S. J. and Alonso, C. V.: Turbulent flow and bed pressure within headcut scour holes due to plane reattached jets, J. Hydraul. Res., 44, 510–521, 2006.

Bennett, S. J., Robinson, K. M., Simon, A., and Hanson, G. J.: Stable knickpoints formed in cohesive sediment, Proceedings, Joint Conference on Water Resource Engineering and Water Resources Planning and Management, Minneapolis, Minnesota, United States, 30 July–2 August, 10 pp. 2000.

Brooks, P.: Experimental study of erosional cyclic steps, MS thesis, University of Minnesota, 63 pp., 2001.

Brush, L. M. and Wolman, M. G.: Knickpoint behavior in noncohesive material: A laboratory study, Geol. Soc. Am. Bull., 71, 59–74, 1960.

Cantelli, A., Paola, C., and Parker, G.: Experiments on upstream-migrating erosional narrowing and widening of an incisional channel caused by dam removal, Water Resour. Res., 40, 1–12, doi:10.1029/2003WR002940, 2004.

Cantelli, A., Wong, M., Parker, G., and Paola, C.: Numerical model linking bed and bank evolution of incisional channel created by dam removal, Water Resour. Res., 43, W07436, doi:10.1029/2006WR005621, 2008.

Crosby, B. T. and Whipple, K. X.: Knickpoint initiation and distribution within fluvial networks: 236 waterfalls in the Waipaoa River, North Island, New Zealand, Geomorphology, 82, 16–38, 2006.

Fildani, A., Normark, W., Kostic, S., and Parker, G.: Channel formation flow stripping: large-scale scour features along the Monterey East Channel and their relation to sediment waves, Sedimentology, 53, 1265–1287, 2006.

Gardner, T. W.: Experimental study of knickpoint and longitudinal profile evolution in cohesive, homogeneous material, Geol. Soc. Am. Bull., 94, 664–672, 1983.

Hayakawa, Y. and Matsukura, Y.: Recession rates of waterfalls in Boso Peninsula, Japan, and a predictive equation, Earth Surf. Proc. Land., 28, 675–684, 2003.

Hayakawa, Y. S. and Oguchi, T.: GIS analysis of fluvial knickzone distribution in Japanese mountain watersheds, Geomorphology, 111, 27–37, 2009.

Hasbargen, L. E. and Paola, C.: Landscape instability in an experimental drainage basin, Geology, 28, 1067–1070, 2000.

Holland, W. N. and Pickup, G.: Flume study of knickpoint development in stratified sediment, Geol. Soc. Am. Bull., 87, 76–82, 1976.

Lee, H. Y. and Hwang, S. T.: Migration of a backward-facing step, J. Hydraul. Eng., 120, 693–705, 1994.

Muto, T.: Shoreline autoretreat substantiated in flume experiment, J. Sediment. Res., 71, 246–254, 2001.

Muto, T. and Swenson, J. B.: Large-scale fluvial grade as a non-equilibrium state in linked depositional systems: Theory and experiment, J. Geophys. Res., 110, F03002, doi:10.1029/2005JF000284, 2005a.

Muto, T. and Swenson, J. B.: Controls on alluvial aggradation and degradation during steady fall of relative sea level: Flume experiments, in: River, Coastal and Estuarine Morphodynamics, edited by: Parker, G. and Garcia, M. H., v. 2: London, Talor and Francis, 665–674, 2005b.

Muto, T. and Swenson, J.: Autogenic attainment of large-scale alluvial grade with steady sea-level fall: An analog tank-flume experiment, Geology, 34, 161–164, 2006.

Muto, T., Yamagishi, T., Sekiguchi, T., Yokokawa, M., and Parker, G.: The hydraulic autogenesis of distinct cyclicity in delta foreset bedding: Flume experiments, J. Sediment. Res., 82, 545–558, 2012.

Papanicolaou, A. Wilson, C., Dermisis, D., and Elhakeem, M.: The effects of headcut and knickpoint propagation on bridges in Iowa, Final Report submitted to: Iowa Department of Transportation, Highway Division, Iowa Highway Research Board, 57 pp., 2008.

Parker, G. and Izumi, N.: Purely erosional cyclic and solitary steps created by flow over a cohesive bed, J. Fluid Mech., 419, 203–238, 2000.

Shen, Z., Törnqvist, T. E., Autin, W. J., Mateo, Z. R. P., Straub, K. M., and Mauz, B.: Rapid and widespread response of the Lower Mississippi River to eustatic forcing during the last glacial-interglacial cycle, Geol. Soc. Am. Bull., 124, 690–704, 2012.

Stein, O. R. and LaTray, D. A.: Experiments and modeling of headcut migration in stratified soils, Water Resour. Res., 38, 20-1–20-12, 2002.

Taki, K. and Parker, G.: Transportational cyclic steps created by flow over an erodible bed. Part 1. Experiments, J. Hydraul. Res., 43, 488–501, 2005.

Toniolo, H. and Cantelli, A.: Experiments on upstream-migrating submarine knickpoints, J. Sediment. Res., 77, 772–783, 2007.

Winterwerp, J. C., Bakker, W. T., Mastbergen, D. R., and van Rossum, H.: Hyperconcentrated sand–water mixture flows over erodible bed, J. Hydraul. Eng., 119, 1508–1525, 1992.

Yokokawa, M., Takahashi, Y., Yamamura, H., Kishima, Y., Parker, G., and Izumi, N.: Phase diagram for antidunes and cyclic steps based on suspension index, non-dimensional Chezy resistance coefficient and Froude number, Proceedings, River, Coastal and Estuarine Morphodynamics Conference, Beijing, China, 6–8 September, 7 pp., 2011.

12

Automated landform classification in a rockfall-prone area, Gunung Kelir, Java

G. Samodra[1,2], G. Chen[2], J. Sartohadi[1], D. S. Hadmoko[1], and K. Kasama[2]

[1]Environmental Geography Department, Universitas Gadjah Mada, Yogyakarta, Indonesia
[2]Graduate School of Civil and Structural Engineering, Kyushu University, Fukuoka, Japan

Correspondence to: G. Samodra (guruh.samodra@gmail.com)

Abstract. This paper presents an automated landform classification in a rockfall-prone area. Digital terrain models (DTMs) and a geomorphological inventory of rockfall deposits were the basis of landform classification analysis. Several data layers produced solely from DTMs were slope, plan curvature, stream power index, and shape complexity index; whereas layers produced from DTMs and rockfall modeling were velocity and energy. Unsupervised fuzzy k means was applied to classify the generic landforms into seven classes: interfluve, convex creep slope, fall face, transportational middle slope, colluvial foot slope, lower slope and channel bed. We draped the generic landforms over DTMs and derived a power-law statistical relationship between the volume of the rockfall deposits and number of events associated with different landforms. Cumulative probability density was adopted to estimate the probability density of rockfall volume in four generic landforms, i.e., fall face, transportational middle slope, colluvial foot slope and lower slope. It shows negative power laws with exponents 0.58, 0.73, 0.68, and 0.64 for fall face, transportational middle slope, colluvial foot slope and lower slope, respectively. Different values of the scaling exponents in each landform reflect that geomorphometry influences the volume statistics of rockfall. The methodology introduced in this paper has possibility to be used for preliminary rockfall risk analyses; it reveals that the potential high risk is located in the transportational middle slope and colluvial foot slope.

1 Introduction

In attempts to study and understand landforms, people have tried to map and document landform features since a long time ago. Summerfield (1991) explained that the first attempts of humans to document landforms started in the age of Herodotus (5th century BC) and Aristotle (384–322 BC). It was described in a simple way. In the early stage of mapping, such features of topography were drawn by the hachure method (Gustavsson, 2005). Nowadays, topography is mapped as contour lines or DTMs (digital terrain models). Topographic maps and DTMs are very important for landform classification and geomorphological mapping.

The landform classification which is based on landform genesis (Verstappen, 1983; van Zuidam, 1983) has been widely used in Indonesia. It is suitable for small-scale geomorphological mapping. However, it is necessary to add the other landform information in order to map geomorphological features in the medium to large scales. In a later development of geomorphological mapping in Indonesia, medium- to large-scale geomorphology maps include the information about relief, parent rock, and geomorphological process.

The detailed geomorphological information is very useful in many fields of study and application. It offers a comprehensive discussion related to another aspect. For instance, the study of hazard analysis will be very beneficial if it is analyzed in the context of geomorphology (Panizza, 1996). Here, geomorphometric analysis can be used as a tool for incorporating disaster risk reduction and transfer measures into development planning. This provides basic ideas for planning priorities in promoting a risk management plan and strategy, and evaluating spatial planning policies. Thus, by using geomorphometry as a preliminary tool for risk assessment, the spatial planning manager can make a balance between minimizing risk and promoting some development priorities.

Risk can be defined as "the expected number of lives lost, persons injured, damage to property and disruption of economic activity due to a particular damaging phenomenon for a given area and reference period" (Varnes, 1984). The definition was originally used to describe landslide risk. Later, the terminology was used for all types of mass movements including rockfalls.

The word "rockfall" is often distinguished from more general landslide phenomena due to its typical material, size and failure mechanism. It is defined as rock fragments (Hungr and Evans, 1988) with size from a few cubic decimeters to 10^4 m cubic meters (Levy et al., 2011) that detach from their original position (Crosta and Agliardi, 2003) followed by free falling, bouncing, rolling or sliding (Peila et al., 2007). Rockfall risk can be expressed by the simple product of temporal probability, spatial probability, reach probability, vulnerability and value of the element at risk (Fell et al., 2005; van Westen et al., 2005; Agliardi et al., 2009) as follows:

$$R = \sum_{i=1}^{I}\sum_{j=1}^{J}\sum_{k=1}^{K}\sum_{m=1}^{M} P(L)_{jkm} \cdot P(T|L)_{ij} \cdot P(I|T)_i \cdot V_{ij} \cdot E_i, \quad (1)$$

where $P(L)_{jkm}$ is the temporal probability (exceedance probability) of rockfall in the magnitude scenario (i.e., boulder volume) class j and crossing landform k for different period m; $P(T|L)_{ij}$ is the probability of the rockfall in the volume class j reaching the element at risk i; $P(I|T)_i$ is the temporal spatial probability of the element at risk i, V_{ij} is the vulnerability of the element at risk i to the magnitude class j and E_i is the economic value of the element at risk i.

Based on the Eq. (1), the magnitude and exceedance probability of rockfalls are diverse in time and places. The 9-unit slope model (Dalrymple et al., 1968, i.e., interfluves, seepage slope, convex creep slope, fall face, transportational midslope, colluvial foot slope, alluvial foot slope, channel wall and channel bed) can pose important zones of rockfall processes where energy and velocity are diverse in places. It can be delineated into key information for prioritization of mitigation actions. The information is useful to expose the spatial distribution of elements at risk of potentially high damage from rockfalls. Thus, selection of preventive mitigation measure type, structural protection location, and structural protection dimension should be supported by a rockfall risk assessment based on landform analysis.

Traditionally, landform analysis (delineation and classification procedure) is based on the stereoscopic technique of aerial photo and field investigations. This method is very common in Indonesia. It has been applied for soil mapping, land evaluation analysis, land suitability analysis, spatial planning, and so on. It is also mentioned in Indonesia's national standard document of geomorphological mapping that the technical requirement for geomorphological mapping is an interpretation of remote sensing data combined with field measurements (SNI, 2002). The standard landform classification in Indonesia is based on the ITC (International Institute for Geo-Information Science and Earth Observation) classification system (van Zuidam, 1983). However, the traditional method of landform classification requires simultaneous consideration and synthesis of multiple different criteria (MacMillan and Shary, 2009) and the quality depends on the skill of the interpreter. The developed landform classification has been applied mostly in soil landscape studies. Thus, we try to automatically classify landforms based on the 9-unit slope model, which is appropriate to rockfall analysis. Even though, the 9-unit slope model is significant for a pedogeomorphic process–response system (Conacher and Dalrymple, 1977), it is also relevant for preliminary rockfall risk zoning.

2 Study area

Gunung Kelir is located in the Yogyakarta Province, Indonesia. It lies in the upper part of the Menoreh Dome, which is located in the central part of Java Island (Fig. 1). The area is dominated by a tertiary Miocene Jonggrangan Formation that consists of calcareous sandstone and limestone. Bedded limestone and coralline limestone, which form isolated conical hills, may also be found in the highest area surrounding the study area.

Landforms in Gunung Kelir are a product of the final uplifting of the Complex West Progo Dome in the Pleistocene. The evolution or chronology of the Kulon Progo Dome has been well explained by van Bemmelen (1949). It started with the rising up of the geosyncline of southern Java in the Eocene Epoch. It made the magma of the Gadjah Volcano, consisting of basaltic pyroxene andesites, reach up to the surface. Then, it was followed by the activity of the Idjo Volcano in the south with more acid magma consisting of hornblende-augite andesites and dacite intrusions. After the strong denudation process, exposing the chamber of the Gadjah Volcano, the Menoreh Volcano in the north began to be active. The material consists of hornblende-augite andesites and without lava flow ended by dacitic intrusion and hornblende andesite with the doming up process. Then, in the lower Miocene, the Kulon Progo Dome subsided below sea level and the Jonggrangan Formation was formed by coral reef sedimentation. Finally, the complex of The Progo Dome was uplifted during the Pleistocene. The uplifting caused jointing and large cracks and caused abundant rockfalls and slides to the foot of the Kulon Progo Dome especially in its eastern flank.

The terms Gunung and Kelir come from Javanese Language. Gunung can be translated as mountain and Kelir is a curtain that is used to perform wayang (performance with traditional Javanese shadow puppets). Its toponym describes a 100–200 m high escarpment that has a maximum slope of

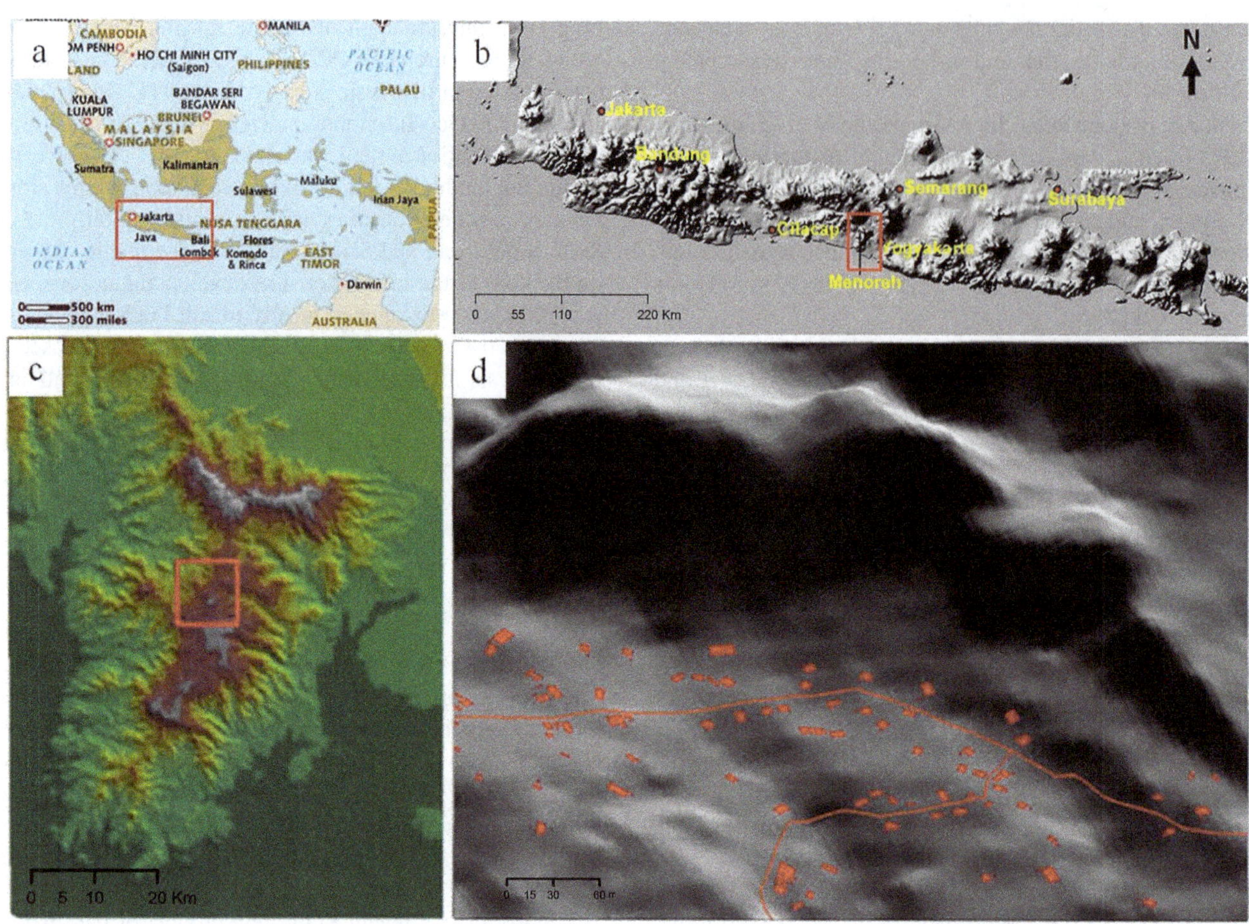

Figure 1. Study area (**a**), geographical position of Java Island (**b**), DTM of Java Island (**c**), DTM of the Kulon Progo Dome (**d**), Gunung Kelir area viewed from the east: red rectangles are elements at risk.

nearly 80°. The complex of Gunung Kelir consists several generic landforms that are prone to rockfall. Its mean slope gradient is 23.14° with the standard deviation of 13.05°. Altitude ranges from 297.75 to 837.5 m. There are 152 buildings exposed as elements at risk on the lower slope of the escarpment (Fig. 1d).

3 Data and methods

Rockfall risk analysis requires assessment of susceptibility and identification of an element at risk. To portray the susceptible area, geomorphological opinion is commonly used to classify landform through interpretation of aerial photos and field surveys. However, the subjectivity of the investigator hinders the application of this method. Therefore, unsupervised landform classification based on the 9-unit slope model is applied in the present study. The main objective of this study is to provide automated landform classification particularly for rockfall analysis. To achieve the primary objective, several works are conducted in this study: (1) fieldwork, (2) DTM preprocessing, (3) DTM processing, (4) rockfall modeling, (5) landform classification based on fuzzy k means, and (6) rockfall volume statistics.

Fieldwork was intended to identify rockfall boulders and elements at risk. A field inventory of fallen rockfall boulders of different size was done to obtain the spatial distribution and dimension of rockfall deposition. The dimension and potential source of rockfalls were determined to simulate rockfall trajectory, velocity, and energy. The buildings on the lower slope of the escarpment were also plotted in order to obtain the spatial distribution of elements at risk. Finally, DGPS (differential global positioning system) profiling was conducted to improve the performance of DTM.

The objective of DTM preprocessing was to improve the quality of DTM-derived products. We applied DTM preprocessing proposed by Hengl et al. (2004) including reduction of padi terraces, reduction of outliers, incorporation of water bodies, and reduction of errors by error propagation. Padi terraces are usually caused by the interpolation method and are located in a closed contour where all the surrounding pixels were assigned the same elevation value. The 5 m resolution of DTM was produced by interpolation, using the ILWIS (Integrated Land and Watershed Management Information

Table 1. Coefficient restitution of surface type.

Surface types	R_N	R_T
Sandstone face	0.53	0.9
Vegetated soil slope	0.28	0.78
Soft soil, some vegetation	0.30	0.3
Limestone face	0.31	0.71
Talus cover with vegetation	0.32	0.8

System) linear interpolation method, from a 1 : 25 000 topographical map of 1999 with a contour interval of 12.5 m and elevation data from DGPS profiling. DTM processing generated several morphometric and hydrological variables such as slope, plan curvature, SPI (stream power index) and SCI (shape complexity index) (Fig. 2). DTM-derived products were processed in ILWIS software with several scripts available in Hengl et al. (2009).

The other morphometric variables were rockfall velocity and energy. These were processed by RockFall Analyst as an extension of ArcGIS (Lan et al., 2007). It included modeling of rockfall trajectory by a kinematic algorithm and raster neighborhood analysis to determine their velocity and energy of rockfalls. Rockfall velocity and energy analyses are needed information about slope geometry and other parameters such as mass, initial velocity, coefficient of restitution, friction angle and minimum velocity offset. There were two coefficients of restitution, i.e., normal restitution (R_N) and tangential restitution (R_T), employed in the model (Table 1). Normal restitution acts in a direction perpendicular to the slope surface and tangential restitution acts in a direction parallel to the surface during each impact of the incoming velocity of the rocks. Velocities change because of the energy loss defined by both. We determined normal restitution and tangential restitution with a geological map presenting elasticity of the surface material and a land use map presenting vegetation cover and surface roughness, respectively. Slope geometry was derived from a corrected DTM. The other parameters were derived from secondary data and field data. For example, the coefficient friction angle was derived from a literature review and mass was determined from the dimension of boulders derived from field measurements data.

The landform elements were derived, as the 9-unit slope model, by using the unsupervised fuzzy k means classification (Burrough et al., 2000) as

$$\mu_{ic} = \frac{\left[(\mathrm{d}_{ic})^2\right]^{-1/(q-1)}}{\sum_{c'=1}^{k}\left[(\mathrm{d}_{ic})^2\right]^{-1/(q-1)}}, \qquad (2)$$

where μ is the membership of ith object to the cth cluster, d is the distance function, which is used to measure the similarity or dissimilarity between two individual observations, and q is the amount of fuzziness or overlap ($q = 1.5$). Supervised k means classification was written and applied in ILWIS script with an additional class center for each morphometric variable (Table 2). The 9-unit slope model was modified by excluding the alluvial toe slope and seepage slope for the final landform classification. Channel wall was also modified as lower slope. Since the study area is located in the upper part of the Kulon Progo Dome, the depositional process of alluvium does not work in such an area. Seepage slope was merged with interfluves because both are more related to a pedogeomorphic process rather than a gravitational process.

The observed volume of rockfall and cumulative distribution in four generic landforms were plotted in a log–log chart. Hungr et al. (1999) and Dussauge et al. (2003) investigated the frequency–volume distribution of rockfalls. They found that rockfall volumes follow a power-law distribution with relatively similar exponent value. The observed cumulative volume distribution was adjusted by a power-law distribution as follows:

$$N_R = r V_R^{-b}, \qquad (3)$$

where N_R is the number of events greater than V_R, V_R is the rockfall volume and b is a constant parameter (cumulative power-law scaling exponent). Linear regression was adopted to estimate the b value.

Similarly to Eq. (3), Dussauge et al. (2003) and Malamud et al. (2004) show that magnitude–frequency distribution of rockfall events in a given volume class j followed a power-law distribution and can be described as

$$\mathrm{Log}N(V) = N_0 + b\log V, \qquad (4)$$

where $N(V)$ is the cumulative annual frequency of rockfall events exceeding a given volume V, N_0 is the total annual number of rockfall events, and b is the power-law exponent.

4 Results and discussion

4.1 Landform classification

Geomorphometry defined as quantitative landform analysis (Pike et al., 2008) was initially applied for the assessment and mitigation of natural hazards (Pike, 1988). Van Dijke and van Westen (1990), for example, introduced rockfall hazard assessment based on geomorphological analysis. Later, Iwahashi et al. (2001) analyzed slope movements based on landform analysis. Both utilized DTMs derived from interpolation of 1 : 25 000 contour maps to analyze the geomorphological hazard. Nowadays, the interpolation of contour maps is still useful to create medium-scale mapping when better resolution DTMs are not available. However, the reduction of error in interpolation of contour mapping is needed to obtain a plausible geomorphological feature.

The result of DTM preprocessing shows that padi terraces still exist where the sampling points of elevation data are unavailable. In addition, "flattening" topography can also be found on slopes of less than 2 %. The remaining padi terraces

Table 2. Class centers for each morphometric variable (SD, standard deviation; PlanC, plan curvature; SPI, stream power index; SCI, shape complexity index).

Landforms	Slope (%)	PlanC	SPI	SCI	Energy (kJ)	Velocity ($m\,s^{-1}$)
Interfluve	0	0	1.0	0	0	0
Convex creep slope	6.0	5.0	3.0	5.0	0.5	0.2
Fall face	40.0	−2.0	50.0	5.5	800.0	20.0
Transportational mid. slope	10.0	−1.0	30.0	7.2	1800.0	30.0
Colluvial foot slope	4.0	2.0	15.0	5.0	400.0	10.0
Lower slope	5.0	2.0	75.0	5.0	0	0
Channel bed	5.0	−5.0	400.0	3.0	0	0
SD/variation	5.79	4.30	158.1	1.4	138.9	3.0

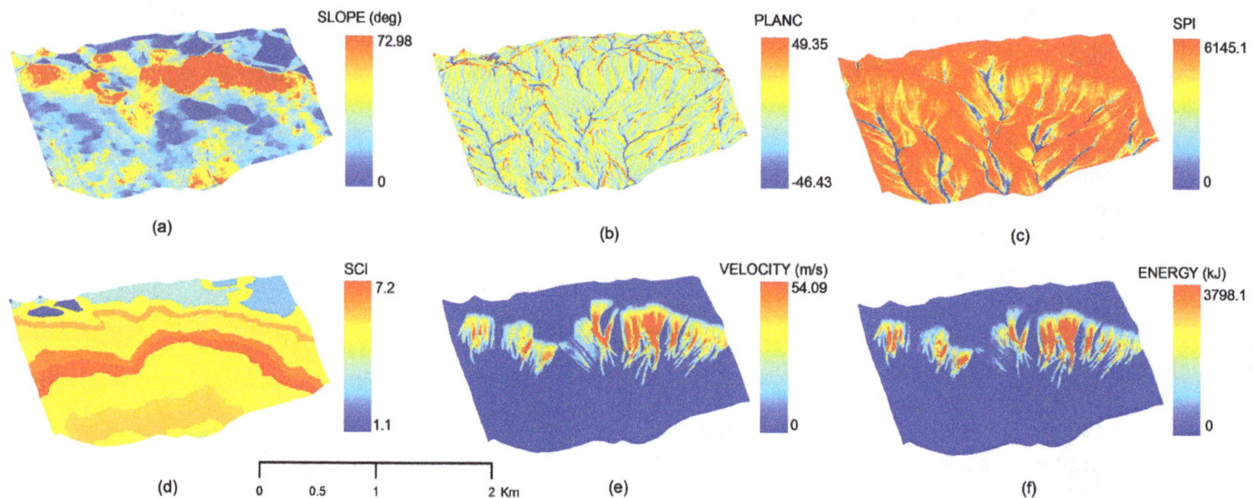

Figure 2. Morphometric variables: **(a)** slope, **(b)** plan curvature, **(c)** stream power index, **(d)** shape complexity index, **(e)** rockfall velocity, and **(f)** rockfall energy.

mostly occur in the transportational middle slope and the flattening phenomenon mostly occurs in the interfluves. Both errors influence the plausibility of slope (Fig. 2a), but do not influence much the final classification of landform elements.

Prior to data analysis, a fundamental decision should be made in relation to the number of landform class and the selection of morphometric variables to be used. The final classification of landform elements should represent an appropriate semantic description related to rockfall processes. A modified 9-slope model was used to represent conceptual entities of rockfall deposition in each slope segment. Convex creep slopes represent a potential rockfall source. Considering that its position is adjacent to a fall face, convex creep slopes and the upper part of fall face are the most likely rockfall sources. A big boulder, which eventually falls, could be part of a convex creep slope and part of a fall face. A fall face represents the Gunung Kelir escarpment, which is dominated by slope > 60° and falling processes. Velocity increases significantly in the fall face and reaches a maximum in the transportational middle slope. In the transportational middle slope, velocity starts to decrease during the contact between boulder and surface. Bouncing, rolling and sliding are dominant in a transportational middle slope. Some high-velocity and high-energy boulders may continue their movement to a colluvial foot slope. This depends on the local surface and the presence of an obstacle that can stop the movement of boulders.

When selecting morphometric variables one should also consider rockfall processes, besides morphology of the landscape. They should reflect the movement and deposition of rockfall boulders. Prior to the selection of morphometric variables, knowledge of rockfall processes in relation to generic landforms should be utilized. Experience and former knowledge are involved during the selection of morphometric variables.

Derivation of morphometric variables through DTM processing was divided into two parts, i.e., morphometric variables derived from RockFall analyst (velocity, energy) and from the ILWIS script (slope, plan curvature, shape complexity index, stream power index). Rockfall velocity and energy are secondary derivatives of a DTM (Lan et al., 2007). The first derivatives (i.e., slope angle and aspect angle) were

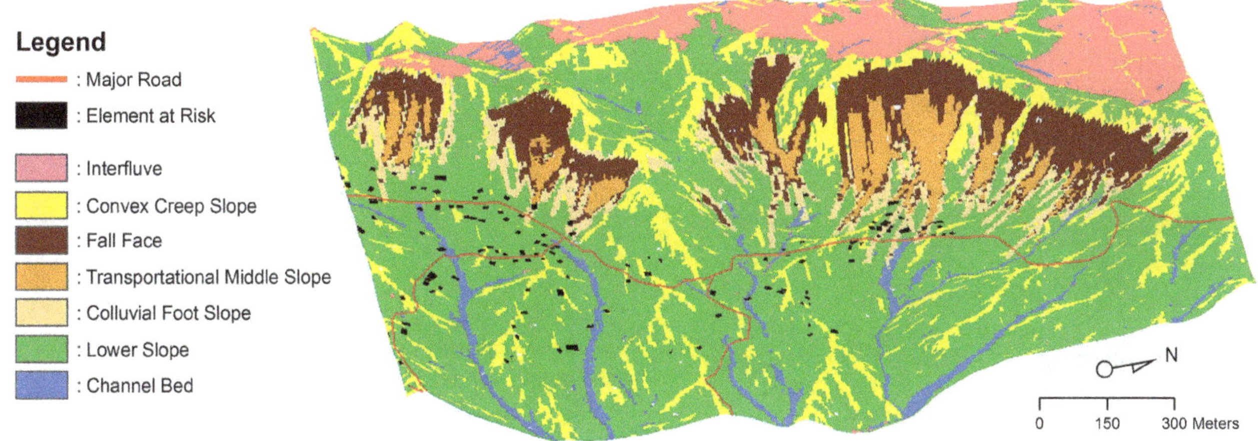

Figure 3. Generic landforms in Gunung Kelir.

employed to compute the rockfall trajectory. Then, rockfall trajectory was used to model the rockfall velocity and rockfall energy by using neighborhood and geostatistical analysis. Velocity and energy of rockfall, as a result of gravitational slope phenomena, may be spatially correlated. Those which are closer tend to be more alike than those that are farther apart. The spatial autocorrelation can be performed with geostatistical techniques.

The highest velocity occurs in the transportational middle slope. Velocity gradually increases in the fall face and decreases in the colluvial foot slope. Since the energy is also calculated from rockfall velocity, the spatial distribution pattern of energy is very similar to the rockfall velocity. Both velocity and energy of rockfall are mostly influenced by slope geometry, coefficient of restitution, and friction angle. The first change of a pixel into zero velocity and energy of its neighborhood operation is determined at the end of boulder movements, meaning that the rockfall boulders are deposited on this site.

Plan curvature and stream power index influence the pattern of the convex creep slope and the channel bed. Shape complexity index, sliced using an equal interval of 25 m, was measured as the outline complexity of a 2-D object. It was calculated using the perimeter to boundary ratio of the sliced feature. SCI indicates how oval a feature is. A low value of SCI represents how simple and compact a feature is. SCI predominantly influences the spatial distribution of the interfluves, which have a low value of around 1, meaning that interfluves are more oval, while convex creep slope and fall face are more longitudinal. Its effect on the other landforms is not apparent because the value of the shape complexity index in the lower slope is relatively homogeneous, i.e., 4–5.

The generic landform result will depend on how well morphometric variables are selected to perform automated landform classification. It represents how well a morphometric variable can describe the specific process working on a landform element. Its spatial dependency influences the application of automated landform mapping in different places and different geomorphological process.

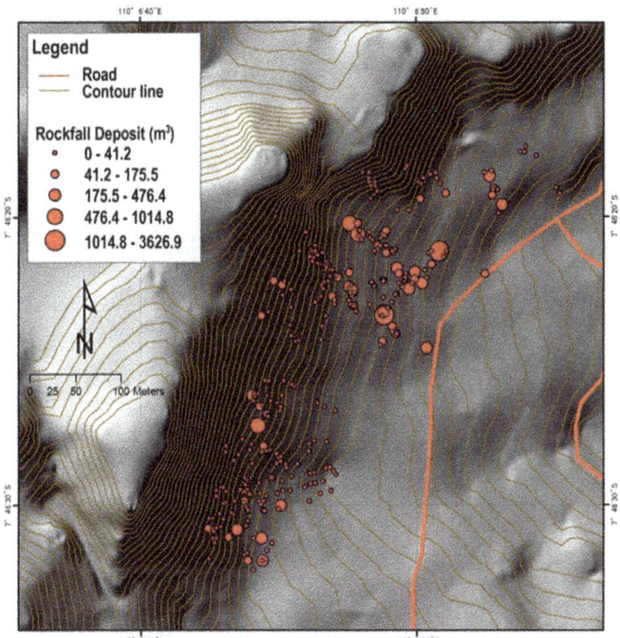

Figure 4. Distribution of rockfall boulders in Gunung Kelir obtained from geomorphological survey.

The final classification result (Fig. 3) was draped over a DTM. The volume of statistical rockfall deposits was employed to evaluate the coincidence between landform classification and rockfall frequency–magnitude. Since landform classification considers surface form and process, we argue that landform classification in a rockfall-prone area exhibits scale-specificity (Evans, 2003). The magnitude (volume) and frequency of boulder deposits may have a specific scale related to each generic landform.

Table 3. Characteristic of rockfall volume distribution in Gunung Kelir.

Generic landform	Area, km^2	N_{events}	V_{total}, m^3	V_{range}, m^3	V_{fit}, m^3	N_{fit}	b_{lr}	R^2	Error margin*
Fall face	0.11	53	513.49	18×10^{-4}–1.0×10^2	2–1.0×10^2	28	0.58	0.98	0.046
Transportational middle slope	0.06	211	9627.59	39×10^{-4}–3.6×10^3	11–3.6×10^3	63	0.73	0.99	0.022
Colluvial foot slope	0.1	199	6287.16	37×10^{-4}–4.8×10^2	10.5–4.8×10^2	70	0.68	0.99	0.019
Lower slope	4.18	58	5004.30	21×10^{-4}–3.6×10^3	11–3.6×10^3	21	0.64	0.97	0.071

* Assumes a 95 % level of confidence.

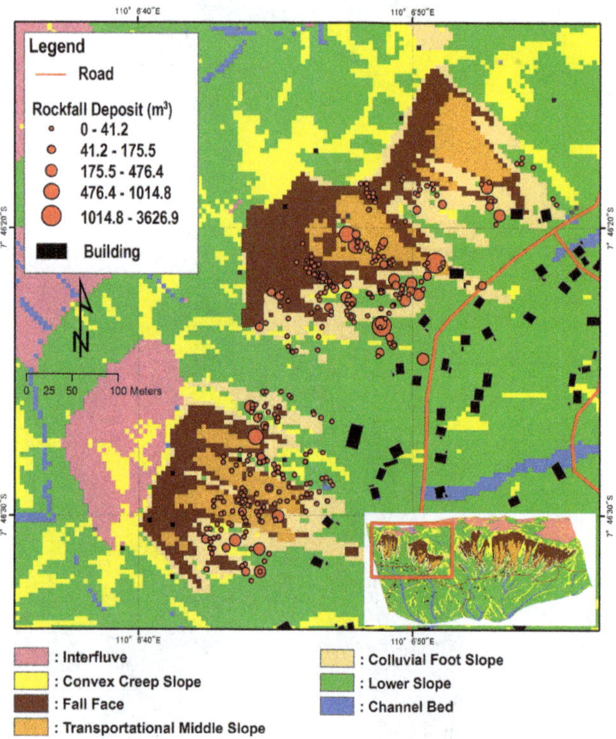

Figure 5. Distribution of rockfall boulders associated with elements at risk and generic landforms.

4.2 Rockfall statistics and landform

The 521 rockfall deposits in our geomorphological inventory range in size from 18×10^{-4} to 3.6×10^3 (Fig. 4). Those are deposited spatially in four main landforms, i.e., fall face, transportational middle slope, colluvial foot slope and lower slope (Fig. 5). Rockfalls deposited on the fall face are mostly found in the southern part of the Gunung Kelir area. There are 47 rockfall boulders in the southern part. The southern fall face has a gentler slope and softer rock than the northern part. Gully erosion can be found in this place due to weathering and erosion. Small volumes of rockfall are mostly deposited in gullies; those are stopped and trapped due to local surface affected by weathering and erosion. However, 12.5 m contours cannot draw this phenomenon. A better resolution of DTMs may be useful to show gully erosion.

The rockfall statistics observed based on the main landforms corresponding to rockfall deposition, i.e., fall face, transportational middle slope, colluvial foot slope and lower slope, indicate that the observed distributions for 53, 211, 199, and 58 events larger than 2, 11, 10, and 11 m^3 are well fitted by power laws with b of 0.58, 0.73, 0.68, and 0.64, respectively. The power-law distribution is well fitted to explain 55, 30, 36 and 38 % of the boulder deposits population in the fall face, transportational middle slope, colluvial foot slope and lower slope, respectively. The model is not well fitted to explain small size rockfall deposits due to the rollover phenomenon. It needs many reports obtained from complete historical rockfall data in many places with different physical characteristics to obtain "general" agreement in assessing rockfall distribution. In many places, such complete historical data are absent.

However, several authors agreed that volume distribution of rockfall follows a power-law distribution. There is still lack knowledge of the b value due to the absence of complete inventory data. Dussauge et al. (2003) argue that the variation of the b value is due to the variability of cliff dimension, area scale, lithology, geometrical and mechanical parameters of rockfalls. Hungr et al. (1999) proposed that jointed rock ($b = 0.65$–0.70) has a higher b value than massive rock ($b = 0.40$–0.43). Gunung Kelir is a subvertical cliff dominated by calcareous limestone. It has $b = 0.58$–0.73 for a volume larger than 2 and 10 m^3. It shows that this study may also confirm the b value for jointed rock proposed by Hungr et al. (1999). Lithology and surface material play important roles for rockfall volume distribution; they influence the bouncing velocity during the impact between rockfall boulder and surface material. Soft rock tends to reduce the energy and decrease the velocity of rockfall.

Landform also influences the value of a scaling exponent. Fall face has the smallest b value, which also indicates that lower frequency of smaller events is more dominant in the fall face. Whereas, higher frequency of greater events is more dominant in the transportational middle slope and colluvial foot slope, which shows the gradation pattern of rockfall deposition around generic landforms. This may correspond to the morphometric condition. The shape and characteristics of the surface, i.e., morphometric variables, determine how a rockfall was deposited. Initially, we considered that the distribution pattern along the x axis and y axis was influenced by the number of measurements in the rockfall boulder data sets in each landform. The distribution pattern of the fall face seems similar to the lower slope while the transportational middle slope is similar to the colluvial foot slope (Table 3). However, the trend only occurs in the volumes $< 2\,m^3$ for the fall face and lower slope, and $< 80\,m^3$ for the transportational middle slope and colluvial foot slope. As stated by Brunetti et al. (2008), we also consider that the distribution pattern is not influenced by the number of measurements in the data set.

All landforms exhibit a rollover of the frequency of rockfall boulders. It is similar to the rollover identified by Dussauge et al. (2003), Hungr et al. (1999), and Guthrie et al. (2004). Roll over occurs in the volume size of around $3\,m^3$ for the fall face, $11\,m^3$ for lower slope, and $6\,m^3$ for colluvial foot slope and transportational middle slope. Since the rockfall process is more related to the deposition zone rather than failure zone, a "rollover" to frequencies should be addressed to the process during the impact between rockfall boulder and surface. The rockfall boulders were deposited on the site when the local surface decreased the volume and energy to zero velocity. This can be influenced by a soft surface condition and or an obstacle that can interfere with the movement of a boulder.

The gradation pattern of rockfall deposition may be addressed to scale specificity (Evans, 2003). The volume of the individual rockfall deposit in the fall face spans 5 orders of magnitude. The landforms that have higher orders of magnitude are lower slope, colluvial slope and transportational middle slope, respectively. Careful attention should be addressed to the maximum individual boulder deposited in the lower slope. Figure 6 shows that the volume of rockfall deposits in the lower slope spans 6 orders magnitude. However, it indicates a long missing gap between the largest boulder ($3626.97\,m^3$) and the second largest boulder ($372.84\,m^3$). The local surface parameter may influence this problem. We consider that the maximum order magnitude on the lower slope is rather similar with the colluvial foot slope (around $400\,m^3$). The likelihood for the deposition of greater rockfall volume can be defined. The higher magnitude of rockfall is more likely to be deposited on the transportational middle slope rather than on the colluvial foot slope, transportational middle slope or lower slope. This information is important for rockfall risk analyses.

4.3 Implication for preliminary rockfall risk analysis

In the past, many people used to consider that natural hazards should be approached from the engineering fields. However, both structural and nonstructural mitigation should be included in natural hazard mitigation comprising geomorphological, geographical, and geological approaches (Oya, 2001). Specific geomorphology features may pose a specific hazard. The most susceptible places, in order, for rockfall susceptibility in the Gunung Kelir area are the fall face,

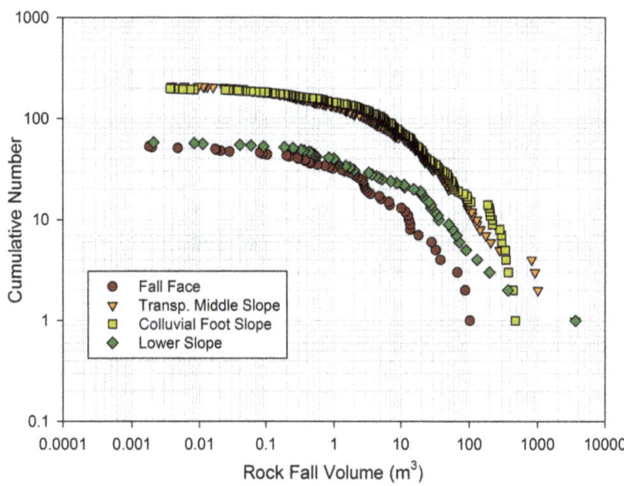

Figure 6. Cumulative frequency curves of rockfall volume.

transportational middle slope, colluvial foot slope and lower slope, respectively, each of which exhibits scale specificity.

Automated landform analysis and rockfall statistics can estimate the likelihood of rockfall magnitude in a specific landform. Each generic landform indicates its degree of susceptibility to rockfall events. The magnitude–frequency relation of rockfalls can be calculated to estimate the annual frequency of rockfall events in each generic landform. It can be defined with reference to specific event magnitude class in a specific generic landform

Preliminary rockfall analysis can be delivered by evaluating elements at risk located in the place susceptible to rockfall hazards. There are 3 buildings located on the transportational middle slope and 14 buildings located on the colluvial foot slope. This is useful information on which to base prioritization action for countermeasure policies and design. Geomorphologic analysis should be taken into account to locate structural measures (e.g., barriers, embankments, rock sheds) in suitable locations. It will improve cost efficiency by optimizing budget and design. The information of buildings located on landforms classified with a high hazard can also be an input to the prioritization of an evacuation procedure. Therefore, the prioritization of mitigation action based on geomorphometric analysis can meet the technical suitability and the effectiveness of selected mitigation options.

5 Conclusions

The application of geomorphometry can be an alternative tool to minimize the subjectivity of Indonesia's standard landform classification applied in disaster risk reduction. Our models explain 55, 30, 36 and 38 % of the boulder deposits population. Rockfall protection through structural measures and land use planning should take into account landform analysis.

However, the original classification of 9-unit slope model should be modified if it is applied in different places. It should consider the origin's effect on the specific landforms. The final classification of landform elements, i.e., interfluves, convex creep slope, fall face, transportational middle slope, colluvial foot slope, slope and channel bed, is different with the original classification of the 9-unit slope model. The considerations to merge and exclude some landforms were based on the experience and the judgement of researchers. The proposed methodology applied in the rockfall-prone area should be tested in different areas that have a similar genesis. Further studies should also explain the effects of scale and spatial dependency on the landform classification.

Acknowledgements. The authors are very grateful to Ian Evans, Tomislav Hengl, and an anonymous reviewer for their detailed and constructive review. We also thank to our colleagues in the Environmental Geography Department, Universitas Gadjah Mada, for their support in conducting the field survey.

Edited by: T. Hengl

References

Agliardi, F., Crosta, G. B., and Frattini, P.: Integrating rockfall risk assessment and countermeasure design by 3D modelling techniques, Nat. Hazards Earth Syst. Sci., 9, 1059–1073, doi:10.5194/nhess-9-1059-2009, 2009.

Burrough, P. A., van Gaans, P. F. M., and MacMillan, R. A.: High resolution landform classification using fuzzy *k* means, Fuzzy Sets Syst. , 113, 37–52, 2000.

Conacher, A. J. and Dalrymple, J. B.: The nine unit land surface model: an approach to pedogeomorphic research, Geoderma, 18, 1–154, 1977.

Crosta, G. B. and Agliardi, F.: A methodology for physically based rockfall hazard assessment, Nat. Hazards Earth Syst. Sci., 3, 407–422, doi:10.5194/nhess-3-407-2003, 2003.

Dalrymple, J. B., Blong, R. J., and Conacher, A. J.: A hypothetical nine unit landsurface model, Zeitschrift für Geomorphologie, 12, 60–76, 1968.

Dussauge, C., Grasso, J., and Helmstetter, A.: Statistical analysis of rockfall volume distributions: implications for rockfall dynamics, J. Geophys. Res., 108, 2286, doi:10.1029/2001JB000650, 2003.

Evans, I. S.: scale specific landforms and aspects of the land surface, in: concepts and modelling in geomorphology: international perspective, edited by: Evans, I. S., Dikau, R., Tokunaga, E., Ohmori, H., Hirano, M., TERRAPUB, Tokyo, 61–84, 2003.

Fell, R., Ho, K. K. S., Lacasse, S., and Leroi, E. A: framework for landslide risk assessment and management, in: Landslide Risk Management, edited by: Hungr, O., Fell, R., Couture, R., Eberhardt, E., Taylor and Francis, London, 3–26, 2005.

Guthrie, R. H. and Evans, S. G.: Analysis of landslide frequencies and characteristics in a natural system, Coastal British Columbia, Earth Surf. Proc. Land., 29, 1321–1339, 2004.

Hengl, T., Gruber, S., and Shrestha, D. P.: Reduction of errors in digital terrain parameters used in soil–landscape modelling, Internat. J. Appl. Earth Observat. Geoinform., 5, 97–112, 2004.

Hengl, T., Maathuis, B. H. P., and Wang, L.: Geomorphometry in ILWIS, in: Geomorphometry: concepts, software, applications, edited by: Hengl, T. and Reuter, H. I., Developments in soil science, Elsevier, Amsterdam, vol. 33, 497–525, 2009.

Hungr, O. and Evans, S. G.: Engineering evaluation of fragmental rockfall hazards, Proceedings 5th International Symposium on Landslides, Lausanne, Switzerland, 1, 685–690, 1988.

Hungr, O., Evans, S. G., and Hazzard, J.: Magnitude and frequency of rock falls and rock slides along the main transportation corridors of southwestern British Columbia, Can. Geotech. J., 36, 224–238, 1999.

Iwahashi, J. and Pike, R. J.: Automated classifications of topography from DEMs by an unsupervised nested-means algorithm and a three-part geometric signature, Geomorphology, 86, 409–440, 2007.

Lan, H., Martin, C. D., and Lim, C. H.: RockFall analyst: a GIS extension for three-dimensional and spatially distributed rockfall hazard modelling, Comput. Geosci., 33, 262–279, 2007.

Levy, C., Jongmans, D., and Baillet, L.: Analysis of seismic signals recorded on a prone-to-fall rock column (Vercors massif, French Alps), Geophys. J. Internat., 186, 296–310, 2011.

MacMillan, R. A. and Shary, P. A.: Landforms and landform elements in geomorphometry, in: Geomorphometry: Concepts, Software, Applications, edited by: Hengl, T. and Reuter, H. I., Developments in Soil Science, Elsevier, Amsterdam, vol. 33, 227–254, 2009.

Malamud, B. D., Turcotte, D. L., Guzzetti, F., and Reichenbach, P.: Landslide inventories and their statistical properties, Earth Surf. Proc. Land., 29, 687–711, 2004.

Oya, M.: Applied geomorphology for mitigation of natural hazard. Kluwer Academic Publisher, Dordrecht, the Netherlands, 2001.

Panizza, M.: Environmental geomorphology, Elsevier, Amsterdam, the Netherlands, 1996.

Peila, D., Oggeri, C., and Castiglia, C.: Ground reinforced embankments for rockfall protection: design and evaluation of full scale tests, Landslides, 4, 255–265, 2007.

Pike, R. J.: The geometric signature: quantifying landslide-terrain types from Digital Elevation Models, Mathemat. Geol., 20, 491–511, 1988.

Pike, R. J., Evans, I., and Hengl, T.: Geomorphometry: a brief guide, in: Geomorphometry: concepts, software, applications, edited by: Hengl, T. and Reuter, H. I., Developments in soil science, Elsevier, Amsterdam, 3–30, vol. 33, 2008.

SNI: Standar Nasional Indonesia (Indonesia National Standard), Penyusunan peta geomorfologi (geomorphology mapping), Badan Standarisasi Nasional, 2002.

Summerfield, M. A.: Global geomorphology, Addison Wesley Longman Limited, Harlow, Essex, England, 1991.

van Bemmelen, R. W.: The geology of indonesia, Vol. II, Government Printing Office the Hague, 1949.

van Dijke, J. J. and van Westen, C. J.: Rockfall hazard: a geomorphologic application of neighbourhood analysis with Ilwi, ITC Journal, 1, 40–44, 1990.

van Westen, C. J., Asch van, T. W. J., and Soeters, R.: Landslide hazard and risk zonation-why is it still so difficult?, Bull. Eng. Geol. Env., 65, 167–184, doi:10.1007/s10064-005-0023-0, 2005

van Zuidam, R. A.: Guide to geomorphological aerial photographic interpretation and mapping, ITC, Enschede, the Netherlands, 1983.

Varnes, D. J.: Landslide hazard zonation: a review of principles and practice, United Nations Educational, Scientific and Cultural Organization (UNESCO), Paris, 1984.

Verstappen, H. T.: Applied geomorphology, Elsevier Science Publisher Co., Amsterdam, 1983.

13

Intertidal finger bars at El Puntal, Bay of Santander, Spain: observation and forcing analysis

E. Pellón, R. Garnier, and R. Medina

Environmental Hydraulics Institute (IH Cantabria), Universidad de Cantabria, Santander, Spain

Correspondence to: E. Pellón (pellone@unican.es) and R. Garnier (garnierr@unican.es)

Abstract. A system of 15 small-scale finger bars has been observed, by using video imagery, between 23 June 2008 and 2 June 2010. The bar system is located in the intertidal zone of the swell-protected beaches of El Puntal Spit, in the Bay of Santander (northern coast of Spain). The bars appear on a planar beach (slope = 0.015) with fine, uniform sand ($D_{50} = 0.27$ mm) and extend 600 m alongshore. The cross-shore span of the bars is determined by the tidal horizontal excursion (between 70 and 130 m). They have an oblique orientation with respect to the low-tide shoreline; specifically, they are down-current-oriented with respect to the dominant sand transport computed (mean angle of 26° from the shore normal). Their mean wavelength is 26 m and their amplitude varies between 10 and 20 cm. The full system slowly migrates to the east (sand transport direction) with a mean speed of 0.06 m day^{-1}, a maximum speed in winter (up to 0.15 m day^{-1}) and a minimum speed in summer. An episode of merging has been identified as bars with larger wavelength seem to migrate more slowly than shorter bars. The wind blows predominantly from the west, generating waves that transport sediment across the bars during high-tide periods. This is the main candidate to explain the eastward migration of the system. In particular, the wind can generate waves of up to 20 cm (root-mean-squared wave height) over a fetch that can reach 4.5 km at high tide. The astronomical tide seems to be important in the bar dynamics, as the tidal level changes the fetch and also determines the time of exposure of the bars to the surf-zone waves and currents. Furthermore, the river discharge could act as input of suspended sediment in the bar system and play a role in the bar dynamics.

1 Introduction

Transverse bars are morphological features attached to the shore that appear with a noticeable rhythmicity along the coast of sandy beaches. They have been identified in many types of environments and have been observed with a wide range of characteristics; therefore a classification of the existing bar systems is necessary. This is not straightforward since these features can be classified using many criteria such as their geometry (length scale, orientation with respect to the shoreline), their dynamics (formation time, migration) or their hydro-morphological environment. Alternatively, classification can be made based on the physical processes governing their formation and their dynamics, although these are sometimes not well understood.

The most documented and observed transverse bar type is probably the "TBR" ("transverse bar and rip") described by Wright and Short (1984), which imposes a cuspate shape on the shoreline, sometimes called mega-cusps (Thornton et al., 2007). They sometimes appear with an oblique orientation with respect to the shoreline (Lafon et al., 2002; Castelle et al., 2007). The TBR are typically linked to outer morphological patterns; specifically, they form due to the onshore migration of a crescentic bar (Ranasinghe et al., 2004; Garnier et al., 2008). They are generally found on intermediate wave-dominated beaches of open coasts and they have wavelengths (distance between two bars) of 100–500 m, and are associated with the presence of rip currents flowing offshore between two bars. Remarkably, the study of Goodfellow and Stephenson (2005) shows that these systems can also appear, at smaller scales, in lower-energy environments (40 km limited fetch).

Table 1. Transverse bar types and main characteristics.

Type	Beach type	Mean wave height (m)[a]	Bar wave length (m)[a]	Cross-shore span (m)[a]	Bar orientation	Migration rate[b] (m day^{-1})	Reference of observed bars
TBR (transverse bars and rips)	Intermediate wave-dominated beaches	> 0.5	100–500	< 150	Normal, oblique	5[c]	Wright and Short (1984) Lafon et al. (2002) Ranasinghe et al. (2004) Goodfellow and Stephenson (2005)[d] Castelle et al. (2007) Thornton et al. (2007)
Large-scale finger bars	Low-energy beaches, wide (~ 1 km) with gentle slope (0.002)	< 0.5	~ 100	~ 1000	Normal or slightly oblique	1	Niederoda and Tanner (1970) Gelfenbaum and Brooks (2003) Levoy et al. (2013)
Finger bars of intermediate beaches	Intermediate wave-dominated beaches	> 0.5	50–100	< 100	Oblique up-current-oriented	40	Konicki and Holman (2000) Ribas and Kroon (2007) Ribas et al. (2014)
Small-scale low-energy finger bars	Very fetch-limited (< 10 km)	< 0.1	< 50	< 100	Oblique down-current-oriented	Lack of data	Falqués (1989) Bruner and Smosna (1989) Nordstrom and Jackson (2012) Present study

[a] Typical observed values.
[b] The values given for the migration rates are the maximum alongshore velocities detected.
[c] Some studies have detected much larger (~ 50 m day^{-1}) alongshore migration rates of crescentic bars (van Enckevort et al., 2004) and mega-cusps (Galal and Takewaka, 2008), but these systems are not clearly coupled with TBR.
[d] These authors identify smaller scale TBR in a low-energy environment.

Here we will focus on "(transverse) finger bars", which differ from the TBR because they do not emerge from offshore bathymetric features but are assumed to form "alone". Moreover, they are not necessarily associated with rip currents. Regarding their geometry, the main difference with the TBR is that the finger bars are long-crested, i.e. their cross-shore extent is generally larger than their wavelength. We identify three types of finger bars (Table 1).

1. The first type of finger bar was identified by Niedoroda and Tanner (1970). We will refer to them as "large-scale finger bars" because of their large cross-shore span (~ 1 km). Their wavelength is ~ 100 m and they appear in low-energy environments (mean wave height < 0.5 m) on very wide (~ 1 km) beaches with a gentle slope (0.002). They are oriented almost perpendicular to the shore or with a slight obliquity, in both micro- and macro-tidal environments (Gelfenbaum and Brooks, 2003; Levoy et al., 2013).

2. Although finger bars are often associated with very low wave energy (Wijnberg and Kroon, 2002), a second type of finger bar can be observed in intermediate morphological beach states (Konicki and Holman, 2000; Ribas and Kroon, 2007; Ribas et al., 2014). They coexist, at a smaller wavelength (typically 50–100 m), with other rhythmic morphologies present in the surf zone, such as with TBR and with crescentic bars. One of the particularities of these "finger bars of intermediate beaches" is that they have an oblique up-current orientation with respect to the mean alongshore current (Ribas et al., 2007).

3. Finally, a third type of finger bar, the "small-scale low-energy finger bars", appears for very low wave energy in fetch-limited environments (fetch < 10 km) with wavelengths of ~ 10 m and a cross-shore span (10–100 m) that depends on the horizontal tidal excursion (Bruner and Smosna, 1989; Garnier et al., 2012). These bars are not strictly normal to the shore (Falqués, 1989; Nordstrom and Jackson, 2012) but seem to be down-current-oriented with respect to the dominant sand transport (Bruner and Smosna, 1989), which is opposite to the finger bars of intermediate beaches.

The processes of generation and evolution of finger bars are probably different depending on their type, and, in particular, their orientation. It is thought that finger bars generally migrate in the direction of sediment transport, but transport direction is not always identified, possibly due to the lack of field data. The theoretical modelling studies of Ribas et al. (2003) and Garnier et al. (2006) have shown different mechanisms to explain the dynamics of up- and down-current-oriented bars by considering forcing due to waves. Ribas et al. (2012) successfully applied their model to finger bars of an intermediate beach, based on continuous observations obtained from video imagery. However, the dynamics of finger bars appearing in low-energy environments is poorly understood, especially concerning the small-scale low-energy finger bars because (1) the forcing is difficult to determine, with forces due to wind, waves and tidal currents all similar in magnitude in very limited fetch environments, and (2) there has been no continuous, long-term survey of such systems. Some observation studies on large-scale finger bars have measured mean migration rates of less than $2\,\mathrm{m\,month^{-1}}$ (Gelfenbaum and Brooks, 2003; Levoy et al., 2013) and maximum speeds of $1\,\mathrm{m\,day^{-1}}$ (Levoy et al., 2013). For the case of small-scale low-energy finger bars, only the preliminary study of Garnier et al. (2012) has given information on the dynamics of such systems, but the migration rates detected are overestimated due to strong noise in the data.

The objective of this contribution is to gain insight into the dynamics of small-scale low-energy transverse bars by performing a continuous survey of finger bars detected in the Bay of Santander, Spain, and by analysing the possible forcing mechanisms. These finger bars are located in the intertidal zone, and the survey is performed by using video images at low tide. Section 2 presents the field site and the data set obtained by video imagery. Section 3 describes the characteristics and the dynamics of the bar system. Section 4 reports the forcing analysis based essentially on wind data. The conclusions are listed in Sect. 5.

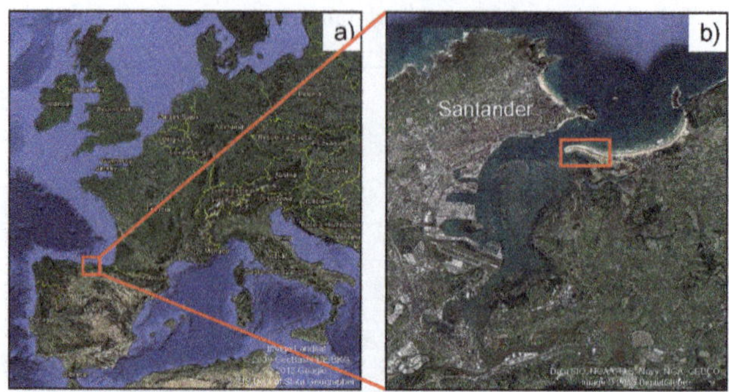

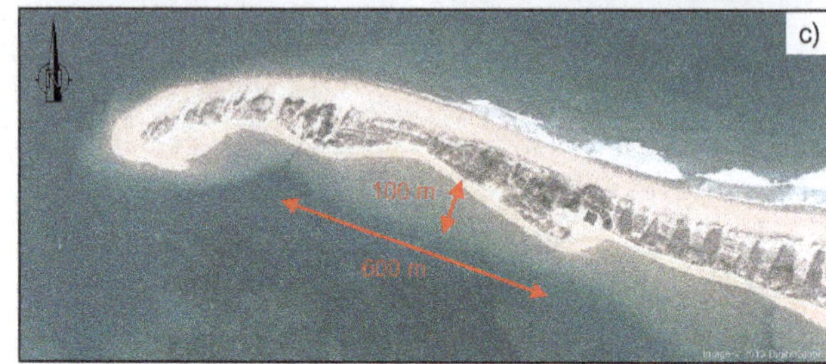

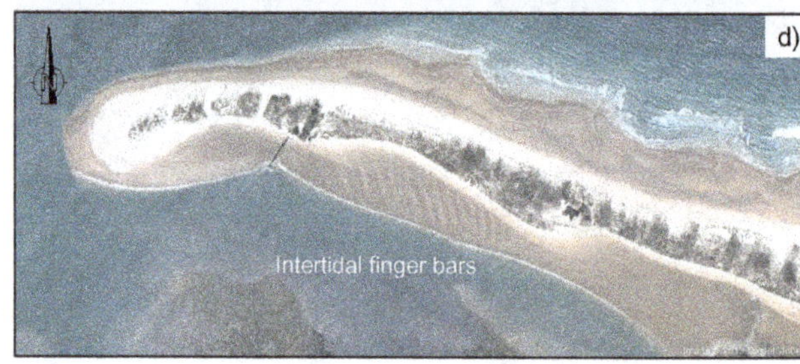

Figure 1. **(a)** Location of Santander, **(b)** map of the bay, **(c)** El Puntal at high tide, and **(d)** El Puntal at low tide. Images from Google Earth; Landsat; © 2009 GeoBasis-DE/BKG; © 2013 Google; US Dept of State Geographer; Data SIO, NOAA, U.S. Navy, NGA, GEBCO; © 2013 DigitalGlobe.

2 Field site and video imagery

2.1 Study site

El Puntal Spit is part of the natural closure of the Bay of Santander (Fig. 1). This bay is one of the largest estuaries on the northern coast of Spain (Cantabrian Sea). The closure of the bay is composed of two natural formations, the Magdalena Peninsula to the north-west, and El Puntal Spit to the north-east. This spit is a sand accumulation which extends from east to west over approximately 2.5 km. Historically, more than 50 % of the surface of this bay has been filled in, reducing the tidal prism and changing the morphological equilibrium of El Puntal (Losada et al., 1991), which tends to extend westward. However, for navigation purposes (Medina et al., 2007), the entrance channel is periodically dredged, and thus the west end of El Puntal is maintained artificially.

There are numerous studies on El Puntal analysing the morphodynamics of the northern face and the west end (Losada et al., 1992; Kroon et al., 2007; Requejo et al., 2008; Medellín et al., 2008, 2009; Gutiérrez et al., 2011), but the lower-energy southern face remains unstudied. The incoming swell from the Cantabrian Sea only reaches the northern face of the spit (Medellín et al., 2008). The southern protected beaches of El Puntal are part of the bay and are located in a low-energy mesotidal environment. The maximum range of the semidiurnal tide is 5 m. Recent hydrodynamic studies (Bidegain et al., 2013) have reported an ebb-oriented mean annual flow of up to $0.1\,\mathrm{m\,s^{-1}}$ in the channel to the south of El Puntal. This flow is mainly driven by the (ebb-dominated) tidal current and by the flow from the Miera River, which enters the bay at the east end of the El Puntal Spit. In the shallower areas the mean flow is much weaker and wind effects can become predominant (Bidegain et al., 2013), especially if we take into account the waves that can be generated over a fetch of up to 4.5 km from the south-west direction. The fetch is highly variable over a tidal cycle due to the numerous intertidal shoals in the bay (Fig. 1b), which can reduce the maximum fetch to 200 m at low tide.

The finger bar system is located in the intertidal zone of the beach on the southern side of the spit. Aerial images show a system of 15 well-developed finger bars that is fully submerged at high tide (Fig. 1c) and fully emerged at low tide (Figs. 1d and 2a). At mid-tide the coastline exhibits a cuspate shape (Fig. 2) and processes of wave refraction and wave breaking are observed (Fig. 2c).

The alongshore extent of the bar system is less than 600 m and the mean bar wavelength is about 25 m. The cross-shore extent of the bars is controlled by the horizontal tidal excursion and is larger in the middle of the domain (130 m) than on the lateral sides (70 m). The bars are almost parallel and have an oblique orientation with respect to the low-tide coastline. The bar angle with respect to the low-tide shore normal is about 25° toward the southeast (where 0° would correspond to shore normal, transverse bars). However, the western end of the system is more irregular, with slight changes in bar orientation and bifurcations (Fig. 1d).

The intertidal beach where the bars appear is planar with a constant slope of approximately 0.015. The offshore boundary of the bars is delimited by a steep slope that ends in the subtidal channel. Sediment sampling has shown the same grain size on bars and troughs with $D_{50} = 0.27$ mm.

2.2 Video imagery and bar detection

In the last few decades, video monitoring systems have been increasingly used to study the shoreline around the world (Holman et al., 1993). To obtain geometric data of the bar system, the images of the Horus video imagery system were used (Medina et al., 2007). This system is composed of four cameras located on the roof of Hotel Real, 91 m a.s.l. and 1.5 km from the study area (Fig. 3a). The Horus station was established in 2008 and takes images every 10 min. In the present study only camera 2 was used (Fig. 3b). The pixel resolution on the study area is variable on the alongshore direction, with values from 4.5 to 6.6 m pixel^{-1}. In the cross-shore direction the resolution is around 0.5 m pixel^{-1}. One daily image of the bar system has been selected at low tide between 23 June 2008 and 2 June 2010, which is the longest period found without long interruptions in the image database. All the interruptions were of less than 6 consecutive days and were due to technical problems (27 days) and poor meteorological conditions (fog 18 days, strong wind 3 days and bad sharpness 85 days). The geometry of the bars was extracted on 577 days, which is 81 % of the time.

Each bar has been digitised manually by selecting three points along the trough at the upper, middle and lowest end of each bar (Fig. 4). Three points were found to describe the position and orientation sufficiently. These digitised data were rectified using seven geographic control points (GCP), established for the Horus system, thus giving geographic coordinates of each digitised bar.

Figure 2. Photos at **(a, b)** low tide and with **(c, d)** rising tide. Pictures taken from the east end of the study area **(a, c)**, and from the west end **(b, d)**. Capture date: 25 February 2012.

Figure 3. Horus video system. **(a)** Cameras on the roof of Hotel Real. **(b)** Image taken by camera 2.

The data processed by Garnier et al. (2012) have been reanalysed in order to correct an apparent periodic movement due to sun shadows in the bars. The amplitude of this periodic movement is of the order of the pixel resolution, and it has been found that its period is related to the capture times. This apparent movement seems to be a systematic error linked to the different sun positions at low tide during the fortnightly cycle of neap-spring tides, which causes different shadows due to the bars and different light reflections in the wet areas. This light shadowing/reflection also occurs for fixed structures present in the surrounding areas. This allowed us to partially correct this spurious movement.

For further analysis, a local, approximately shore-parallel coordinate system has been defined with the alongshore, y axis at 113° from north (Fig. 4). Within the new coordinate system, the mean shoreline position now is approximately parallel to the y axis during most of the tidal cycle. To better understand the behaviour of the finger bars, the digitised bars are then subsampled at $x1 = 45, x2 = 65, x3 = 85$, and $x4 = 105$ m over the length of the alongshore domain ($y1$–$y4$ axes; see Fig. 4). These positions have been chosen to give complete cross-shore coverage of the bars, and each one is representative of one level of the beach profile (Fig. 5c).

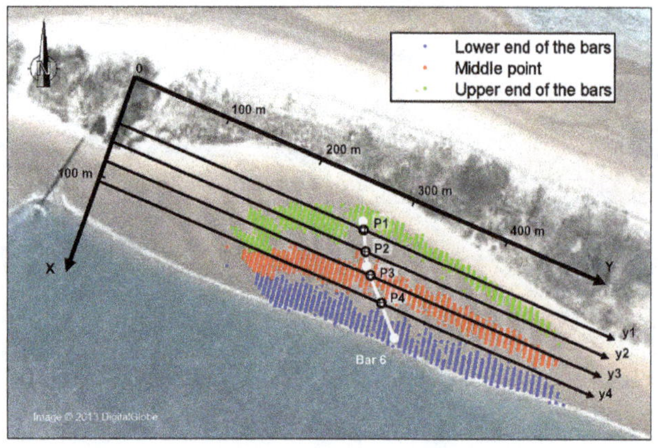

Figure 4. Coordinate system and bar digitisation. The x and y axes stand for the cross-shore and the alongshore direction respectively. The colour points represent the digitised data (each bar is represented by three points); blue, red and green are the lowest, the middle and the upper points of the bars respectively. The bar positions (P1–P4) are defined along the $y1$–$y4$ axes, at $x1 = 45$, $x2 = 65$, $x3 = 85$, and $x4 = 105$ m respectively (see positions of bar 6, in white). Image from Google Earth, © 2013 DigitalGlobe.

2.3 3-D geometry

The Horus system captures one image of the study area every 10 min. This means that the path of the shoreline can be observed along the tidal cycle with high frequency. To extract information about the 3-D geometry of the finger bar system, a reconstruction of the intertidal bathymetry of the study area has been performed by mapping the shoreline from every image. This must be done on a day with perfect conditions, as the meteorological conditions and image sharpness need to be excellent. Furthermore, the tide should have the highest range possible, allowing the extraction of a large intertidal region, and this must occur during daylight hours.

On good days, the shoreline is digitised and rectified on each image. To obtain the bathymetry we assume that the sea level measured at the tide gauge of Santander (less than 2 km away) is the same as the level of the shoreline in the study area. The tide level (with respect to the local Santander Harbour datum, Z) at the time of each image is associated with the rectified shoreline from that image, obtaining an intertidal bathymetry.

2.4 Piecewise regression of the bar movement

The method proposed here to find the time-dependent migration rates is based on piecewise regressions. This allows us to focus on the medium-term movements rather than on the daily fluctuations. The time series of the bar position for each bar at each cross-shore position has been decomposed into segments of variable length. The segment length has been set in order to minimise the error between the piecewise segment and the measured positions. After this decomposition, each bar is represented by several segments of different lengths (the segment k has a length of T_k). For each segment, we can therefore obtain the approximate bar migration rate V_k, which is the migration rate of this bar during the time interval T_k.

Considering that, at a time t, N segments are obtained (where N is the number of bars of the system at this time t, multiplied by 4, which is the number of cross-shore positions studied), the time-dependent migration rate of the bar system V_{m} (which is the average of the speeds, at this time t, of all the bars on all the cross-shore positions) is computed as

$$V_{\mathrm{m}}(t) = \sum_{i=1}^{N(t)} \frac{\hat{V}_i(t)}{N(t)}, \text{ where } \hat{V}_i(t) = \begin{cases} V_{\mathrm{k}}, & \text{if } t \in T_{\mathrm{k}} \\ 0, & \text{otherwise}. \end{cases} \quad (1)$$

3 Bar characteristics and dynamics

3.1 Bathymetry reconstruction

A bathymetry reconstruction has been done on 12 days with excellent meteorological conditions. Figure 5a shows the bathymetry obtained for 24 June 2008, the day with the best image quality. Cross-shore profiles of this bathymetry (Fig. 5c) show that the bars only appear on the region of the intertidal beach profile which has constant slope of 0.015. The extraction of alongshore profiles from these bathymetries allows us to measure the amplitude of the bars, which oscillates between 10 and 20 cm. These profiles also show the asymmetry of the bars (Fig. 5b) with steeper slopes on the lee sides (relative to the migration direction), in agreement with previous studies (Gelfenbaum and Brooks, 2003).

3.2 Bar dynamics

During the 2-year study period the position and geometry of 15 bars were digitised daily. Figure 6 shows the position of the bar system along the $y3$ axis once the digitised data have been corrected, rectified and transformed to the described coordinate system (Fig. 4). Taking into account that the pixel resolution on the study area is of about 5 m pixel^{-1} in the alongshore direction, the small oscillations visible in Fig. 6 are not deeply analysed, as they could be either physical or measurement errors. The bar system is persistent in time, appearing in all the observed images with similar geometric characteristics, but the entire system slowly migrates to the east. As a result of the eastward migration a new bar becomes visible at the western end of the study area (bar 1, Fig. 6). Although aerial images and the migration of the system suggest that the bars are formed at the west of the study area, the formation area is not included in the present results as it is hidden by the dune (Fig. 5a). At the eastern end of the area, the last bar decays and slowly disappears. In addition, during the study period, only one episode of merging of two bars into one has been detected, on 28 March 2009 (bars 5–6, Fig. 6).

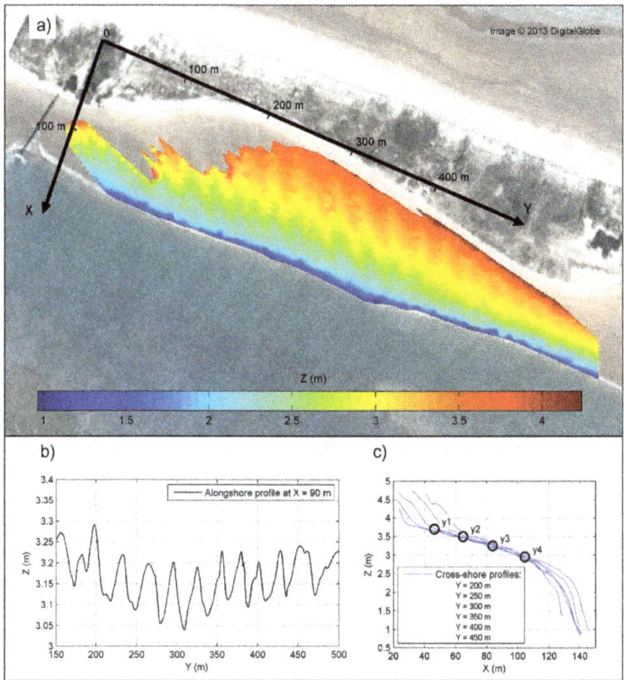

Figure 5. **(a)** Bathymetry reconstruction with videoed shoreline positions during rising tide (24 June 2008). The north–west area without data is the shadowed area by the dune, from the point of view of the camera. **(b)** Alongshore profile of the bed level. **(c)** Cross-shore profiles of the bed level and cross-shore positions of the $y1$–$y4$ axes. Image from Google Earth, © 2013 DigitalGlobe.

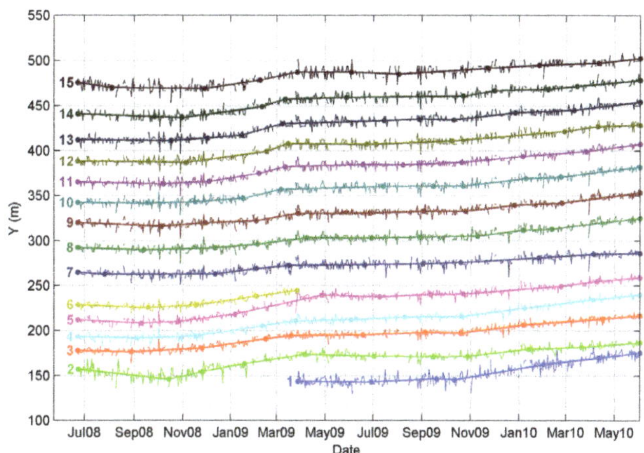

Figure 6. Evolution of the bar system. Time series of the bar position along the $y3$ axis. The thin, discontinuous lines represent the measured position. The thick segments represent the piecewise regression of the measured position. The number to the left of each line indicates the bar number.

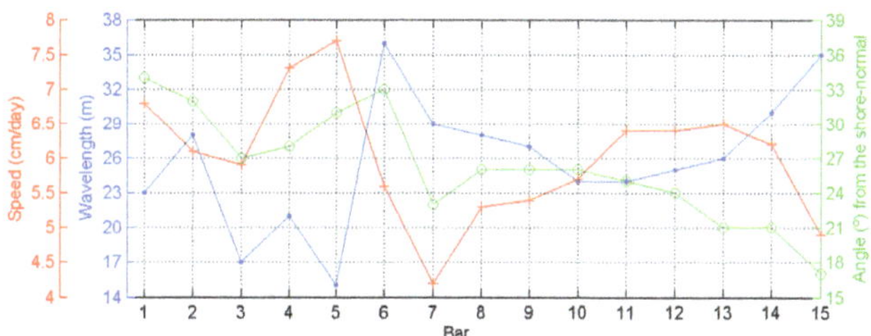

Figure 7. Time-averaged wavelength, angle and migration rate of each bar. The bar angles are measured from the shore normal to the east. Positive values of the bar speeds represent movements of the bars to the east.

3.2.1 Time-averaged characteristics

The digitised and rectified data allow the daily measurement of the bar wavelength. The bar wavelength is computed as the difference between the positions (on the y_i axis) of two consecutive troughs. For each bar, the wavelength has been averaged over the complete study period (Fig. 7). The wavelength is approximately constant for each bar during the study period (standard deviation, σ, around 4 m for all bars), but varies between bars, with a minimum of 15 m and a maximum of 36 m. The mean wavelength of the whole bar system is 25.8 m.

Similarly, the variability of the mean bar angle is low, with σ around 5° for each bar. The mean angle of the system, measured from the x axis, is 26.4°, with a maximum angle of 34° at the western end, decreasing to a minimum of 17° at the eastern end (Fig. 7). The bars are not straight in plan view, so their angle has also been studied by splitting the bars into two parts, the upper (onshore) half and the lower (offshore) half. The upper part of all the bars has a lower angle with the shore normal (mean of the whole system of 23°), while the lower part has higher angles (mean of 31°).

The time series of bar position is almost continuous and allows us to compute the time-averaged migration rate of the system, which is obtained by linear regression. The mean speed of each bar (for the whole study period) is shown in Fig. 7. All the bars of the system slowly migrate to the east, with a mean speed of $6\,\mathrm{cm\,day^{-1}}$ (approximately one wavelength per year). The maximum migration rate is obtained for the bar with the shortest wavelength ($8\,\mathrm{cm\,day^{-1}}$, bar 5) that merges with the next bar, which is longer and slower (bar 6). In general, the longer the wavelength, the slower the migration rate. This is in agreement with previous studies on transverse bars (Garnier et al., 2006).

There are noticeable differences in the dynamics and in the characteristics of the first five bars (western bars) compared with the eastern bars. The western bars (close to the formation zone) are more irregular in shape, with a larger mean angle (5° larger), a smaller wavelength (20 m mean) and a corresponding higher migration rate. The eastern bars are well defined and remarkably parallel. Their cross-shore span decreases as they approach the decaying zone.

3.2.2 Time-dependent migration rates

Each bar signal has been decomposed into 10 segments by means of the piecewise regression described in Sect. 2.4 (Fig. 6). It was found that 10 is the best number to represent the medium-term movement of the bar and to filter the daily fluctuations. As we are analysing 2 years of data, the mean segment length is 70 days.

The migration rate of the bar system V_m, computed with (Eq. 1), is not constant in time with maximum migration rates in winter (Fig. 8). The maximum speeds, about 0.15 m day^{-1}, are reached during the first winter studied (2009), while during the second winter (2010) the maximum speeds are lower than 0.1 m day^{-1}. During summer the system migration is slower, with negative speeds for summer 2008, and migration rates lower than 0.01 m day^{-1} for summer 2009. The negative speeds (i.e. migration to the west) found in summer 2008 can be due to limitations in the computation of V_m. Specifically, the accuracy of the piecewise regression is expected to be lower at the beginning and end of the time series, due to the lack of previous/subsequent data. The negative migration rate is obtained for the first segment of the bars only; therefore this result may not be realistic.

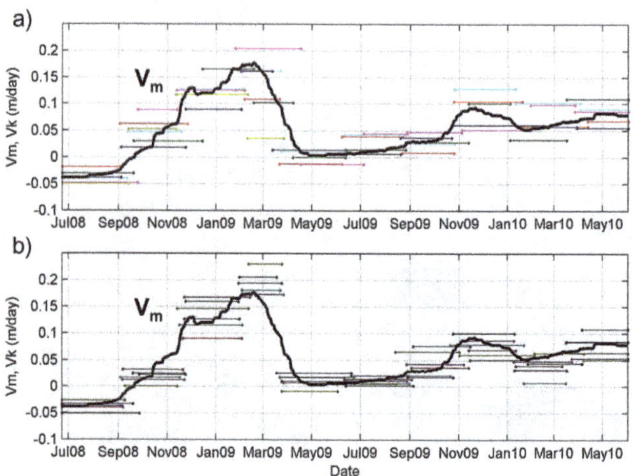

Figure 8. V_m (thick black line), time-dependent migration rate of the entire bar system (average of all coloured lines). V_k (colour lines), individual bar migration rate (the colours correspond to Fig. 6). **(a)** V_k for bars 3–8. **(b)** V_k for bars 9–14.

4 Forcing analysis

4.1 Forcing candidates

The migration to the east of the bar system indicates a dominant forcing coming from the west. The wind data have been extracted from the SeaWind (Menéndez et al., 2011) reanalysis database. Figure 9a shows the wind rose, and the time series of the wind speed is displayed in Fig. 9c. The predominant wind is from the west, reaching values of up to 25 m s^{-1}. The wind from the east is also frequent but less energetic, with speeds lower than 15 m s^{-1}. The mean wind speed is 5 m s^{-1}.

Other studies on transverse bars (Ribas et al., 2003) suggest that waves are the main forcing that controls their dynamics. The study area is protected from the incoming swell (Medellín et al., 2008) and the waves that can act on the bar system are generated locally. According to estuarine studies these wind waves can have a significant effect in the sediment transport (Green et al., 1997). Here, wind waves are generated over a maximum fetch of 4.5 km (from the south-west of the study area). Toward the south and south-east the fetch is reduced by the proximity of land.

During the survey period, the tidal range oscillates between 1 and 5 m (Fig. 9d). Maximum values of the tidal current in the channel (offshore of the bar system) occur during spring tides, with values of up to 0.25 m s^{-1}. In the channel the mean (residual) flow is ebb-oriented, but the residual tidal current is small in the intertidal areas. Computations performed (not shown) with an H2D model (Bárcena et al., 2012) show that the maximum residual current (obtained dur-

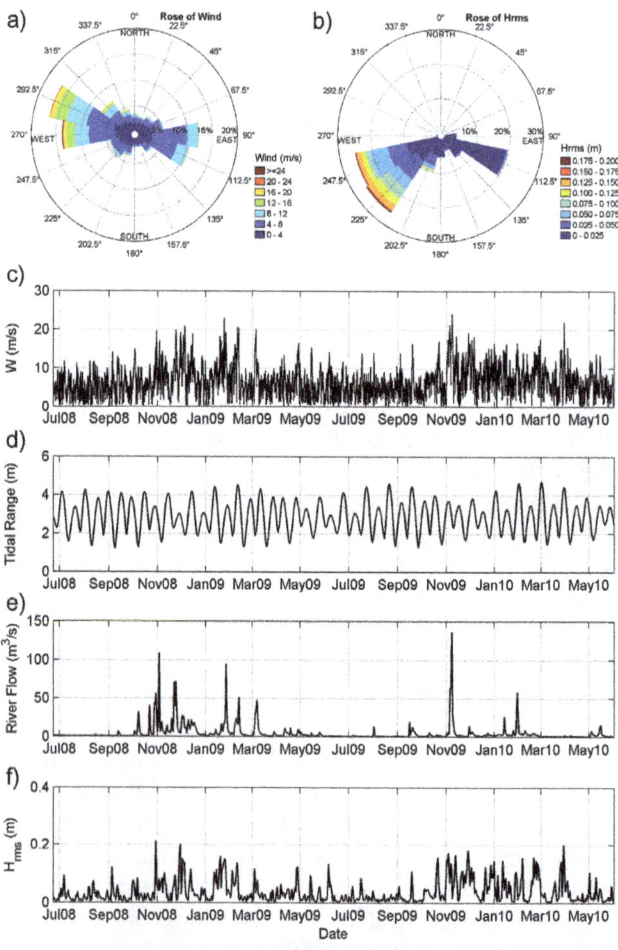

Figure 9. **(a)** Wind rose. **(b)** Wave rose. **(c–f)** Time series of the **(c)** wind speed W, and the daily averaged **(d)** tidal range, **(e)** river flow rate and **(f)** root-mean-squared wave height of the wind waves H_{rms}.

ing spring tides) is lower than $0.01\ \mathrm{m\,s^{-1}}$ in the study area. Although the residual current is small, the tide can have an effect on the bar dynamics because tidal currents can cause sediment stirring (which is stronger during mid tides) and because of the changes in water level. Firstly, the fetch is strongly dependent on the water level (Green et al., 1997) according to the emersion and submersion of the numerous intertidal shoals during the tidal cycle, and this is taken into account in the wave computations (see Sect. 4.2.1). Secondly, the changes in tidal level affect the time of bar submersion (that is larger during neap tides) and the volume of sand that can be transported (larger if high tide coincides with strong winds/waves). This will be taken into account in the sediment transport computations by including the tidal correction factor (see further explanations in Sect. 4.3.2).

Hydrodynamic studies of the Santander Bay have highlighted the effect of the water discharge produced by the Miera River (to the east of the study area) in the annual mean current magnitude in the bay (Bidegain et al., 2013). Time series of the daily averaged river flow rate are shown in Fig. 9e. Bidegain et al. (2013) have shown that, although the effect of the river is strong in the channel (ebb-oriented flow), the current produced close to the bar system is weak. However, the river discharge can play a role in the bar dynamics as it is linked to a strong sediment supply, which can act as an input of suspended sediment to the bar system.

4.2 Wind acting on water surface

4.2.1 Wave computation

The wind waves over the system have been simulated from the wind speed and direction by using the SWAN model (Booij et al., 1999). In the computations, changes in tidal level affecting the fetch have been included. The time series of the wind waves has been obtained with an interpolation technique based on radial basis functions (RBF), a scheme which is convenient for scattered and multivariate data (Camus et al., 2011). Results of the root-mean-squared (rms) wave height H_{rms} of the waves approaching the bar system are displayed in Fig. 9b (wave rose) and in Fig. 9f (time series of the daily averaged H_{rms}). The waves arrive from the west-south-west and south-west 65 % of the time, with a mean H_{rms} of 5 cm and a period of 1.5 s. During the westward windstorms the waves can extend to 20 cm from the west-south-west, with a period of 3 s. The other 35 % of the time the waves come from the east-south-east, with H_{rms} lower than 7 cm and periods below 1.7 s. The mean H_{rms} from this sector is less than 2 cm with a period of 1.2 s.

4.2.2 Wind stress vs wind-wave stress forcing

The previous studies on transverse bars, where the waves appear to be the main forcing, usually focus on wave parameters to relate the dynamics of the bars with the incident wave forcing, for example the alongshore component of the wave energy flux (e.g. Castelle et al., 2007; Price and Ruessink, 2011) or the wave radiation stress (e.g. Ribas and Kroon, 2007).

Here, the effect of the local wind is also analysed by computing the alongshore component of the wind shear stress acting on the water surface (Figs. 10a and 11a), defined as (Dean and Dalrymple, 1991)

$$T_y = -\rho C_{\mathrm{f}} W^2 \cos\theta_{\mathrm{w}}, \qquad (2)$$

where ρ is the water density ($\rho = 1025\ \mathrm{kg\,m^{-3}}$); C_{f} is the friction coefficient, equal to 1.2×10^{-6}; W is the wind speed; and θ_{w} is the incoming wind angle (from the shore normal).

In order to compare the relative effect of the wind stress and of the wave radiation stress, we define the alongshore wave stress, $S_y = S_{xy}/X_{\mathrm{b}}$ (Figs. 10b and 11b). S_{xy} is the alongshore component of the wave radiation stress (Longuet-Higgins and Stewart, 1964) and X_{b} is the surf-zone width. By considering $X_{\mathrm{b}} = H_{\mathrm{rms}}/(\beta\,\gamma_{\mathrm{b}})$, we obtain

$$S_y = \frac{\rho g}{16}\frac{H_{\mathrm{rms}}}{\beta\gamma_{\mathrm{b}}}\sin\theta\cos\theta, \qquad (3)$$

where g is the gravitational acceleration ($g = 9.81\ \mathrm{m\,s^{-2}}$), γ_{b} is the breaking coefficient for irregular waves ($\gamma_{\mathrm{b}} = 0.42$, Thornton and Guza, 1983), β is the beach slope ($\beta = 0.015$) and θ is the offshore wave angle (from the shore normal). S_y is an approximation of the term $\partial S_{xy}/\partial y$ in the alongshore momentum balance equation, a term that is equivalent to T_y in the same equation.

Figure 11a and b show the seasonal variability of S_y and T_y respectively. The comparison of both figures shows that both forcings are of the same order of magnitude and can therefore play a role in the bar dynamics, although S_y is twice as large as T_y. Only the wind stress seasonal analysis shows more highly energetic conditions in winter 2009 than in winter 2010, in accordance with the results of the migration rate. However, the wind stress shows more highly energetic conditions in autumn than in winter, while the migration rate shows lower values in autumn 2009 than in winter 2009. The wave stress seasonal analysis shows lower differences between autumns and winters, with larger values still being in the autumn.

4.3 Sediment transport

The relationship between the bar migration and the alongshore component of the sediment transport is also investigated. We use a formulation based on the Soulsby and Van Rijn formula (Soulsby, 1997). This formulation has been used in modelling studies to explain the formation of different kinds of transverse bars (Ribas et al., 2012; Garnier et al., 2006).

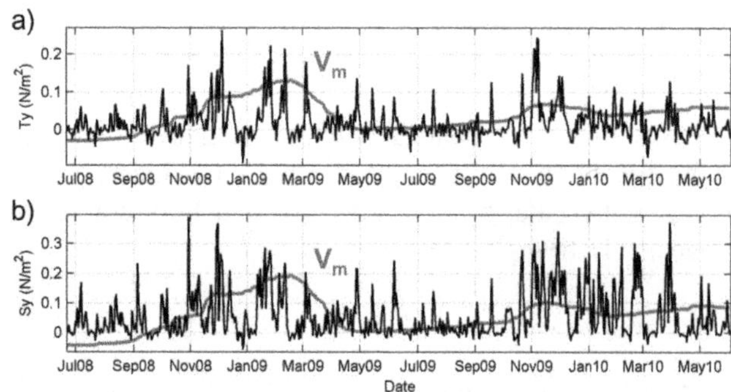

Figure 10. Time series of the daily averaged **(a)** alongshore wind stress (T_y, black) and **(b)** alongshore wave stress (S_y, black). The grey lines represent the behaviour of the bar migration rate $V_{\rm m}$ that has been rescaled with the above variables.

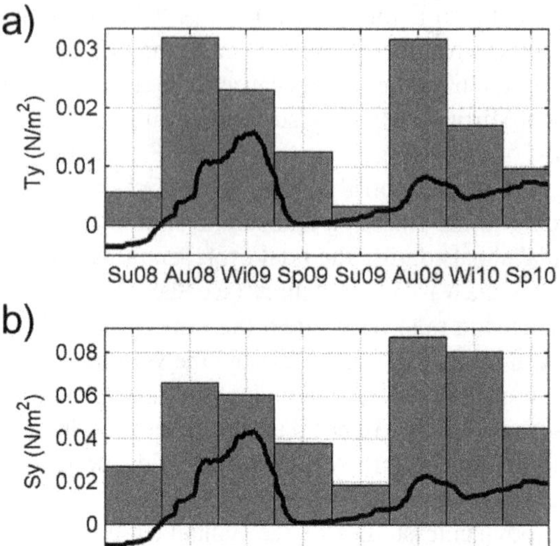

Figure 11. Seasonal variability of **(a)** alongshore wind stress (T_y) and **(b)** alongshore wave stress (S_y). The black lines represent the behaviour of the bar migration rate $V_{\rm m}$ that has been rescaled with the above variables. The bottom axes indicate the seasons, from summer 2008 to spring 2010.

4.3.1 Soulsby and Van Rijn formula

Here, we assume that the general formulation of the alongshore component of the sediment transport is given by

$$q = \alpha \left(V_{\rm wave} + V_{\rm wind} \right), \qquad (4)$$

where α is the sediment stirring function, $V_{\rm wave}$ is the alongshore component of the wave- and depth-averaged current driven by the wind waves, and $V_{\rm wind}$ is the alongshore component of the depth-averaged current driven by the local wind.

The alongshore current generated by the wind waves is approximated from the formula presented by Komar and Inman (1970):

$$V_{\rm wave} = 1.17 \left(g H_{\rm rms} \right)^{0.5} \sin\theta_{\rm b} \cos\theta_{\rm b}, \qquad (5)$$

where $\theta_{\rm b}$ is the wave angle at breaking. By using Snell's law and the dispersion relationship, Eq. (5) has been evaluated at the breaking depth, defined as $H_{\rm rms}/\gamma_{\rm b}$ ($\gamma_{\rm b} = 0.42$) from the incident wave angle computed with the SWAN model (Sect. 4.1).

The alongshore current generated by the wind is computed by assuming the alongshore momentum balance between the wind stress and the bottom friction in the case of a quadratic friction law:

$$V_{\rm wind} = \pm \left| \frac{T_y}{\rho c_{\rm d}} \right|^{0.5}, \qquad (6)$$

where $c_{\rm d}$ is the hydrodynamic drag coefficient set as $c_{\rm d} = 0.005$.

The stirring function in Eq. (4) is approximated with the Soulsby and Van Rijn formula (Soulsby, 1997) as

$$\alpha = \begin{cases} A_{\rm S} \left(U_{\rm eq} - U_{\rm crit} \right)^{2.4} & \text{if } U_{\rm eq} > U_{\rm crit} \\ 0 & \text{otherwise}, \end{cases} \qquad (7)$$

where $A_{\rm S}$ is a coefficient that represents the suspended load and the bed load transport and $U_{\rm crit}$ is the critical velocity above which the sediment can be transported. $A_{\rm S}$ and $U_{\rm crit}$ depend essentially on the sediment characteristics and on the water depth (for more details see Soulsby, 1997; Garnier et al., 2006). The equivalent stirring velocity is defined as

$$U_{\rm eq} = \left(U_{\rm wind}^2 + V_{\rm wave}^2 + \frac{0.018}{C_{\rm d}} U_{\rm b}^2 \right)^{0.5}, \qquad (8)$$

where $U_{\rm b}$ is the wave orbital velocity amplitude at the bottom (computed at wave breaking), $C_{\rm d}$ is the morphodynamic drag coefficient computed with the formula of Soulsby (1997) and $U_{\rm wind}$ is the modulus of the current generated by the wind:

$$U_{\rm wind} = \left(\frac{C_{\rm f}}{c_{\rm d}} W^2 \right)^{0.5}. \qquad (9)$$

4.3.2 Tidal correction factor

Although the tidal level variations have been included to compute the incoming wave time series, the sediment transport formula defined in Sect. 4.3.1 does not take into account that the bars can be emerged, and are therefore inactive, during part of a tidal cycle. If strong winds and high waves (despite the limited fetch) coincide with the time of emersion, they will have no effect and the effective sediment transport should be zero. Furthermore, the time of submersion depends on the tidal range. During neap tides the bar system is affected by the marine dynamics almost 100 % of the time because the full emersion of the bars occurs only when the tide is at its lowest level (during a short time period). However, during spring tides, the active time period is reduced because the tide falls lower and the bars stay emerged for a longer time.

These effects have been quantified by means of a tidal correction factor (α_t), ranging from 0 to 1, which evaluates how exposed the bars are due to the stirring resulting from the hydrodynamics. The corrected transport formula is then computed as

$$q^t = \alpha_t q; \quad (10)$$

α_t varies every hour, depending on the surf-zone width (X_b) and on the tidal level (η_t). It is computed by using the following formula (see Fig. 12):

$$\alpha_t = \begin{cases} 0 & \text{if} \quad Z_3 < \eta_t \\ \dfrac{Z_3 - \eta_t}{Z_3 - Z_2}\alpha_{t,\max} & \text{if} \quad Z_2 < \eta_t < Z_3 \\ \alpha_{t,\max} & \text{if} \quad Z_1 < \eta_t < Z_2 \\ \dfrac{\eta_t - Z_0}{Z_1 - Z_0}\alpha_{t,\max} & \text{if} \quad Z_0 < \eta_t < Z_1 \\ 0 & \text{if} \quad \eta_t < Z_0, \end{cases} \quad (11)$$

where $\alpha_{t,\max}$ quantifies the ratio between the surf-zone width (corresponding to the H_{rms}) and the cross-shore span of the bars, representing the percentage of the bars that could be within the active surf zone,

$$\alpha_{t,\max} = \frac{X_b}{L} = \frac{H_{rms}}{\beta\gamma_b L}, \quad (12)$$

where L is the mean cross-shore span of the bars ($L = 100$ m) and X_b is the width of the active surf zone (Fig. 12). Z_0, Z_1, Z_2 and Z_3 are defined as (see Fig. 12)

$$\begin{cases} Z_0 & = 2.5 \text{ m} = \text{level of the bar lower end} \\ Z_1(t) & = Z_0 + h^*(t) = 2.5 + \gamma_b^{-1} H_{rms}(t) \\ Z_2 & = 3.7 \text{ m} = \text{level of the bar upper end} \\ Z_3(t) & = Z_2 + h^*(t) = 3.7 + \gamma_b^{-1} H_{rms}(t). \end{cases} \quad (13)$$

Z_0 and Z_2 (levels of the bar lower end and upper end) are constant and determined from the 3-D-geometry (Fig. 5). Z_1 and Z_3 depend on the active depth h^* defined as ($h^* = H_{rms}/\gamma_b$).

To better understand these formulas, let us consider a day with constant wave height. The tidal correction factor is maximum ($\alpha_t = \alpha_{t,\max}$) when the maximum depth at the bars is larger than the active depth ($\eta_t \geq Z_1$) and when the sea level does not exceed the upper end of the bars ($\eta_t \leq Z_2$). This means that the complete surf-zone width is located over the bars. Furthermore, the sediment transport over the bars vanishes if the sea level does not reach the lower end of the bars ($\eta_t \leq Z_0$, i.e. too shallow) and if the minimum water depth at the bars is larger than the active depth ($\eta_t \geq Z_3$, i.e. too deep).

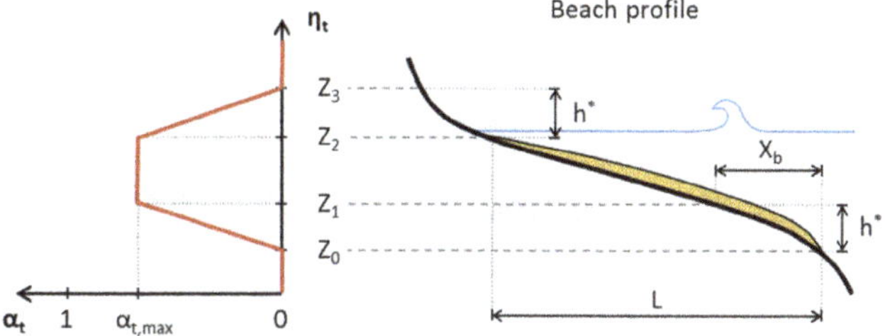

Figure 12. Parameters for the calculation of the tidal correction factor (α_t), which depends on the tidal level (η_t); the mean cross-shore span of the bars (L); the active depth (h^*); the surf-zone width (X_b); and the level of the bar lower end (Z_0), the bar upper end (Z_2) and the levels Z_1 and Z_3, which vary with the wave height (H_{rms}).

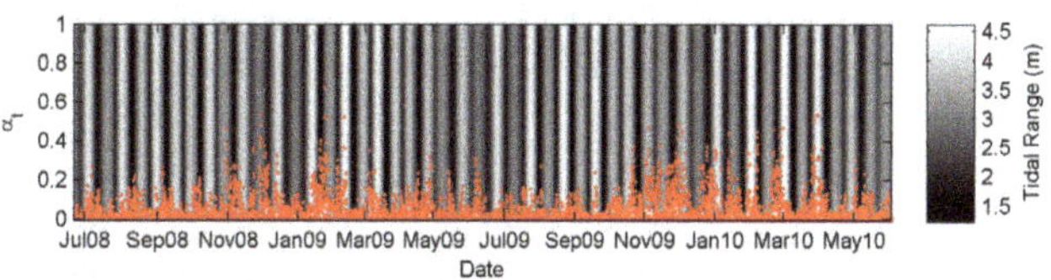

Figure 13. Time series of: the tidal correction factor α_t (red dots); and the tidal range (greyscale, vertical bars).

From our observations, the tidal factor never reaches 1 (Fig. 13). $\alpha_t = 1$ could occur for strong stormy conditions if the surf-zone width is as large as the bar width L ($H_{rms} > 0.5$), and if it coincides with high tide ($\eta_t = Z_2$). Figure 13 shows that α_t reaches its maximum values during neap tides, as was expected, generally as the tidal level is close to Z_2 a larger part of the day. During spring tides, the time during which the tidal level is between Z_3 and Z_0 is highly reduced. Consequently, the tidal factor is generally minimum.

4.3.3 Results

In order to analyse the correlation between q and the bar migration rate V_m, the time-dependent sediment transport rate must be computed. First, we integrate the sediment transport over the time intervals T_k, for each segment that characterises the bar movement (as explained in Sect. 3.2.2), and then we apply an equation equivalent to Eq. (1) for that sediment transport data. The obtained time series of the time-dependent sediment transport is shown in Fig. 14.

Figure 14a, which displays the sediment transport without the tidal correction, shows that the sediment transport is weaker in spring 2009 than in spring 2010, corresponding to a smaller migration rate. However, the seasonal average of q shows similar values for autumn–winter 2009 and autumn–winter 2010, while the migration rate results show lower values during 2010. The correlation coefficient obtained is $r = 0.75$.

The addition of the tidal factor improves the results (Fig. 14b) increasing the correlation coefficient to $r = 0.8$. All correlations obtained are highly significant ($p < 0.001$).

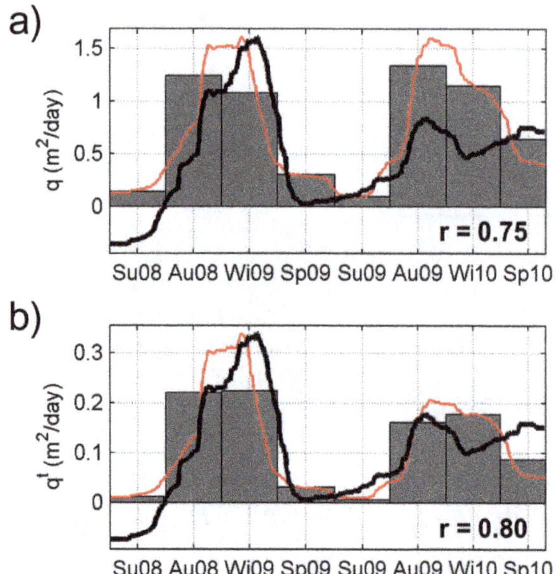

Figure 14. Sediment transport evaluation. Analysis of **(a)** alongshore component of sediment transport q (without tidal correction), and **(b)** q^{t} (with tidal correction). The grey areas show the seasonally averaged transport and the red lines show the time-dependent sediment transport time series (obtained by averaging over the T_k intervals). The black lines represent the behaviour of the bar migration rate V_{m} that has been rescaled with the above variables. The correlation coefficient of both lines is shown in the bottom right corner. The bottom axes indicate the seasons, from summer 2008 to spring 2010.

The seasonal analysis shows higher values of sediment transport during autumn–winter 2009 than autumn–winter 2010, corresponding with higher time-dependent migration rates during autumn–winter 2009. Again, the sediment transport computed in spring 2009 is lower than in spring 2010, corresponding well with the smaller migration rates measured in spring 2009. The time-dependent sediment transport time series of q^{t} (Fig. 14b) follows the main shape of the measured time-dependent migration rate. The bigger differences are found at the beginning and end of the study period, where the time-dependent migration rates are less reliable, as explained in Sect. 3.2.2. In particular, none of the formulas used managed to predict the negative (westward) migration reported during summer 2008, but, as previously mentioned, these negative migration rates may be not realistic.

Figure 9e shows the flow rate of Miera River. This flow rate is greater during winter 2009 than winter 2010, so the faster migration rate of the bars during this period could be influenced by the river discharge, possibly because it is a source of sediment. However, tests performed by including additional sediment stirring due to the river flow do not show improvement in the results.

5 Conclusions

A small-scale finger bar system has been identified on the intertidal zone of the swell-protected beach of El Puntal Spit in the Bay of Santander (northern coast of Spain). The beach is characterised by a constant slope of 0.015 and by uniform sand with $D_{50} = 0.27$ mm. This system appears on the flat intertidal region, which extends over 600 m on the alongshore direction and between 70 and 130 m on the cross-shore direction (the cross-shore span is determined by the tidal horizontal excursion).

A system of 15 bars was observed by using the Horus video imaging system during 2 years (between 23 June 2008 and 2 June 2010). The bar system has been digitised from daily images at low tide. The data set is almost continuous, with good quality data 81 % of the time and a maximum continuous period of time without data of no more than 6 days.

The geometric characteristics of the system are almost constant in time. The mean wavelength of the bar system is 26 m and the bar amplitude is between 10 and 20 cm. Moreover, the bars have an oblique orientation with respect to the low-tide shoreline, with a mean angle of 26° to the east from the shore normal. We noticed differences in the geometry along the domain: the western bars (first half) are more irregular and have smaller wavelength than the eastern bars (second half).

The full system slowly migrates to the east (against the ebb flow) with a mean speed of 6 cm day^{-1} that varies between bars. In general, larger wavelength bars migrate more slowly, in agreement with previous studies (Garnier et al., 2006). An episode of merging of two bars was observed on 28 March 2009: the bar with the smallest wavelength is faster and merges with the next bar. Bars form on the western end of the system, migrate east and then decay at the eastern end.

A detailed analysis of the bar motion, from a piecewise regression of the bar positions, has shown that bars migrate more quickly in winter than in summer, with maximum migration rates obtained in winter 2009 (0.15 m day^{-1}). Some negative speeds (migration to the west) have been computed (during summer 2008), but this result could be an effect of the limitations of the piecewise regression at the beginning and end of the time series.

The primary forcing mechanism that is acting on the bar dynamics is the wind over the water surface. Offshore of the bar system, the mean (annual) flow is ebb-oriented (to the west), because of the Miera River discharge and the astronomical tide. However, in the intertidal zone their effects on the mean flow vanish. There, wind shear stress and wind waves generated over a fetch of up to 4.5 km at high tide seem to determine the direction of the alongshore transport.

Although the residual tidal current is weak, the tide seems to be important in the bar dynamics as the tidal range changes the mean (daily) fetch as well as the time of exposure of the bars to the marine dynamics. Furthermore, the river discharge could act as input of suspended sediment in the bar system and play a role in the bar dynamics.

The correlation between the bar migration and the alongshore component of the sediment transport has been analysed by using the Soulsby and Van Rijn formulation. The inclusion of a tidal correction factor, simulating that the active time depends on the tidal level and the wave height, improves the results.

Finally, the bar system is persistent and neither formation nor destruction events of the entire system have been observed. Further studies are necessary to understand the formation processes and the full dynamics of these small-scale finger bars. In situ measurements of the hydrodynamics and sediment concentrations, as well as numerical morphological modelling, are essential in order to deepen our understanding. The bar system here has an oblique down-current orientation with respect to the migration direction and has similar characteristics and dynamics to the system described by previous theoretical (modelling) studies that consider the forcing due to waves only (e.g. Garnier et al., 2006). However, in our estuarine environment, the dynamics are more complex as the actions of different forcings are of the same order of magnitude.

Acknowledgements. The authors thank Puertos del Estado (Spanish Government) for providing tide-gauge data. The work of R. Garnier is supported by the Spanish Government through the "Juan de la Cierva" programme. This research is part of the ANIMO (BIA2012-36822) project, which is funded by the Spanish Government. The authors thank the editor, G. Coco, and the referees, F. Ribas and E. Gallagher, for their useful comments.

Edited by: G. Coco

References

Bárcena, J. F., García, A., García, J., Álvarez, C., and Revilla, J. A.: Surface analysis of free surface and velocity to changes in river flow and tidal amplitude on a shallow mesotidal estuary: an application in Suances Estuary (Nothern Spain), J. Hydrol., 14, 301–318, 2012.

Bidegain, G., Bárcena, J. F., García, A., and Juanes, J. A.: LARVAHS: predicting clam larval dispersal and recruitment using habitat suitability-based particle tracking mode, Ecol. Model., 268, 78–92, 2013.

Booij, N., Ris, R. C., and Holthuijsen, L. H.: A third-generation wave model for coastal regions, Part I, Model description and validation, J. Geophys. Res., 104, 7649–7666, 1999.

Bruner, K. R. and Smosna, R. A.: The movement and stabilization of beach sand on transverse bars, Assateague Island, Virginia, J. Coastal Res., 5, 593–601, 1989.

Camus, P., Méndez, F. J., and Medina, R.: A hybrid efficient method to downscale wave climate to coastal areas, Coast. Eng., 58, 851–862, 2011.

Castelle, B., Bonneton, P., Dupuis, H., and Senechal, N.: Double bar beach dynamics on the high-energy meso-macrotidal French Aquitanian coast: a review, Mar. Geol., 245, 141–159, 2007.

Dean, R. G. and Dalrymple, R. A.: Water wave mechanics for engineers and scientists, World Scientific, 157–158, 1991.

Falqués, A.: Formación de topografía rítmica en el Delta del Ebro, Revista de Geofísica, 45, 143–156, 1989 (in Spanish).

Galal, E. M. and Takewaka, S.: Longshore migration of shoreline mega-cusps observed with x-band radar, Coast. Eng. J., 50, 247–276, 2008.

Garnier, R., Calvete, D., Falqués, A., and Caballeria, M.: Generation and nonlinear evolution of shore-oblique/transverse sand bars, J. Fluid Mech., 567, 327–360, 2006.

Garnier, R., Calvete, D., Falqués, A., and Dodd, N.: Modelling the formation and the long-term behavior of rip channel systems from the deformation of a longshore bar, J. Geophys. Res., 113, C07053, doi:10.1029/2007JC004632, 2008.

Garnier, R., Medina, R., Pellón, E., Falqués, A., and Turki, I.: Intertidal finger bars at El Puntal spit, bay of Santander, Spain, in: Proceedings of the 33rd Conference on Coastal Engineering, ASCE, Santander, Spain, 1–8, 2012.

Gelfenbaum, G. and Brooks, G. R.: The morphology and migration of transverse bars off the west-central Florida coast, Mar. Geol., 200, 273–289, 2003.

Goodfellow, B. W. and Stephenson, W. J.: Beach morphodynamics in a strong-wind bay: a low-energy environment?, Mar. Geol., 214, 101–116, 2005.

Green, M. O., Black, K. P., and Amos, C. L.: Control of estuarine sediment dynamics by interactions between currents and waves at several scales, Mar. Geol., 144, 97–116, 1997.

Gutiérrez, O., González, M., and Medina, R.: A methodology to study beach morphodynamics based on self-organizing maps and digital images, in: Proceedings of the Coastal Sediments 2011, World Scientific, 2453–2464, 2011.

Holman, R. A., Sallenger Jr., A. H., Lippmann, T. C. D., and Haines, J. W.: The application of video image processing to the study of nearshore processes, Oceanography, 6, 78–85, 1993.

Komar, P. and Inman, D.: Longshore sand transport on beaches, J. Geophys. Res., 75, 5514–5527, 1970.

Konicki, K. M. and Holman, R. A.: The statistics and kinematics of transverse sand bars on an open coast, Mar. Geol., 169, 69–101, 2000.

Kroon, A., Davidson, M. A., Aarninkhof, S. G. J., Archetti, R., Armaroli, C., González, M., Medri, S., Osorio, A., Aagaard, T., Holman, R. A., and Spanhoff, R.: Application of remote sensing video systems to coastline management problems, Coast. Eng., 54, 493–505, 2007.

Lafon, V., Dupuis, H., Howa, H., and Froidefond, J. M.: Determining ridge and runnel longshore migration rate using spot imagery, Oceanol. Acta, 25, 149–158, 2002.

Levoy, F., Anthony, E. J., Monfort, O., Robin, N., and Bretel, P.: Formation and migration of transverse bars along a tidal sandy coast deduced from multi-temporal Lidar datasets, Mar. Geol., 342, 39–52, 2013.

Longuet-Higgins, M. S. and Stewart, R. W.: Radiation stresses in water waves: a physical discussion with applications, Deep-Sea Res., 11, 529–562, 1964.

Losada, M. A., Medina, R., Vidal, C., and Roldan, A.: Historical evolution and morphological analysis of "El Puntal" Spit, Santander (Spain), J. Coastal Res., 7, 711–722, 1991.

Losada, M. A., Medina, R., Vidal, C., and Losada, I. J.: Temporal and spatial cross-shore distributions of sediment at "El Puntal" spit, Santander, Spain, Coast. Eng., 1992, Proceedings of the Twenty-Third International Conference, 2251–2264, 1992.

Medellín, G., Medina, R., Falqués, A., and González, M.: Coastline sand waves on a low-energy beach at "El Puntal" spit, Spain, Mar. Geol., 250, 143–156, 2008.

Medellín, G., Falqués, A., Medina, R., and González, M.: Coastline sand waves on a low-energy beach at El Puntal spit, Spain: linear stability analysis, J. Geophys. Res., 114, C03022, doi:10.1029/2007JC004426, 2009.

Medina, R., Marino-Tapia, I., Osorio, A., Davidson, M., and Martín, F. L.: Management of dynamic navigational channels using video techniques, Coast. Eng., 54, 523–537, 2007.

Menéndez, M., Tomás, A., Camus, P., García-Díez, M., Fita, L., Fernández, J., Méndez, F. J., and Losada, I. J.: A methodology to evaluate regional-scale offshore wind energy resources, OCEANS'11 IEEE Santander Conference, Spain, 1–9, 2011.

Niedoroda, A. W. and Tanner, W. F.: Preliminary study of transverse bars, Mar. Geol., 9, 41–62, 1970.

Nordstrom, K. F. and Jackson, N. L.: Physical processes and landforms on beaches in short fetch environments in estuaries, small lakes and reservoirs: a review, Earth-Sci. Rev., 111, 232–247, 2012.

Price, T. D. and Ruessink, B. G.: State dynamics of a double sandbar system, Cont. Shelf. Res., 31, 659–674, 2011.

Ranasinghe, R., Symonds, G., Black, K., and Holman, R.: Morphodynamics of intermediate beaches: a video imaging and numerical modelling study, Coast. Eng., 51, 629–655, 2004.

Requejo, S., Medina, R., and González, M.: Development of a medium-long term beach evolution model, Coast. Eng., 55, 1074–1088, 2008.

Ribas, F. and Kroon, A.: Characteristics and dynamics of surfzone transverse finger bars, J. Geophys. Res., 112, F03028, doi:10.1029/2006JF000685, 2007.

Ribas, F., Falqués, A., and Montoto, A.: Nearshore oblique sand bars, J. Geophys. Res., 108, C43119, doi:10.1029/2001JC000985, 2003.

Ribas, F., de Swart, H. E., Calvete, D., and Falqués, A.: Modeling and analyzing observed transverse sand bars in the surf zone, J. Geophys. Res., 117, F02013, doi:10.1029/2011JF002158, 2012.

Ribas, F., ten Doeschate, A., de Swart, H., Ruessink, G., and Calvete, D.: Observations and modelling of surf-zone transverse finger bars at the Gold Coast, Australia, Ocean Dynam., doi:10.1007/s10236-014-0719-4, in press, 2014.

Soulsby, R. L.: Dynamics of Marine Sands, Thomas Telford, London, UK, 1997.

Thornton, B. and Guza, R. T.: Transformation of wave height distribution, J. Geophys. Res., 88, 5925–5938, 1983.

Thornton, E. B., MacMahan, J., and Sallenger Jr., A. H.: Rip currents, mega-cusps, and eroding dunes, Mar. Geol., 240, 151–167, 2007.

van Enckevort, I. M. J., Ruessink, B. G., Coco, G., Suzuki, K., Turner, I. L., Plant, N. G., and Holman, R. A.: Observations of nearshore crescentic sandbars, J. Geophys. Res., 109, C06028, doi:10.1029/2003JC002214, 2004.

Wijnberg, K. M. and Kroon, A.: Barred beaches, Geomorphology, 48, 103–120, 2002.

Wright, L. D. and Short, A. D.: Morphodynamic variability of surf zones and beaches: a synthesis, Mar. Geol., 56, 93–118, 1984.

Permissions

All chapters in this book were first published in ESD, by Copernicus Publications; hereby published with permission under the Creative Commons Attribution License or equivalent. Every chapter published in this book has been scrutinized by our experts. Their significance has been extensively debated. The topics covered herein carry significant findings which will fuel the growth of the discipline. They may even be implemented as practical applications or may be referred to as a beginning point for another development.

The contributors of this book come from diverse backgrounds, making this book a truly international effort. This book will bring forth new frontiers with its revolutionizing research information and detailed analysis of the nascent developments around the world.

We would like to thank all the contributing authors for lending their expertise to make the book truly unique. They have played a crucial role in the development of this book. Without their invaluable contributions this book wouldn't have been possible. They have made vital efforts to compile up to date information on the varied aspects of this subject to make this book a valuable addition to the collection of many professionals and students.

This book was conceptualized with the vision of imparting up-to-date information and advanced data in this field. To ensure the same, a matchless editorial board was set up. Every individual on the board went through rigorous rounds of assessment to prove their worth. After which they invested a large part of their time researching and compiling the most relevant data for our readers.

The editorial board has been involved in producing this book since its inception. They have spent rigorous hours researching and exploring the diverse topics which have resulted in the successful publishing of this book. They have passed on their knowledge of decades through this book. To expedite this challenging task, the publisher supported the team at every step. A small team of assistant editors was also appointed to further simplify the editing procedure and attain best results for the readers.

Apart from the editorial board, the designing team has also invested a significant amount of their time in understanding the subject and creating the most relevant covers. They scrutinized every image to scout for the most suitable representation of the subject and create an appropriate cover for the book.

The publishing team has been an ardent support to the editorial, designing and production team. Their endless efforts to recruit the best for this project, has resulted in the accomplishment of this book. They are a veteran in the field of academics and their pool of knowledge is as vast as their experience in printing. Their expertise and guidance has proved useful at every step. Their uncompromising quality standards have made this book an exceptional effort. Their encouragement from time to time has been an inspiration for everyone.

The publisher and the editorial board hope that this book will prove to be a valuable piece of knowledge for researchers, students, practitioners and scholars across the globe.

List of Contributors

F. U. M. Heimann
WSL Swiss Federal Institute for Forest, Snow and Landscape Research, 8903 Birmensdorf, Switzerland
Department of Environmental System Sciences, ETH Zurich, 8092 Zurich, Switzerland

D. Rickenmann
WSL Swiss Federal Institute for Forest, Snow and Landscape Research, 8903 Birmensdorf, Switzerland

J. M. Turowski
Helmholtz Centre Potsdam, GFZ German Research Centre for Geosciences, Telegrafenberg, 14473 Potsdam, Germany

J. W. Kirchner
Department of Environmental System Sciences, ETH Zurich, 8092 Zurich, Switzerland
WSL Swiss Federal Institute for Forest, Snow and Landscape Research, 8903 Birmensdorf, Switzerland

M. Böckli
WSL Swiss Federal Institute for Forest, Snow and Landscape Research, 8903 Birmensdorf, Switzerland

A. Badoux
WSL Swiss Federal Institute for Forest, Snow and Landscape Research, 8903 Birmensdorf, Switzerland

J. M. Turowski
Helmholtz Centre Potsdam, GFZ German Research Centre for Geosciences, Telegrafenberg, 14473 Potsdam, Germany
WSL Swiss Federal Institute for Forest, Snow and Landscape Research, 8903 Birmensdorf, Switzerland

A. C. Cunningham
School of Geosciences, University of the Witwatersrand, Johannesburg, South Africa
Centre for Archaeological Science, School of Earth and Environmental Sciences, University of Wollongong, Wollongong, Australia

J. Wallinga
Soil Geography and Landscape group&Netherlands Centre for Luminescence dating, Wageningen University, Wageningen, the Netherlands

N. Hobo
Soil Geography and Landscape group&Netherlands Centre for Luminescence dating, Wageningen University, Wageningen, the Netherlands
Alterra, Wageningen University and Research Centre, Wageningen, the Netherlands
Department of Physical Geography, Utrecht University, Utrecht, the Netherlands

A. J. Versendaal
Soil Geography and Landscape group&Netherlands Centre for Luminescence dating, Wageningen University, Wageningen, the Netherlands

B. Makaske
Soil Geography and Landscape group&Netherlands Centre for Luminescence dating, Wageningen University, Wageningen, the Netherlands
Alterra, Wageningen University and Research Centre, Wageningen, the Netherlands

H. Middelkoop
Department of Physical Geography, Utrecht University, Utrecht, the Netherlands

M. Fox
Institute of Geology, Swiss Federal Institute of Technology, ETH Zürich, Switzerland

F. Herman
Institute of Geology, Swiss Federal Institute of Technology, ETH Zürich, Switzerland
Institute of Earth Science, University of Lausanne, Switzerland

S. D. Willett
Institute of Geology, Swiss Federal Institute of Technology, ETH Zürich, Switzerland

D. A. May
Institute of Geology, Swiss Federal Institute of Technology, ETH Zürich, Switzerland

N. Stark
Dalhousie University, Department of Oceanography, Halifax, Canada
Virginia Tech, Department of Civil and Environmental Engineering, Blacksburg, VA, USA

A. E. Hay
Dalhousie University, Department of Oceanography, Halifax, Canada

R. Cheel
Dalhousie University, Department of Oceanography, Halifax, Canada

C. B. Lake
Dalhousie University, Department of Civil Engineering, Halifax, Canada
Virginia Tech, Department of Civil and Environmental Engineering, Blacksburg, VA, USA

C. B. Phillips
Earth and Environmental Science, University of Pennsylvania, Hayden Hall, 240 South 33rd st., Philadelphia, PA 19104, USA

D. J. Jerolmack
Earth and Environmental Science, University of Pennsylvania, Hayden Hall, 240 South 33rd st., Philadelphia, PA 19104, USA

M. Scherler
Departement of Geosciences, University of Fribourg, Chemin du Musée 4, 1700 Fribourg, Switzerland
Swiss Federal Research Institute WSL, Zürcherstrasse 111, 8903 Birmensdorf, Switzerland

S. Schneider
Departement of Geosciences, University of Fribourg, Chemin du Musée 4, 1700 Fribourg, Switzerland

M. Hoelzle
Departement of Geosciences, University of Fribourg, Chemin du Musée 4, 1700 Fribourg, Switzerland

C. Hauck
Departement of Geosciences, University of Fribourg, Chemin du Musée 4, 1700 Fribourg, Switzerland

M. Attal
School of GeoSciences, Univ. of Edinburgh, Drummond Street, Edinburgh, EH8 9XP, UK

S. M. Mudd
School of GeoSciences, Univ. of Edinburgh, Drummond Street, Edinburgh, EH8 9XP, UK

M. D. Hurst
School of GeoSciences, Univ. of Edinburgh, Drummond Street, Edinburgh, EH8 9XP, UK
British Geological Survey, Keyworth, Nottingham, NG12 5GG, UK

B. Weinman
Department of Soil, Water, and Climate, University of Minnesota, 439 Borlaug Hall, 1991 Upper Buford Circle, St. Paul, MN 55108-6028, USA

K. Yoo
Department of Soil, Water, and Climate, University of Minnesota, 439 Borlaug Hall, 1991 Upper Buford Circle, St. Paul, MN 55108-6028, USA

M. Naylor
School of GeoSciences, Univ. of Edinburgh, Drummond Street, Edinburgh, EH8 9XP, UK

D. M. Krzeminska
Water Resources Section, Faculty of Civil Engineering and Geosciences, Delft University of Technology, Delft, Netherlands

T. A. Bogaard
Water Resources Section, Faculty of Civil Engineering and Geosciences, Delft University of Technology, Delft, Netherlands

T.-H. Debieche
Université d'Avignon et des Pays de Vaucluse, EMMAH UMR1114 INRA-UAPV, 33 rue Louis Pasteur, 84000 Avignon, France
Water and Environment Team, Geological Engineering Laboratory, Jijel University, P.O. Box 98, 18000 Jijel, Algeria

F. Cervi
Université d'Avignon et des Pays de Vaucluse, EMMAH UMR1114 INRA-UAPV, 33 rue Louis Pasteur, 84000 Avignon, France
Dipartimento di Ingegneria Civile, Chimica, Ambientale e dei Materiali (DICAM) Università di Bologna, Viale Risorgimento 2, 40136, Bologna, Italy

V. Marc
Université d'Avignon et des Pays de Vaucluse, EMMAH UMR1114 INRA-UAPV, 33 rue Louis Pasteur, 84000 Avignon, France

J.-P. Malet
Institut de Physique du Globe de Strasbourg, CNRS UMR7516, Université de Strasbourg, Ecole et Observatoire des Sciences de la Terre, 5 rue Descartes, 67084 Strasbourg, France

R. M. Headley
Geosciences Department, University of Wisconsin-Parkside, Kenosha, WI, USA

T. A. Ehlers
Department of Geosciences, University of Tübingen, Tübingen, Germany

A. Cantelli
Shell International Exploration and Production, Houston, Texas, USA

T. Muto
Graduate School of Fisheries Science and Environmental Studies, Nagasaki University, 1–14 Bunkyomachi, Nagasaki 852-8521, Japan

G. Samodra
Environmental Geography Department, Universitas Gadjah Mada, Yogyakarta, Indonesia
Graduate School of Civil and Structural Engineering, Kyushu University, Fukuoka, Japan

G. Chen
Graduate School of Civil and Structural Engineering, Kyushu University, Fukuoka, Japan

J. Sartohadi
Graduate School of Civil and Structural Engineering, Kyushu University, Fukuoka, Japan

D. S. Hadmoko
Graduate School of Civil and Structural Engineering, Kyushu University, Fukuoka, Japan

K. Kasama
Graduate School of Civil and Structural Engineering, Kyushu University, Fukuoka, Japan

E. Pellón
Environmental Hydraulics Institute (IH Cantabria), Universidad de Cantabria, Santander, Spain

R. Garnier
Environmental Hydraulics Institute (IH Cantabria), Universidad de Cantabria, Santander, Spain

R. Medina
Environmental Hydraulics Institute (IH Cantabria), Universidad de Cantabria, Santander, Spain

www.ingramcontent.com/pod-product-compliance
Lightning Source LLC
Chambersburg PA
CBHW050438200326
41458CB00014B/4984
9781682860816